international AIR POWER REVIEW

AIRtime Publishing

United States of America • United Kingdom

international® AIR POWER REVIEW

Published quarterly by AIRtime Publishing Inc.
120 East Avenue, Norwalk, CT 06851
Tel (203) 838-7979 • Fax (203) 838-7344

© 2005 AIRtime Publishing Inc.
Sea Harrier cutaway © Mike Badrocke

Photos and other illustrations are the copyright of their respective owners
Created for AIRtime Publishing by Aeromedia Communications Ltd

Softbound Edition ISSN 1473-9917 / ISBN 1-880588-83-8
Hardcover Deluxe Casebound Edition ISBN 1-880588-32-3

Publisher
   Mel Williams

Editor
   David Donald   e-mail: airpower@btinternet.com

Assistant Editor
   Daniel J. March

Sub Editor
   Karen Leverington

US Desk
   Tom Kaminski

Russia/CIS Desk
   Piotr Butowski, Zaur Eylanbekov

Europe and Rest of World Desk
   John Fricker, Jon Lake

Correspondents
   Argentina: Santiago Rivas and Juan Carlos Cicalesi
   Australia: Nigel Pittaway
   Belgium: Dirk Lamarque
   Brazil: Claudio Lucchesi
   Bulgaria: Alexander Mladenov
   Canada: Jeff Rankin-Lowe
   France: Henri-Pierre Grolleau
   Germany: Marnix Sap
   India: Pushpindar Singh
   Israel: Shlomo Aloni
   Italy: Luigino Caliaro
   Japan: Yoshitomo Aoki
   Netherlands: Tieme Festner
   Romania: Danut Vlad
   Spain: Salvador Mafé Huertas
   USA: Rick Burgess, Brad Elward, Mark Farmer (North Pacific region), Peter Mersky, Bill Sweetman

Artists
   Mike Badrocke, Zaur Eylanbekov, Keith Fretwell,
   Grant Race, John Weal

Controller
   Linda DeAngelis

Sales Director
   Jill Brooks

Origination by Chroma Graphics and Universal Graphics, Singapore
Printed in Singapore by KHL Printing

All rights reserved. No part of this publication may be copied, reproduced, stored electronically or transmitted in any manner or in any form whatsoever without the written permission of the publishers and copyright holders.

International Air Power Review is published quarterly in two editions (Softbound and Deluxe Casebound) and is available by subscription or as single volumes. Please see details opposite.

Acknowledgments
The author of the DHS article would like to thank the following individuals for their assistance: CDR Doug Stephan and ENS Paul Smith USCG Aviation Program Office, Ms. Jolie Shifflet and LT Ron Mench US Coast Guard Media Relations. Garrison Courtney ICE/PA, Gary Bracken AMO Public Affairs, Zachary Mann, Vince Bond CBP Public Affairs, Dave Aycock, AMO San Diego Air & Marine Branch, Martin Vaughn, Brough McDonald, AMO Tucson Air Branch, Carl Robinson, Attilla 'A.T.' Guyris, AMO Riverside Air Unit, Robert Viador, Luis Bencosme, Marty Wade, AMO Miami Air & Marine Branch, Mario Villarreal OBP Public Affairs, Victor Colon (Miami Sector OBP), Ed Duda (Buffalo Sector OBP), Greg Mish (Buffalo Sector OBP), Lonny Schweitzer (Grand Forks Sector OBP), Joseph Guiliano (Blaine Sector OBP), David Bernard, Joe Mulhurn (Havre Sector OBP), Special thanks to Mark Johnson Patrol Agent-in-Charge OBP Tucson Sector Air Operations

Thanks to RADM James Kelly, CAPTs Joseph Aucoin, Garry Mace, CDRs Jay Bynum, Paul Foster, Dave Mayo, Doug McGowen, Tim McMahon, Mike Wettlaufer, LCDRs Bill Blacker, Brian Bronk, Des Connolly, Steven DeSantis, Douglass Spool, LTs Joe Espiritu, Andy Green, AW1 Michael Martin, and the personnel of CVW-5 for their help with the Air Wing Five feature.

The author of the Sea Harrier article would like to thank Commander Tim Eastaugh, Commander Chips Lawler and all 899 NAS personnel for their kind help. Many thanks also to Dale Donovan, IO(CC)1, Headquarters Strike Command, and Lieutenant Commander David Constant, Sea Harrier Force Public Relation Officer. Special thanks to pilot Lieutenant Jamie Haggo for the superb flight.

Contact and Ordering Information
(hours: 10 am - 4.30 pm EST, Mon-Fri)
addresses, telephone and fax numbers
International Air Power Review, P.O. Box 5074, Westport, CT 06881, USA
   Tel (203) 838-7979 • Fax (203) 838-7344
   Toll free within USA and Canada: 1 800 359-3003
   Toll free from the United Kingdom, Belgium, Denmark, France, Germany,
      Holland, Ireland, Italy, Luxembourg, Norway, Portugal, Sweden and
      Switzerland (5-6 hours ahead): 00 800 7573-7573
   Toll free from Finland (6 hours ahead): 990 800 7573-7573
   Toll free from Australia (13-15 hours ahead): 0011 800 7573-7573
   Toll free from New Zealand (17 hours ahead): 00 800 7573-7573
   Toll free from Japan (14 hours ahead): 001 800 7573-7573

website
   www.airtimepublishing.com
e-mails
   airpower@airtimepublishing.com
   inquiries@airtimepublishing.com

Subscription & Back Volume Rates
One-year subscription (4 quarterly volumes),
inclusive of ship. & hdlg./ post. & pack.:
   Softbound Edition
   USA $59.95, UK £48, Europe EUR 88, Canada Cdn $99,
   Rest of World US $119 (air)

   Deluxe Casebound Edition
   USA $79.95, UK £68, Europe EUR 120, Canada Cdn $132,
   Rest of World US $155 (air)

Two-year subscription (8 quarterly volumes),
inclusive of ship. & hdlg./ post. & pack.:
   Softbound Edition
   USA $112, UK £92, Europe EUR 169, Canada Cdn $187,
   Rest of World US $219 (air)

   Deluxe Casebound Edition
   USA $149, UK £130, Europe EUR 230, Canada Cdn $246,
   Rest of World US $295 (air)

Single/back volumes by mail (each):
   Softbound Edition
   USA $19.95, UK £14.95, Europe EUR 22
   (incl. ship. & hdlg./ post. & pack.)
   Deluxe Casebound Edition
   US $24.95, UK £18.95, Europe EUR 30
   (incl. ship. & hdlg./ post. & pack.)

Prices are subject to change without notice. Canadian residents please add GST. Connecticut residents please add sales tax.

Volume Sixteen

CONTENTS

AIR POWER INTELLIGENCE
Programme Update 4
Project Development 4
Upgrades and Modifications 8
Procurement and Deliveries 9
Air Arm Review 13
John Fricker and Tom Kaminski

NEWS FEATURE
Iraq: airpower in the counter-insurgency war 16
Tim Ripley

DEBRIEF
F-16 EPAF M3 upgrade 30
Danny van der Molen
Eurocopter Tiger for AAAC 31
Nigel Pittaway
Team 60: Sweden's ambassadors 32
Robert Hewson

PHOTO REPORT
Cruzex 2004 34

FOCUS AIRCRAFT
HALE/MALE Unmanned Air Vehicles 40
Part 2: 21st Century warfighters
The second instalment of this feature looks at the operations of the USAF's Global Hawk and Predator fleets, and describes the current generation of HALE/MALE UAVs – and what might lie in the future.
Bill Sweetman

SPECIAL FEATURE
Sea Harrier Farewell 60
Henri-Pierre Grolleau

SPECIAL REPORT
CVW-5: at home at Atsugi 70
Ted Carlson

AIR POWER ANALYSIS
US Department of Homeland Security
 US Customs and Border Protection 78
  Office of Border Patrol 78
  Office of Air & Marine Operations 82
 United States Coast Guard 86
  Direct-reporting units 88
  LANTAREA/MDZA 88
  PACAREA/MDZP 88
  US Coast Guard aircraft 90
Tom Kaminski

SPECIAL FEATURE
Tucson Testers – the AATC 94
Andy Wolfe

VARIANT FILE
Boeing 707 Military variants
Part 1: Military 707s and conversions 98
Jon Lake

PIONEERS & PROTOTYPES
VTOL Flat-risers:
Lockheed XV-4 and Ryan XV-5 118
Daniel J. March

WARPLANE CLASSIC
Messerschmitt Bf 110 128
Dr Alfred Price

AIR COMBAT
RAF Liberators at war
Part 2: Overseas operations 156
Jon Lake

INDEX 174

MAJOR FEATURES PLANNED FOR VOLUME SEVENTEEN
Focus Aircraft: SEPECAT Jaguar, **Warplane Classic:** Douglas A-1 Skyraider, **Special Report:** Holloman's QF-4s, **Variant Briefing:** Boeing 707 military versions Pt 2, **Unit report:** HX-21 Rotary-wing Test Squadron, **Technical Briefing:** Swiss Hornets, **Air Combat:** F-84 in Korea

Air Power Intelligence

PROGRAMME UPDATE

A-10C – upgraded Warthog

Lockheed Martin Systems Integration and the USAF unveiled the first example of the A-10C Thunderbolt II at Eglin AFB, Florida, on 25 January 2005. Assigned to the 46th Test Wing, serial 81-0989 is the first of 357 examples that will be updated under the USAF's $360 million Precision Engagement Capability modification programme by 2009. The programme involves 207 aircraft for the USAF, 50 for the AFRC and 100 for the ANG. Flown for the first time on 20 January, the prototype will support flight tests through the end of 2005. Weapons tests, which will be carried out by the 40th Flight Test Squadron (FLTS) under the auspices of Eglin's Air Armament Center (AAC), are planned for the second half of 2005.

Awarded in February 2001, the programme is a spiral development that will install a digital stores management system (DSMS), integrated capabilities for delivering precision-guided weapons, targeting pod integration, increased DC electrical power and joint-service battlefield interface via digital datalink.

Known as Spiral #1, the first phase integrates a MIL-STD-1760 databus, Joint Direct Attack Munition (JDAM), Wind Corrected Munition Dispenser (WCMD), Litening and Sniper targeting pods, and doubles the output of the DC power system. The DSMS provides the A-10C with a central interface control unit (CICU) and a pair of multi-function colour displays (MFCD). This equipment replaces the A-10A's armament control panel (ACP), interstation control unit (ICU) and television monitor. In addition, the A-10A control stick and throttle quadrant are replaced by improved hands-on throttle and stick (HOTAS) equipment. Although the A-10C will be capable of carrying up to six precision weapons, its normal configuration will comprise four stores. The incorporation of the new weapons will allow the USAF to task its premier ground attack aircraft against higher value targets than is currently possible. Under Spiral #2 the A-10C will be equipped with a new digital datalink and the Army joint tactical radio set (JTRS) with the enhanced position location reporting system (EPLRS).

The precision engagement modification is the first of several modifications intended to extend the Warthog's service life until 2028. Numerous follow-on projects include replacement of General Electric TF34-GE-100 engines with the more powerful TF34-GE-101 variant. The engine development effort will likely begin in 2006 and production incorporation is planned for FY09. Additionally, as part of the 'Hog-up' programme, the aircraft are receiving new outer wing panels and other modifications intended to extend service life from 8,000 to 16,000 hours. They are also receiving a new Integrated Flight and Fire Control Computer (IFFCC) developed by BAE Systems. The new computer's automated, continuously-computed weapon delivery function improves accuracy and reduces pilot workload. It also has a digital terrain system that uses terrain elevation data for ground collision avoidance and to aid in weapon delivery, and supports the incorporation of the precision engagement capability.

The contractor will initially modify 13 aircraft under the $74 million engineering manufacturing and development (EMD) contract and was expected to have received a production contract by March 2005. Lockheed Martin will subsequently deliver upgrade kits to the Ogden Air Logistics Center at Hill AFB, Utah, for installation. Each kit will take about 45 days to install, and the initial batch of A-10Cs should be fielded during the first half of 2006.

PROJECT DEVELOPMENT

International

A400 construction begins

Production of the first major airframe component for the prototype A400M military transport was formally launched on 26 January at the Varel facility of Airbus Germany. Representatives of the seven customer nations joined government officials and Airbus Military management members to witness milling of the first lower-fuselage frame.

Enforcing a rigorous timetable, first metal cut for the A400M came only 18 months after programme launch, in May 2003. A400M first flight is planned in 2008, with deliveries of 180 from 2009 to seven European NATO nations, comprising Belgium (7); Germany (60); France (50); Luxembourg (1); Spain (27); Turkey (10); and UK (25). South Africa is also applying to join the consortium and interest has been reported from Malaysia, to replace ageing C-130Hs.

Greece

Greece joins M-346 programme

The Hellenic aerospace industry plans to become a primary partner with Aermacchi in the latter's M-346 advanced transonic twin-turbofan lead-in fighter-trainer (LIFT) trainer programme, from a Memorandum of Understanding (MoU) signed by the Greek Defence Ministry in January. Currently in its early industrialisation phase, with the first of two prototypes flying since 15 July 2004, continuing towards type certification, the M-346 programme is fully supported by the Italian Government. Aermacchi is now identifying other potential European partners, for possible programme participation.

Hungary

Gripen roll-out and first flight

Formal roll-out took place at Saab's Linköping factory in Sweden on 25 January of the first of 14 Hungarian JAS 39EBS-HU Gripens for the 15-year lease-to-buy contract signed in early 2003. After extensive ground tests, the aircraft was first flown on 16 February by a Saab test-pilot, before completing further functional checks of Hungary-specific systems.

Production is also well advanced of another five Hungarian air force (LeRP) Gripens, for delivery from March 2006. After final assembly, the Gripens will undergo ground and flight acceptance tests by Saab and official FMV pilots. In January, the first five LeRP pilots started their year-long JAS 39 conversion training with Flygvapnet's F 7 at Såtenäs.

After graduation, they will return to Hungary as Gripen instructor pilots, although initial LeRP conversions will start in Sweden in mid-2006. First LeRP Gripen unit will be the 59th Fighter Wing, at Kecskemét, from March 2006.

AMRAAM for Gripen

A $25,389,904 US Letter of Offer and Acceptance (LOA) for acquisition of 40 Raytheon AIM-120C-5 AMRAAMs and associated equipment, was signed in mid-December by the Hungarian MoD. The contract is conditional on successful airframe/AMRAAM integration certification with new Gripens.

Work on AMRAAM/Gripen integration is already underway in Sweden, from a previous Raytheon/Saab/BAE deal, plus a Swedish Materiel Administration (FMV) March 2004 order for two AIM-120C-5s. The Gripen is currently certified only for earlier AIM-120B AMRAAMs, so more integration, related mainly to software, is required, prior to planned 2006-07 AIM-120C deliveries to Hungary. This will allow the LeRP to meet its NATO Quick Reaction Alert (QRA) tasks from January 2009.

The AMRAAM offer was part of a larger US Gripen weapons package comprising stockpiled AIM-9M Sidewinder AAMs, newly manufactured GBU-10/12/16 laser guidance kits for Mk 82/83/84 bombs, and Raytheon AGM-65H/K Maverick ASMs. These, however, were apparently beyond Hungary's budget. Hungarian MoD tender invitations were expected earlier this year, to fulfil LeRP Gripen requirements for short-range AAMs and air-to-ground weapons, for mid-2005 selection. Hungary favours the Bodenseewerk Gerätetechnik/Saab Dynamics IRIS-T with a helmet-cued sight, but funding considerations may limit LeRP short-range AAM procurement to stockpiled German or Swedish AIM-9L Sidewinders.

India

Su-30 programme progress

Recent major milestones in India's Sukhoi Su-30K/MKI procurement and production programmes included roll-out on 28 November of the first of 140 Su-30MKIs being built by Hindustan Aeronautics Ltd (HAL) at its Ozar (Nasik) air base facility. This preceded the 26 December delivery of the 50th and last Su-30 to various standards from Irkut production, comprising 38 Su-30Ks, including 12 with canard foreplanes, and the first dozen

Eurofighter Typhoon

In London, Defence Secretary Geoff Hoon said in December that Tranche 2 contracts will provide the RAF with another 89 aircraft costing £4.3 billion ($8.17 billion), representing a £48.3 million ($91.8 million) programme unit cost. With manufacturing plants in Germany and Spain, EADS is receiving Tranche 2 Typhoon contracts worth Euro 4.3 billion ($5.84 billion).

Associated weapons systems will be extended to include precision-guided missiles for additional limited ground-attack roles, for which RAF Typhoons will use Raytheon Paveway IVs with a 500-lb (227-kg) GBU-12-based Mk 82 warhead, and LITENING III targeting pods. Tranche 2 Eurojet EJ200 turbofans will incorporate new digital engine control and monitoring units (DECMUs). As a 36 per cent shareholder in Eurojet, the Rolls-Royce share of Tranche 2 component production for 519 EJ200 turbofans is worth over £750 million ($1.425 billion). Rolls-Royce is now also starting to assemble, test and deliver 195 RAF EJ200s at about 40 annually until 2010.

Except for the UK, other Eurofighter partners' Typhoons will include IRIS-T close-combat advanced AAMs in their weapons fit, following a 20 December German Federal Office for Military Technology and Procurement (BWB) signature with Nuremberg-based Diehl BGT Defence, of a Euro 1 billion ($1.36 billion) production contract. Diehl is prime contractor in the six-nation German-led IRIS-T programme, following original European Sidewinder AAM production under US licence.

Recent German press claims that Luftwaffe Eurofighters have only limited combat readiness after soaring R&D costs were rejected in January by the German Defence Ministry, which maintained that all development tasks were on schedule. The Ministry said that multi-national Development Aircraft have already proved Eurofighter's capabilities, including gun firing, air-refuelling, carriage of external fuel tanks, and air-to-air missile launches.

Full Typhoon day/night air-refuelling clearance was achieved late last year, after eight successful trial sorties completed by Alenia with the instrumented production IPA2 aircraft, from the AMI's Pratica di Mare air base, near Rome. The initial night sortie concentrated mainly on operating with only position lights from the AMI Boeing KC-707 tanker, and its trailing basket. The tests also validated the decision to exclude illumination of the refuelling probe, which presented no operating problems. Series production aircraft have all these capabilities, but release of relevant clearance to service pilots is a progressively staged process. This will include the DASS electronic self-defence system, due for full implementation by December 2006. The initial stage of DASS clearance was granted by the NATO-Eurofighter Agency in December 2004, when the first aircraft to this standard entered Spanish air force (EdA) service.

This view shows the Case White hangar at BAE Systems Warton, from where the RAF's first two units (Nos 17 and 29 Squadrons) have been operating eight Typhoon T.Mk 1s for initial training and operational evaluation duties. No. 17 moved to Coningsby in April.

Eight two-seat Eurofighters were then in Luftwaffe hands, one being used exclusively for maintenance training. The others are being operated in evaluations of operational issues, and their service release will gradually be upgraded to full production standards. EADS Military Aircraft delivered the first single-seat German production Typhoon to the Luftwaffe on 14 February, as one of more than 30 similar aircraft then received by Germany, Great Britain, Italy and Spain.

The EdA received its first single-seat Eurofighter from EADS Spain's Getafe site on 29 December 2004, as the first in the new standard configuration, featuring improved equipment performance, data-networking Link 16 MIDS, and enhanced software. The Italian air force (AMI) received its first production single-seat Typhoon in early 2005, as did the RAF. Eurofighter Typhoon s/n IS002 made its initial 65-minute flight by Alenia Defence chief test-pilot Commander Maurizio Cheli in December at the company's Caselle plant near Turin.

Su-30MKIs with added thrust-vectoring and advanced mission system avionics.

OAO Irkut Scientific and Production Corporation (NPK Irkut) president Aleksei Fedorov then announced that Indo-Russian Su-30 programme contracts totalled nearly $5 billion. He said that the first Indian contract, signed in November 1996, for an initial 40 Irkutsk Aviation Plant-built Su-30s, and worth nearly $1.5 billion, was extended in 1998 for another 10 Su-30s. All the earlier IAF Su-30Ks are scheduled for HAL upgrades to full Su-30MKI standards.

Interestingly, he referred to HAL's Su-30MKI licensed production programme, officially quoted in Delhi as costing Rs 22,122.78 crore, or $4.8 billion, as involving 140 aircraft, and not the reduction to 120 quoted in some recent reports. Rosoboronexport is also quoting 140 Su-30MKIs as HAL's production target, plus manufacturing Su-30 components for export customers, for completion by 2017.

Irkut listed Su-30MKI avionics as including the most advanced developments from Russia, India, France, Israel and the UK, integrated by the Ramenskoye Design Bureau (OKB). The Su-30MKI's NIIP Tikhomirov N011 Bars (Panther) radar has a phased-array antenna, allowing simultaneous tracking and attack of several air and surface targets. HAL also licence-produces some avionics, together with the Su-30MKI's thrust-vectoring Saturn/UMPO AL-31FP turbofans – the first to enter world operational service.

OAO Saturn Scientific Production Association (NPO Saturn) General Director Yuri Lastochkin has quoted new service life figures for these IAF engines and their rotating nozzle assemblies, comprising 1,000 and 500 flying hours, respectively, between major overhauls. Initial IAF Su-30MKI experience has reportedly indicated difficulties in reaching even these modest targets, and overall AL-31FP life is currently quoted at only 2,000 hours.

The Su-30s are among the main users of the IAF's Ilyushin Il-78MK tanker/transports, of which the sixth and last was delivered from the Tashkent Chkalov Aviation Production Association (TAPOiCh) in December, from a late 2001 $152 million contract. The IAF received its first Il-78 in March 2003, and air-refuelling has become a standard procedure to extend its operations.

Russia

Fifth-generation fighter

In a recent Moscow press interview, RFAF C-in-C General of the Army Vladimir Mikhailov claimed that flight development of the prototype of Russia's PAK FA fifth-generation fighter was planned to start in 2007. He said that Sukhoi computer modelling and design studies for the "completely new" PAK FA were well advanced, and in a recent visit to the company, he had told General Director Mikhail Aslanovich Pogosyan that the first flight must be in 2007, even if it was on 31 December.

Gripen developments: on 16 February Saab flew the first aircraft destined for Hungary (above). On 24 March the Gripen flew for the first time with the SaabTech-developed SPK 39 reconnaissance pod, intended for Flygvapen service. The pod was carried by JAS 39B prototype 39800 (below), which is retained by Saab at Linköping for development and demonstration purposes.

Air Power Intelligence

ZJ518/PA-02 is the second aircraft to be reworked to Nimrod MRA.Mk 4 standard. Recent announcements suggest that the number of conversions may be restricted to 12, down from the original figure of 18.

South Africa

First Denel Hawk flies
The first of 23 BAE Systems Hawk Mk 120 lead-in fighter trainers assembled in South Africa by Denel Aviation made its initial flight on 13 January from Johannesburg International Airport. It followed the sole UK-assembled Hawk in the SAAF contract, which has been in South Africa for some time, and is currently based at the Air Force's Bredasdorp Test Flight & Development Centre. It is now completing flight-tests of navigation and combat training systems designed and developed by the local Advanced Technologies and Engineering (ATE) company.

Denel's first Hawk was initially flown by South African industry test-pilot Dave Stock, together with BAE Systems test-pilot Gordon McClymont, on an 80-minute general handling sortie. Denel, which manufactures Hawk aerostructures, including stabilators and air brakes, is assembling 23 SAAF Hawks, on schedule for mid-

The USAF's YAL-1A Airborne Laser aircraft recently completed airworthiness testing at Edwards AFB, California. The conclusion of these tests means that the programme can proceed with integration and flight testing of the aircraft's complex mission systems. Initial integration testing involves the system's beam control system and will run through the summer, when the YAL-1A will again be taken out of service to allow the laser to be installed.

year initial deliveries at two per month.

With Armscor, ATE and Rolls-Royce, Denel is receiving much of the $8.7 billion of South Africa's new Hawk programme-associated economic benefits in defence, aerospace and civil sectors. SAAF C-in-C Lt Gen. Roelf Beukes recently visited the Hawk 120 final assembly line, and congratulated Denel and BAE Systems on their work progress.

South Africa joins A400M
The Airbus Military A400M multi-role mission transport aircraft programme gained an additional partner in December, when South Africa joined this already multi-national project. Pretoria's membership is accompanied by funding allocations for acquiring eight to 14 A400Ms between 2010-14. Costs for eight aircraft are quoted as Euro 837 million ($1.084 billion), or $135.5 million each.

Negotiations are progressing with Airbus Military to determine terms of agreement for South African participation, for which Denel and Aerosud are also discussing industrial partnership contracts. Previously, 180 A400Ms had been ordered in May 2003 by seven European NATO nations, to fly from 2008, with deliveries from 2009.

Current SAAF transport and airlift capability is achieved by survivors of 12 C-130B/F Hercules acquired from 1959, and now nearing retirement. The SAAF operates only three of nine remaining C-130s, although four are undergoing local Denel upgrades for continued operation, and two are receiving deep-level maintenance. They will need replacing by about 2010, and already over the past three years Pretoria has had to spend over R100 million ($16.3 million) for privately-owned airlift capabilities to deploy personnel, resources and materiel for African peace-keeping operations. The A400M will allow South Africa to upgrade its airlift capability, while helping to strengthen its aeronautical industry.

South Korea

RoKAF C-in-C flies T-50
A 51-minute flight by Republic of Korea air force (RoKAF) chief-of-staff General Lee Han-ho in one of Korea Aerospace Industries (KAI)'s four development T-50/A-50 Golden Eagle advanced/lead-in fighter trainers from Sacheon air base on 5 January was the type's 759th test-sortie. Accompanying KAI test-pilot Lt Col C.H. Lee, General Lee, as an experienced fighter- and test-pilot, attained a maximum speed of Mach 1.05, at up to 40,000 ft (12192 m), manoeuvres up to 5 g, and 77° maximum angle-of-attack in a high-alpha recovery demonstration.

The T-50 was designed and is being built jointly with Lockheed Martin, and Gen. Lee said it performed and handled much like the F-16. Production of the first 25 T-50s is now well advanced, and deliveries of another 69 will start in 2008. The RoKAF will receive its first production T-50s later this year, to begin operations in early 2006.

Three T-50 flight-test aircraft achieved a monthly peak of 70 sorties in December 2004, to complete about 70 per cent of overall development plans. On 27 and 29 December the T-50 reached its maximum design load factors of plus 8 g and minus 3 g, respectively. Earlier, on 10 December, the T-50 achieved its maximum operating speed of 675 kt (1250 km/h) calibrated airspeed, representing the aircraft's highest dynamic air pressure to date, and corresponding to 815 kt (1509 km/h) TAS and Mach 1.3. T-50 design limit speeds are 730 kt (1352 km/h) CAS and Mach 1.5.

United Kingdom

Hawk 128 contract signed
In December, BAE Systems was awarded an initial £158.5 million ($298 million) MoD Design and Development Contract (DDC), for the RAF's new Hawk 128 advanced jet-trainer (AJT). The MoD will work closely with BAE to manage design of the digital avionics architecture, introduce a modern glass-cockpit environment, and deliver two trials aircraft from BAE's Brough factory in Yorkshire, to support development and test-flying programmes.

Announced on 30 July 2003, the AJT contract will involve RAF receipt of 20 Hawk Mk 128 trainers, with options for another 24, for pilots selected to fly the Tornado GR.Mk 4, Harrier GR.Mk 7/9, Typhoon and, later, the F-35 Joint Strike Fighter. The total AJT programme is worth £3.5 billion ($6.58 billion), which includes £800 million ($1.5 billion) acquisition costs, and the expense of operating, supporting and maintaining the aircraft over 25 years. Exact aircraft numbers, delivery schedule and In-Service Date will all be finalised with the planned early 2006 main investment decision.

United States

New team to market C-27J
L-3 Communications has teamed with Alenia Aeronautica to form a new joint venture called Global Military Aircraft Systems (GMAS). The new team will market and support Alenia's C-27J tactical transport aircraft, which is viewed as a prime contender for the US Army's upcoming Future Cargo Aircraft (FCA) programme. The service plans to purchase as many as 128 FCAs for the active-duty Army and the Army National Guard. Alenia had previously teamed with Lockheed Martin for the C-27J development. GMAS will be located in Greenville, Texas, and will comprise L-3 Communications Integrated Systems and Alenia North America, which is a subsidiary of Alenia Aeronautica. Chrysler Technologies, which is one of L-3's legacy companies, was the prime contractor for the earlier C-27A project and delivered 10 examples of Alenia's earlier G-222 to the USAF in the early 1990s. No longer in service with the USAF, those aircraft are operated by the US Department of State in support of counter-narcotics operations.

Presidential helicopter selected
On 28 January 2005, Lockheed Martin Systems Integration and its Team US101 was named as the winner of the US Navy's Presidential helicopter replacement (VXX) competition to build the new 'Marine One' helicopter. It was awarded a $1.7 billion contract associated with the programme's systems development and demonstration (SDD) phase. The company's US101, which is a development of the AgustaWestland EH101 medium-lift helicopter, will replace the current fleet of VH-3D and VH-60N VIP helicopters assigned to Marine Corps Helicopter Squadron HMX-1.

Team US101 comprises Lockheed Martin, responsible for overall systems integration, AgustaWestland, Bell Helicopter and General Electric. These companies are responsible for aircraft design, aircraft production and construction of the helicopter's CT7-8E turboshaft engines. Although specific components will be built in Italy and

Aero India 2005

The show held at Bangalore in February showcased India's growing aerospace industry, and also some of its recent acquisitions. All three flying ADA Tejas Light Combat Aircraft flew together on the opening day, and the second technology demonstrator (left) performed a daily routine. The three aircraft were painted specially for the show with green and saffron wings. Also much in evidence was the HAL Dhruv Advanced Light Helicopter. Two Dhruvs of No. 201 Army Helicopter Squadron at Nasik Roads flew a daily assault demonstration (above), while Navy examples were also present. The Dhruv is also now in Coast Guard service.

Both prototypes of the HAL HJT-36 Intermediate Jet Trainer were on show, this being the second aircraft which first flew on 26 March 2004. The prototypes are powered by Snecma Larzac 04H20 engines, although production aircraft will use the NPO Saturn Al-55I.

A debutante at Aero India 2005 was the Kamov Ka-31 AEW helicopter now in service with INAS 339. Here it demonstrates the rotating radar antenna in the deployed position. The Indian Navy is examining the E-2C Hawkeye to expand its AEW capability.

India has traditionally imported combat equipment from both East and West. Illustrating this mix is an Ilyushin Il-78MKI refuelling a pair of Dassault Mirage 2000s. Indian 'Mainstays' also support Jaguars and Su-30s, while upgraded MiG-27s may also receive probes.

the UK, Bell Helicopter will carry out final assembly in Amarillo, Texas. The completed helicopters will subsequently be flown to Lockheed Martin's facility in Owego, New York, where systems integration and final delivery will take place.

Under the SDD phase, which runs through 2011, Team US101 will build eight helicopters comprising three test aircraft and five pilot production examples. The test aircraft and two of the latter will be built to the so-called 'increment-one' configuration, and will have increased range. The initial pair of production helicopters will achieve initial operational capability (IOC) in 2009. The final three aircraft and the 15 examples that follow will be delivered in 'increment-two' configuration. The total cost for the initial research and development phase is estimated to be $3.6 billion, while production of the remaining aircraft is expected to cost an additional $2.5 billion. Each of the helicopters is currently expected to cost $110 million.

Super Hornet radar delivered

Raytheon's Space and Airborne Systems (SAS) recently delivered the first low-rate initial production (LRIP) AN/APG-79 Active Electronically Scanned Array (AESA) radar to Boeing Integrated Defense Systems in St Louis, Missouri. The radar set is the first of 415 that will be installed in the F/A-18E/F Super Hornet and is part of the Super Hornet's Block II upgrades. The latest Super Hornet variant, which also features advanced displays and new navigation systems, is expected to reach initial operational capability (IOC) by September 2006.

C-130J news

USAF C-130Js made their operational debut during December 2004 when Air National Guard (ANG) and Air Force Reserve Command (AFRC) examples deployed to Southwest Asia as part of a joint airlift mission. Supported by personnel and aircraft assigned to the Rhode Island ANG's 143rd Airlift Wing and AFRC's 403rd Wing from Keesler AFB, Mississippi, the latter unit logged the C-130J's first USAF combat mission while deployed.

The 418th Flight Test Squadron at Edwards AFB, California, recently cleared the stretched C-130J model's airdrop envelope with the container delivery system (CDS). After conducting eight sorties over a five-day period, the aircraft was approved to airdrop nearly 40,000 lb (18144 kg) of pre-bundled equipment. The 418th had previously cleared the aircraft to deploy a maximum of 80 paratroopers during tests conducted at Fort Bragg, North Carolina.

The USMC's KC-130J aerial refuelling tanker recently passed its second operational evaluation (OPEVAL), and was determined to be operationally effective in both permissive and non-permissive environments. The OPEVAL included 32 sorties, flown against simulated threats at NAWS China Lake, California, and MCAS Cherry Point, North Carolina, which tested the aircraft's on-board electronic countermeasures. As a result of the earlier OPEVAL the KC-130J was already approved to operate in permissive environments.

The US Coast Guard recently supported low-level calibration tests at Robins AFB, Georgia, with two of its five C-130Js. The aircraft supported tests that will eventually certify the C-130J to operate at reduced vertical separation minimums (RVSM) and fly at more fuel-efficient altitudes. An EC-130J from the Pennsylvania ANG's 193rd Special Operations Wing (SOW) also supported the tests. The calibration tests will be followed by additional tests flown at altitudes from 29,000-41,000 ft (8839-12497 m). RVSMs allow a greater number of aircraft to operate in the same airspace simultaneously.

F-35A nears final assembly

Lockheed Martin will begin final assembly of the first F-35A Joint Strike Fighter at its Fort Worth, Texas, facility when its four major subassemblies are mated later this spring. Construction of the aircraft's centre fuselage is underway at Northrop Grumman's facility in Palmdale, California, while BAE Systems is building the aft fuselage and tails. Once the components arrive in Fort Worth they will be joined with Lockheed Martin-built forward fuselage and wings. Assembly of the initial conventional take-off and landing (CTOL) F-35A will be completed at the end of 2005 and the first flight is planned for August 2006.

Armed helicopter unveiled

The US Navy unveiled its MH-60S Armed Helicopter Weapons System (AHWS) at NAS Patuxent River, Maryland, on 8 December 2004. The third block in the spiral development for the MH-60S programme will provide expeditionary strike group commanders with a single platform capable of conducting search and rescue (SAR)/combat SAR (CSAR), maritime interdiction and surface warfare missions. Its configuration includes the AN/AAS-44C EO/IR system, eight AGM-114 Hellfire missiles and crew-operated, laser-sighted 7.62-mm and 0.50-in (12.7-mm) machine-guns. The aircraft is also equipped with ballistic floor armour and an integrated self-defence suite. Air test and evaluation squadron HX-21 will initially perform captive carriage and jettison testing for Hellfire missiles, as well as evaluate the integration of a new digital map system and a digital video recording system. The test programme will run through the summer of 2006, and the system will achieve initial operational capability in the autumn of 2006. Fleet squadrons will receive 126 armed helicopter systems.

Army joins ULB

Boeing has joined with the US Army's Aviation Applied Technology Directorate (AATD) to conduct a $1.6 million project that includes deploying weapons from its Unmanned Little Bird helicopter prototype. A variety of weapons are currently being considered for the programme, including the AGM-114 Hellfire missile, Advanced Precision Kill Weapon System (APKWS) rockets, and the 12.7-mm GAU-19A Gatling gun. Boeing's modified MD 530F helicopter, which is capable of manned and unmanned flight, entered flight-testing in September 2004. The contractor intends to pursue a variety of missions for the Unmanned Little Bird, including precision re-supply, communications relay, ISR and weapons delivery.

On 18 December 2004 the Bell 210 made its first flight at Bristol, Tennessee. The Model 210 is a refurbished UH-1H fuselage with a Honeywell T53-517B engine and the dynamic components from the Model 212. It satisfies the US Army's requirement for a Light Utility Helicopter (LUH) in the TDA (non-combat) units as well as the Army National Guard, and is being offered to Homeland Defense organisations.

UPGRADES AND MODIFICATIONS

Argentina

AT-63 programme resumed
Roll-out on 15 December of the first upgraded AT-63 Pampa armed jet-trainer at Lockheed Martin Aircraft Argentina SA's (LMAASA) Cordoba plant followed reactivation of this original $230 million year-2000 programme, suspended in 2003 through defence budget economies. Two five-year contracts awarded by the Argentine Government to LMAASA in February 2004 revived earlier plans to upgrade 12 existing Argentine air force (FAA) IA-63 Pampa intermediate jet-trainers as AT-63s, with a 3,500-lb (15.57-kN) thrust Honeywell TFE731-2C turbofan, 1553B digital avionics, a glass cockpit, HOTAS, and five stores pylons for additional weapons training and light ground-attack.

Also included is production of six new AT-63s for the FAA, and six more for international sales, plus maintenance and modification services for aircraft, engines and accessories. Flight trials were due to start in mid-2005.

One of the FAA's dozen IA-63s – actually the first prototype – was lost, together with its pilot, in a crash on 12 December near Punta Indio air base in the Buenos Aires region. Earlier, on 19 November, an FAA Bell UH-1H helicopter from VII Brigada Aérea crashed in a mountainous area near San Carlos de Bariloche, killing all four crew members.

Bulgaria

Helicopter enhancement moves
Selection was announced in December by the Bulgarian Defence Ministry of Elbit Systems Ltd, as prime project-contractor, and Lockheed Martin, as preferred bidders to upgrade 12 Mi-24 attack and six Mi-17 transport helicopters to NATO standards. However, in early 2005 this was cancelled due to contractual difficulties.

Defence Minister Svinarov signed a contract on 28 January for procurement of 12 AS 532AL Cougar and six AS 565MB Panther twin-turboshaft helicopters. The 12 Bulgarian air force (BVVS) Cougars will comprise eight for tactical transport, with the first three delivered by late 2006, and four equipped for Combat SAR roles.

The six Panthers for Bulgarian Naval Aviation (ANB) will undertake sea surveillance, ASW/ASuW and SAR missions. All 18 helicopters will be delivered by 2008. Eurocopter has now sold 835 Dauphin/Panther/EC 155 helicopters to 164 customers in 59 countries.

Indonesia

C-130s returned to service
US aid to Indonesia during tsunami relief operations in early 2005 included help in restoring over a dozen of the TNI-AU's 25 Lockheed C-130s to airworthy condition. Only nine C-130s – mostly C-130H/H-30s, apart from nine elderly C-130Bs – had been maintained in TNI-AU service, following long-term US embargoes on arms and military spares sales to Indonesia. Two technicians from Lockheed Martin's Greenville Air Logistics Centre, arrived on 16 January with spares to refurbish five C-130Hs at Jakarta's military airport. They were assisted by crews from the USAF's 517th Airlift Squadron at Elmendorf Air Force Base, Alaska, who arrived on 6 January with four C-130s and 120 personnel. Working with Indonesian technicians, they restored TNI-AU's first five C-130s to flight status and immediate service within about a week of their arrival.

New Zealand

Aircraft upgrades
L-3 Communications announced in December a long-awaited $NZ226 million ($161.4 million) contract award to its Spar Aerospace (L-3 Spar) Canadian subsidiary for a 15-year life-extension programme for the Royal New Zealand air force's five Lockheed C-130Hs. This includes replacement of mechanical, avionics and structural components, plus design and installation of flight-deck communications and navigation improvements. Modifications will begin in early 2006 on the first aircraft at L-3 Spar's facility in Edmonton, Canada, followed by the second to fifth aircraft at Safe Air Ltd's Blenheim facility in New Zealand.

The new contract immediately

Bell is offering an armed version of the Model 407 for the US Army's ARH (Armed Reconnaissance Helicopter) competition, and revealed this mock-up in February 2005. ARH partially replaces the cancelled Comanche programme.

followed NZ Defence Ministry selection in December of L-3's Integrated Systems subsidiary to upgrade mission and nav/com systems for the RNZAF's six Lockheed P-3K maritime-patrol/SAR aircraft. This involves supply by L-3 Communications' WESCAM subsidiary of electro-optical and infra-red (EO/IR) imaging-sensors for the RNZAF's P-3K Project Guardian upgrade programme.

RNZAF P-3Ks, already incorporating structural and some systems upgrades, will receive new mission-management systems and WESCAM MX-20 EO/IR imaging-turrets, plus ground-based support and training, from L-3 Communications' Integrated Systems Division from early 2005.

Two RNZAF Boeing 757-200s operated as passenger transports since replacing two Boeing 727-100s in early May 2003 are being modified and upgraded for additional cargo roles, to FAA Supplementary Type Certificate (STC) standards. Changes costing NZ$100-$200 million ($72.78-145.54 million) will include installation of a cargo door; RB211-535-E4B engine thrust increase; and upgraded civil and military communication, and navigation, surveillance/air traffic management (CNS/ATM) capabilities, from late 2005.

Peru

Russian aircraft upgrade plans
Peruvian government signature was expected late last year with Russia's Rosoboronexport arms agency for an approximately $35 million contract to return a number of FAP support aircraft, grounded from spares shortages, to operational service. Some eight FAP Antonov An-32 twin-turboprop tactical transports and 18 Mil Mi-17 helicopters are reportedly involved in this programme, from large numbers of various other Soviet-sourced types.

Plans to upgrade 17 remaining FAP MiG-29S/SE 'Fulcrum-Cs'; two two-seat MiG-29UB combat trainers; and 10/8 Sukhoi Su-25/UBs bought from Belarus in 1996-98, postponed last year through lack of spares and technical support, may also be revived, given the required funding. Delivered from 1976, the survivors of the FAP's 32/16 Sukhoi Su-20/22M2K 'Fitter-F/J' strike/interceptors and four Su-22UM3 'Fitter-G' two-seat combat trainers are overdue for further upgrades, after interim 1984 modernisation by SEMAN-Peru. One of the Su-22UM3s crashed on 17 December south of Lima, killing both crew-members.

United Kingdom

F.Mk 3 upgrades continue
Despite firm orders and initial deliveries of Eurofighter Typhoons to begin their replacement, further upgrades and technical support for the RAF's 130 or so Panavia Tornado F.Mk 3 air defence fighters are planned from a £25 million ($46.85 million) December MoD Smart Acquisition contract with BAE Systems. The new F.Mk 3 Sustainment Programme (FSP) will integrate the latest air defence weapons, comprising Raytheon's beyond-visual-range AIM-120C-5 AMRAAMs, and MBDA's advanced short-range ASRAAM FOC2 into the Tornado's mission system.

Also included are radar and on-board computer software changes to enhance missile-targeting through improved processing and display systems. The contract should deliver economies of over £14 million ($26.24 million) through rapid prototyping, a combined trials team, and lean engineering. As prime-contractor, BAE Systems is in partnership with QinetiQ, AMS, Raytheon and MBDA, to contribute avionics and other equipment, while overseeing the F.Mk 3 upgrade programme.

United States

C-130 for AMP modification
The first of more than 500 C-130s to be equipped with a new cockpit under the Avionics Modernization Programme (AMP), was inducted at Boeing's Aerospace Support Center in San Antonio, Texas, on 19 January 2005. The aircraft will serve as the prototype for the new avionics system that will provide the USAF, USMC and US Navy aircraft with a standardised cockpit configuration. The prototype is scheduled to fly for the first time with new avionics installed in early 2006. Once the programme has entered the production phase, modification will be carried out at the USAF's Ogden and Warner Robins Air Logistics Centers in Utah and Georgia, as well as at the Boeing facility.

Huey mods
The USAF has announced plans to equip 24 UH-1H helicopters operated by the 23rd Flying Training Squadron (FTS) at Fort Rucker, Alabama, with the Huey II upgrades. The US Army transferred the first of 40 UH-1Hs, which support the USAF's Specialized Undergraduate Pilot Training – Helicopter (SUPT-H) programme, to the service in May 2004. As part of the update the helicopter will be equipped with a Bell 212 nose section, tail and main rotor, new tail boom, a more powerful T53-L-703 turboshaft engine and a fully integrated glass cockpit. US Helicopters, located in Ozark, Alabama, will update the aircraft over a period of four years and the initial example is expected to make its first flight in June 2005. Lockheed Martin will conduct a military qualification in co-operation with the USAF. A new type designation is expected to be assigned to the updated helicopter.

PROCUREMENT AND DELIVERIES

Algeria

EC 225 VIP helicopter delivered
On 22 December Eurocopter officially delivered the first series production EC 225 Super Puma/Cougar executive helicopter to roll off the Marignane assembly lines to the Ministerial Air Liaison Group (GLAM) of the Algerian Republic. This twin-Turboméca Makila 2A turboshaft medium (11-tonne) helicopter has already been chosen by 32 heads of state or governments, since its initial flight in November 2000.

Australia

Airbus MRTT contract signed
RAAF selection last April of five Airbus A330-200 multi-role tanker/transport (MRTT) aircraft to replace its three ageing Boeing 707s in No. 33 Squadron was followed in December by a $1.4 billion Australian Defence Department procurement contract with European Aeronautic Defence and Space Construcciones Aeronauticas SA (EADS CASA) in Spain. The RAAF thus becomes the first customer for the twin General Electric CF6-80E1 turbofan-powered A330 MRTT, which can deliver 65 tonnes of fuel in its air-refuelling role, some 1,000 nm (1853 km) from base, following up to two hours on-station. Fuel can be offloaded through Cobham UK Flight Refuelling Ltd (FRL) 905E underwing hose-and-drogue pods and control systems, plus a rear-fuselage advanced EADS fly-by-wire refuelling boom.

All A330 MRTT fuel is stowed entirely within existing wing and tail tanks, leaving the cabin completely free for up to 272 seats in troop transport roles, and its cargo-hold available for military equipment on pallets or containers. EADS CASA will undertake MRTT conversion of the first Airbus A330-200, with QANTAS modifying the remaining four in Australia, joined by other industrial partners including Australian Aerospace, ADI and GKN.

Already a major Airbus A330 operator, QANTAS has also been chosen by the ADF to provide logistic support for the aircraft, including training, servicing and heavy maintenance, over 20 years. Work performed by Australian industry should exceed $A500 million ($387 million) over service lives.

The A330-200 is the Airbus company's second MRTT conversion, following similar A310 adaptations for the Canadian and German air forces. These also use FRL's 900 series hose-drum pods, as do Swedish Lockheed KC-130Hs and USAF Special Operations MC-130Hs. RAAF A330 MRTT deliveries will start in 2008-09, to refuel Boeing F/A-18s, Boeing 737-700 AEW&C and Joint Strike Fighter aircraft, while retaining a significant strategic airlift capability.

Brazil

New presidential Airbus
Delivery was scheduled on 16 January of an Airbus A319 Corporate Jet ordered from a $56.7 million contract for presidential and government official transport. It replaces a 47-year-old Boeing 707, christened by a former Brazilian president as 'The Junk Heap'.

Chile

More Vipers possible
The Chilean AF is reportedly looking to acquire 20 to 28 used F-16s in order to replace some or all of its Mirage 50M Pantera fighters. The aircraft would be operated alongside 10 new Block 50 F-16C/Ds that will be delivered to Chile in 2005/6.

China

Defence budget increases
For the 15th successive year, China's defence spending plans again involve double-digit growth, and while Beijing's official 2004-05 figures quote a military budget of 211.701 billion yuan ($25.5 billion), US estimates of actual current-year funding are nearer $55 billion. If space and satellite programmes, oceanographic research, and National Oil Corporation security costs are included, the true total could reach $70 billion. Defence expenditures in 2002 and 2003 were 170.778 billion and 190.787 billion yuan, respectively, representing about 1.63 per cent of GDP, with further growth expected this year.

Priority in equipment funding is being given to AF/PLA procurement and local production of the Sukhoi Su-27 and Su-30. Following initial KnAAPO deliveries from 1991-96 of 38 Su-27SKs (locally J-11s), AF/PLA orders followed for an eventual total of 40 IAPO-built two-seat Su-27UBK (JJ-11) trainers, and 76 two-seat Su-30MKK (J-13) multi-role fighters with new digital mission systems from KnAAPO.

Orders for 24 further upgraded Su-30MK2s then followed in January 2003, for delivery from August 2004. Deliveries are also well advanced since 1998 of 200 Shenyang-assembled 'Flankers', of which only the first 80 were Su-27SKs, before production switched to upgraded Su-30MKKs.

Chinese Su-27/30 production does not include the Salyut AL-31FN turbofans, of which contracts worth $900 million for nearly 250 more of these engines were also agreed last December with Rosoboronexport. These AL-31s are required for installation in Chengdu's (CAC) J-10 development of the single turbofan-powered Israeli Lavi, in series Chinese production since 2002. Salyut General Director Yu.S. Eliseyev said that, apart from the J-10 contract, China was interested in acquiring upgraded AL-31F engines for its Sukhoi fighters, costing $1.2-1.4 billion over a six-year period.

Colombia

Light ground-attack aircraft
The Colombian air force (FAC) recently issued requests for proposals relating to long-standing FAC requirements for 24-36 new single-turboprop aircraft, to replace its current force of seven Cessna OA-37B and nine Rockwell OV-10A light ground-attack types. Most of these are currently unserviceable, although the Broncos have been undergoing major overhauls in the US. Some $234 million has been approved for the new light-strike programme, for which at least four bidders were expected to respond. These were reported to include Raytheon-Beech, with the T-6A, recently certified for light combat stores; EMBRAER's EMB-314 Tucano; Pilatus PC-21; and the Korean Aerospace KO-1.

The FAC also has a requirement for SIGINT aircraft, similar to Brazil's specially-equipped EMBRAER EMB-145s. Recent FAC upgrades have included a seventh Douglas AC-47 Dakota gunship, with a new fire-control system, FLIR and NVGs for night-attack roles. Five of these have also undergone Basler PT6A-turboprop conversions, as AC-47T Fantasmas (Phantoms), from a $6.4 million US aid programme. This has totalled $3.7 billion since 1997, compared with Colombia's 2004 defence budget of less than $3 billion.

New equipment for the Iraqi Air Force

On 16 January 2005 the USAF delivered three C-130Es (above) to the Iraqi AF. The aircraft are assigned to 23 Squadron, at Al Muthana AB in Baghdad. An Iraqi C-130 crew undertook its first operational mission on 12 February when it transported the country's Prime Minister on a round trip between Baghdad and As Sulaymaniwah West.

On 17 January a pair of SAMA CH2000 aircraft was delivered to the air force's 70 Squadron in Basrah. Deliveries will continue at a rate of two aircraft per month through the spring of 2005. Kirkush AB is now home to seven Aerocomp Comp Air 7SL light utility aircraft (above) assigned to 3 Squadron. The unit will eventually relocate to Al Muthana as well.

In early February the air force's 2 Squadron received a pair of refurbished UH-1H helicopters (below) from the Jordanian government, which were delivered to Taji AB. A total of 16 Hueys will be delivered by February 2006. The helicopters will be evenly divided between 2 and 4 Squadrons, both at Taji.

Denmark

EH101 deliveries begin
Deliveries are starting of the 14 RDAF EH101s being produced at AgustaWestland's Yeovil factory, and officially named Merlin Joint Supporters late last year by Denmark's Air Materiel Command. The new name reflects the EH101's multi-role capabilities in supporting all three Danish services, additional to its basic SAR and tactical transport roles.

Egypt

More C-130s
Having taken delivery of 23 Lockheed C-130Hs and three stretched C-130H-30s by 1978 and 1990, the Egyptian air force (EAF) is receiving another three (former Danish) C-130Hs and related spares from a new $30.69 million US Foreign Military Sales fixed-price contract announced on 10 December 2004. The aircraft are being refurbished by Lockheed Martin Marietta.

France

First EC 725s delivered
Deliveries started earlier this year to

Air Power Intelligence

Brazilian rarities

Above: The Hawker Siddeley HS-748 (C-91) was still in FAB use in late 2004, this aircraft having received a new grey scheme.

Based at Pirassununga, the Brazilian air force's Clube de Voo a Vela (CVV) is part of the Academia da Força Aérea. It has a number of gliders on strength, with tugs in the shape of the Aero Boero 180 (above) and EMBRAER EMB-201R Ipanema (below). The Aero Boero is an Argentine-produced light aircraft, while the EMB-201R is a dedicated glider-tug version of the locally-produced agricultural aircraft. Another local product, the Aeromot Super Ximango motor-glider (below left), is also used.

the French air force (AdA) of the first of six Eurocopter EC 725 Cougar Mk II+ twin-turboshaft helicopters equipped for combat search and rescue (CSAR). Another eight EC 725s are also being delivered by 2006 for operation by French special forces.

India

IAF considers F-16 procurement

According to newly-appointed IAF C-in-C Air Chief Marshal S.P. Tyagi, Lockheed Martin F-16s may be included in India's planned issue of requests for information for its long-term $3 billion+ programme to acquire 126 new multi-role combat aircraft, as MiG-21 replacements. The IAF is also considering the Dassault Mirage 2000-9, upgraded MiG-29SMT or M2 MRCA, and JAS 39C/D Gripen. Recent removal of long-standing US arms embargoes on both India and Pakistan has broadened the sub-Continent's weapons sources, from earlier almost-total reliance on Soviet, Russian-bloc and Chinese products, with which strong ties will undoubtedly continue.

The new fighter programme, also involving Indian industry participation, is continuing despite progress, albeit slow, with India's own Light Combat Aircraft (LCA) Tejas project, in which two F404-powered technology demonstrators and the similarly-engined first prototype had completed over 360 test sorties by early 2005. Four more similarly-powered prototypes and the first eight production LCAs are due to follow, before 200 single-seat and 40 two-seat IAF versions planned with indigenous GTRE Kaveri turbofans.

In February, Air Chief Marshal Tyagi said that the IAF had signed an initial Rs 4,000 crore ($922 million) contract with HAL for 40 LCAs, including options for 20 two-seaters. Resumption of Kaveri high-altitude operating tests within revised development costs of Rs 2,800 crore ($645.37 million), had been approved, according to Dr K. Ramachandra, Director of India's Gas Turbine Research Establishment (GTRE). They will continue in Russia below the fuselage of a test-bed twin-turbojet Tupolev Tu-16 'Badger'. Production Kaveris would be available by 2009.

Among other Indian military engine news, after extensive evaluations of similar powerplants, selection was announced in late 2004 of the 19.57-kN (4,400-lb) Salyut Lyulka AL-55I turbofan from international tenders for planned production of HAL's HJT-36 Intermediate Jet Trainer (IJT) Kiran replacement. A prototype HJT-36, with a single SNECMA Larzac 04-30 of similar power, has been flying since March 2003 from a Rs 1.8 billion ($41.5 million) R&D contract for 211/24 IAF/Indian Navy procurement requirements. Two AL-55Is are also being proposed for HAL's projected HJT-39 combat air trainer, as a BAE Hawk 132 follow-on.

IAF AEW&C programme

As expected, EMBRAER's announcement in February of a Memorandum Of Understanding (MOU) signed with the Indian Defence R&D Organisation (DRDO), to support a new Airborne Early Warning & Control (AEW&C) development programme using indigenous radar, could involve IAF acquisition of another three EMB-145 Legacy twin-turbofan transports as associated platforms. These would follow 2003 Delhi government orders for five EMB-145s, including four for IAF official transport roles, and the fifth for India's Border Security Force. Equipped with defensive aids sub-systems, including chaff/flare dispensers, the first EMB-145s are expected in India within months.

Last year, the Indian government cleared DRDO proposals to develop and integrate the new AEW&C system with indigenous radars, for a total Rs 18 billion ($415 million) expected programme cost. This resumes a long-term DRDO airborne radar development programme, temporarily halted by the loss of a HAL/Avro 748 test-bed, fitted with a large dorsal radome, and its crew of eight pilots and technicians, on 11 January 1999. The new platforms will be based on the EMB-145 intelligence, surveillance and reconnaissance (ISR) version already ordered for the Brazilian and Greek air forces.

In India, they will supplement an initial three Russian-supplied Il-76-based Beriev A-50EhI AEW&C aircraft costing $1.1 billion, with Israeli Elta EL/M-2075 Phalcon phased-array radar in a fixed dome. IAF deliveries start in 2006, with up to five due by 2008. Pakistan also claims that India is planning to acquire 50 unmanned aerial vehicles (UAVs) and other intelligence-gathering equipment from Israel.

MiG-29K contract details

Defence Minister S.P. Mukherjee recently confirmed details of the contract signed on 20 January 2004 with Rosoboronexport for procurement of 16 carrier-based MiG 29K combat aircraft for the Indian Navy. Costing $740.35 million (unit price, $46.27 million), their delivery will start in June 2007, to provide integral air defence, strike and army support capabilities at extended ranges.

Ireland

More helicopter orders

After competitive evaluation from mid-2004 tenders, Irish Defence Minister Willie O'Dea signed a Euro 49 million ($66.57 million) contract with AgustaWestland in January to acquire four new Air Corps AB139 utility helicopters. These are due from Bell Agusta Aerospace in 2006-07, for IAC operational, VIP transport, air ambulance and training roles. The IAC is also acquiring two more light utility Eurocopter EC 135s for general support roles, from an additional Euro 11 million ($14.94 million) contract.

Italy

C-130J deliveries completed

A stretched Lockheed Martin C-130J-30 which flew into the 46ª Brigata Aerea's Pisa air base on 10 February, was the last of 22 Super Hercules ordered between 1997-2000 for the AMI. This received 10 of the stretched -30s (including the 100th C-130J delivered world-wide), equipping the 50° Gruppo, as well as 12 standard-length C-130s, to replace some 10 remaining early 1970s C-130Hs in the 2° Gruppo.

Six shorter Hercules will be operated as KC-130J tankers, for a new AMI capability. Using only wing and external tanks, the AMI KC-130Js, similar in basic configuration to US Marine Corps tankers, have a 57,500-lb (26082-kg) fuel off-load capability on a 500-nm (926-km) radius mission. The KC-130J can also accommodate a fuselage tank, which adds another 24,392 lb (11064 kg) available fuel off-load.

Tanker unveiled

Boeing rolled out the first of four KC-767A tanker transports, destined for the Italian AF, at its facility in Wichita, Kansas, on 24 February 2005. Based on the 767-200ER commercial transport, the tanker will be delivered in April 2006 at the conclusion of flight-testing and certification. Built at Boeing's Everett, Washington, facility, the 767 was flown to Wichita where the structural modifications were carried out. Powered by a pair of General Electric CF6-80C2 engines, the tanker is equipped with a Boeing air-refuelling boom, remote air-refuelling operator station (RARO), wing-mounted and centreline hose-and-drogue refuelling systems and a refu-

Formed in 1986, Brazil's army aviation currently operates four types of helicopter. For observation and light attack duties the HA-1 (right) is used, an initial 16 Helibras-built HB 350L1s being supplemented by 20 Eurocopter-built AS 550A2 Fennecs. For assault transport the Exercito Brasileiro uses the Eurocopter AS 532UE (HM-3, below right), of which three are in use, the Eurocopter AS 565AA Panther (HM-1, below left, 35 delivered) and the Sikorsky S-70A-6 (HM-2, above). Four Black Hawks have been delivered and another 10 are on order. Fixed-wing equipment comprises EMB-110 Bandeirantes, including some on loan from the air force, and Piper-designed, EMBRAER-built lightplanes.

elling receptacle. The aircraft also features a cargo door and a 'combi' interior that allows passengers or cargo, or both simultaneously on the main deck. Aeronavali, a division of Alenia Aerospace/Finmeccanica will equip the remaining three aircraft with the tanker modifications locally in Italy. The second 767, which flew for the first time on 28 February, will be delivered to the contractor in May 2005.

Jordan

AMRAAMs approved

Despite protests from Israel, the US DoD has given its approval for the sale of 50 AIM-120C AMRAAM air-to-air missiles to Jordan in a deal worth about $39 million. The missiles will equip Jordan's growing fleet of F-16A/B fighters.

Nepal

ALHs ordered

Already operating 11 helicopters of five different types, including five Mil Mi-17s delivered in 2000, the Royal Nepalese Army Air Service is receiving two multi-role HAL Dhruv advanced light helicopters (ALHs), reportedly costing Rs 250 million ($5.76 million) each. Powered by two 801-kW (1,074-shp) Turboméca TM333-2B turboshafts, some 35 ALHs have so far been delivered to Indian armed forces and government agencies since mid-2002. About half of Indian army aviation's planned 120 Dhruvs will incorporate integrated weapons systems for attack missions.

The 13-seat Dhruv, which incorporates about 60 per cent foreign components, has been demonstrated and evaluated as far afield as Chile, where President Ricardo Lagos recently expressed interest in acquiring four or more land and naval versions. The Israeli air force (IDF/AF) also received a single ALH last December, and HAL production is progressing towards one per week. HAL and Turboméca are also co-developing a higher-powered Shakti version of the TM333 for improving hot and high-altitude performance.

Portugal

EH101 deliveries begin

Formal delivery took place at AgustaWestland's Vergiate facility last December of the first of 12 AgustaWestland EH101 Merlin helicopters ordered by the Portuguese government in 2002 for combat SAR role and fishery protection. The FAP is expecting all its EH101s in 2005.

Singapore

Seahawks for naval aviation

Formation of a naval aviation element of the Singapore Republic armed forces was signalled in January, from MINDEF contract signature with Sikorsky to acquire six new S-70B Seahawk helicopters. Equipped with advanced anti-surface/submarine warfare sensors and weapons, the S-70Bs will operate from Singapore's new frigates, following scheduled Seahawk deliveries between 2008-10.

South Korea

E-X AEW programme revived

Announcing revival on 14 February of the suspended $1.9 billion RoKAF E-X programme for four new airborne early warning aircraft, the Ministry of National Defence (MND) said that at least three bidders were sought to maintain a competitive price basis. The RoKAF's original requirements for an aircraft with a maximum speed of over 300 kt (556 km/h), an operating height of 26,500 ft (8077 m), and endurance of at least six hours, with a 360° radar capability to detect objects within a 200-nm (370-km) radius, attracted only two bids.

These comprised a Gulfstream G550, with an Israeli Elta radar, and a Boeing 737 AEW&C, but the MND said that the former's detection capabilities fell short of RoKAF range requirements. Bidders were being invited again in late March, for selection next December after an open competition, and planned procurement of two AEW&C aircraft by 2010 and two more by 2012. Initial funding of about 94 billion won ($89 million) has been allocated in the FY2005 budget for the RoKAF's AEW programme, to end current dependence on USAF services.

Orion fleet expansion plans

RoK Naval Aviation plans to double its fleet of Lockheed P-3C Orion maritime patrol/ASW aircraft from selection announced in December by Seoul's National Defence Ministry of US L-3 Communications to join an industry team headed by Korea Aerospace Industries (KAI). Its task is to complete mission system modernisation and service life extensions of eight P-3Cs, known as 'Lot II'.

These are being procured from US Navy stocks, and with additional ground support systems, will supplement the eight P-3Cs, which were the last Orions to be built, operated by the RoKNA since 1994. Contract details remain to be finalised, but L-3 IS offers experience in upgrading over 300 Lockheed P-3s since the 1960s.

Indigenous helicopters planned

Multi-billion dollar National Security Council plans were approved by the Seoul government on 18 January to limit foreign procurement and implement joint development of indigenous transport/utility and attack helicopter projects. These involve national design, development and production of 300 transport helicopters by 2010, and 200 attack helicopters by 2012, following July 2004 rejection by the RoK's Board of Audit and Inspection of foreign imports on cost grounds.

Priority is being given to replacing dwindling totals of RoK army and air force Bell UH-1Hs and other types, through the proposed Korean Multi-purpose Helicopter (KMH) programme, costing 8 to 13 trillion won ($7.6 to 12.4 billion). KMH development is planned in partnership with foreign manufacturers, for which three companies – Eurocopter, AgustaWestland and Bell – were shortlisted from five initial bidders last July for further negotiations. Progress with the KMH programme will determine whether to follow-on with attack helicopter development.

Spain

More VIP helicopters

EdA's government transport Escuadrón 402, based at Madrid's Cuatro Vientos air base, was reinforced last December by deliveries from Eurocopter España of two AS 532UL Cougars. The two VIP helicopters, which were ordered in June 2003, will join the existing EdA fleet of four Eurocopter AS 332M1s and two AS 332B Super Pumas, for transport of senior officials. The AS 532ULs are receiving full Spanish civil airworthiness-type certification, and their Euro 33 million ($42.75 million) contract includes support services, training and spares.

Thailand

Army delivery

Sikorsky Aircraft delivered a pair of S-70A helicopters to the Royal Thai Army at its facility in Stratford, Connecticut, recently. The delivery, which was funded via a foreign military sales (FMS) contract, increases Thailand's fleet to five Sikorsky Blackhawks.

In a ceremony aboard HTMS Taksin at Sattahip naval base on 7 February 2005, the Royal Thai Navy formally accepted its first Super Lynx 300 for service with 203 Squadron at U-Tapao. The aircraft will primarily operate from the 'Narasuan'-class frigates, but are also capable of operating from all other Royal Thai Navy ships with aviation facilities.

Turkey

S-92 ordered
The Turkish government has placed an order for a single Sikorsky S-92 helicopter that will support head-of-state missions. The order follows Turkmenistan's decision to purchase a pair of S-92 VIP helicopters.

ATR 72 for MR/ASW roles
Selection by Turkey's Undersecretariat for Defence Industries for Turkish Navy Aviation (TCBH) and procurement of 10 twin-PWC PW127E-turboprop ATR 72MP-500 maritime patrol and anti-submarine surveillance aircraft, was announced by Alenia Aeronautica in January. Contract negotiations were then started for the cost-competitive medium-range ATR-72s, equipped with comprehensive Thales mission systems.

These have been selected by several countries, including France, Japan and Turkey. Designed for maximum flexibility, Thales Airborne Maritime and Situation Control System (AMASCOS) meets current and future operational requirements in search-and-rescue, EEZ surveillance, anti-surface and ASW roles. With its modular design, AMASCOS integrates an advanced tactical command sub-system, with a broad range of sensors including Ocean Master Mk II radar, ESM, acoustic and communication systems.

AMASCOS is part of the $400 million September 2002 MELTEM contract, for which Thales is supplying 19 systems to the Turkish Navy and Coast Guard, for installation in nine previously-ordered Airtech CN-235Ms and 10 new-build aircraft, from TAI production.

Ukraine

An-74s to Egypt and Libya
Delivery is planned this year by the Kharkov State Aviation Production Enterprise (KhGAPP), of five twin-turbofan STOL tactical transports to the governments of Egypt and Libya. Egypt will receive three Antonov An-74T-200A cargo aircraft of nine ordered in 2003 for an unspecified agency, with maximum payloads of 10000 kg (22,045 lb) of freight. Libya will take delivery of two An-74TK-200S versions with convertible interiors for up to 52 personnel. Certification tests of the first An-74T-200A were completed on 24 December.

United Arab Emirates

AB139 helicopters ordered
Agusta/Bell Aerospace received a contract award in February from the United Arab Emirates air force (UAEAF) for eight AB139 medium-twin helicopters, costing about $83 million. Six AB319s will be configured for search and rescue operations, with the remaining two assigned to VIP transport. The UAE is the third country to choose the AB139 for SAR applications, following previous governmental orders from Oman and Ireland.

United States

Army ER/MP UAV progress
Northrop Grumman and General Atomics are conducting ground and flight demonstrations of the Hunter II and Warrior UAVs at Fort Huachuca's Libby AAF in Arizona. The demonstrations are in support of a six-week evaluation associated with the US Army's Extended Range/Multi-Purpose (ER/MP) UAV programme. Three Hunter II UAVs that are supporting the effort include one example that is powered by a heavy fuel engine (HFE). Northrop Grumman concluded its initial flight-testing of the Hunter at Cochise College Air Field in Douglas, Arizona, in mid-January. Those tests verified the UAV's endurance, communications and air-to-ground surveillance capabilities, and paved the way for the army evaluation. The twin-boom Hunter II is based on the Heron medium-altitude long-endurance (MALE) UAV built by Israeli Aircraft Industries, but features advanced avionics developed by Northrop Grumman. Based on the RQ-1 Predator, the Warrior, which is powered by a Thielert Centurion heavy fuel engine (HFE), flew for the first time in October 2004. The HFE will allow the UAV's to operate at altitudes above 25,000 ft (7620 m) on either jet or diesel fuel. The evaluations are part of the ER/MP UAV programme's Phase I System Concept Demonstration (SCD) phase and will support the service's selection of a single UAV. A Phase II system development and demonstration (SDD) contract will be awarded to the winning contractor in April 2005.

Helicopter orders
Acting on behalf of the US Navy, the US Army Aviation and Missile Command (AMCOM) recently awarded Sikorsky Aircraft a $180.8 million modification to an existing contract for 15 MH-60S Knighthawk helicopters.

Sikorsky Aircraft recently received two contracts associated with the UH-60 Blackhawk. A $17.2 million modification to an existing contract covers the production of two UH-60Ls and a second contract covered the purchase of eight UH-60Ms at a cost of a $244.7 million. The latter batch was included in the 2005 defense appropriation.

Bell Helicopter was recently awarded a $12.4 million US Army contract covering the purchase of seven TH-67A+ training helicopters that will built at the contractor's facility in Quebec, Canada.

The US Army AMCOM recently awarded Boeing subsidiary McDonnell Douglas Helicopter a $29.5 million modification to an existing contract covering the remanufacture of six AH-64D Aircraft. The helicopters will be delivered by August 2007.

Boeing has received a pair of contracts from the US Army AMCOM and the US Army Aviation Applied Technology Directorate (AATD) associated with the Chinook helicopter. Under the first, which is valued at $549 million, Boeing will build 17 new CH-47F helicopters. The order includes seven aircraft that were funded by the FY03 supplemental defence bill and 10 authorised by the FY05 bill. Deliveries of the new aircraft will begin in September 2006. The second contract covers the remanufacture of 12 CH-47Ds to the MH-47G configuration and is worth $194.4 million. The Army is remanufacturing its existing fleet of 397 CH-47Ds to the CH-47F configuration and plans call for the purchase of at least 55 new production aircraft. Additionally, 34 MH-47D/Es are being updated to MH-47G configuration and there are plans to purchase 27 new examples. The service will eventually operate a fleet of 513 CH-47Fs and MH-47Gs.

Clipper news
The USAF has ordered three additional C-40C transports from Boeing. Based on the 737-700 Boeing Business Jet, the aircraft will enter service with the Air Force Reserve Command's 932nd Airlift Wing (AW)/73rd Airlift Squadron (AS) at Scott AFB, Illinois, in 2007. The 932nd currently operates a similar number of C-9As.

Boeing delivered the eighth C-40A airlifter to the US Naval Reserve on 18 February 2005. The Navy began replacing its fleet of 29 C-9Bs with Clippers in 1997 and has now ordered nine C-40As.

UAVs ordered
The US Army AMCOM recently exercised a contract option with AAI that covers the purchase of eight additional RQ-7 Shadow Tactical UAV (TUAV) systems at a cost of $71.9 million. The second contract is worth $17.4 million and covers a single RQ-7 system. Each TUAV system includes four RQ-7B air vehicles and two ground control stations. The RQ-7B variant was initially fielded by the 572nd Military Intelligence Company (MICO) at Fort Wainwright, Alaska during 2004

More QF-4 targets
The USAF Air Armament Center has awarded BAE Systems a $21 million

Above: The first F-15K for the Republic of Korea AF made its initial flight at Boeing's Lambert Field facility in St Louis, Missouri, on 3 March 2005. Boeing will deliver 40 F-15Ks to the ROKAF, between 2006 and 2008, under a $3.6 billion contract.

Right: The first F/A-22A to wear the 1st Fighter Wing's 'FF' tail code was unveiled during a ceremony at Langley AFB, Virginia, on 11 February 2005. The wing had already received 91-4005 (the fifth Raptor) for use as maintenance trainer.

Air Power Intelligence

A number of the RAF's most 'senior' squadrons are celebrating their 90th anniversaries during 2005. Tornado GR.Mk 4/4A operators Nos 13 (left) and 14 (below) Squadrons have produced commemorative schemes based on their unit markings. No. 13 Squadron was formed on 10 January 1915 and was fighting on the Western Front by October. No. 14 was formed on 3 February and was sent to Egypt and Palestine, a region with which it was closely associated until 1944.

contract that covers the production, delivery and support of 17 Lot 11 QF-4 Full-Scale Aerial Targets (FSATs). BAE Systems has converted 170 F-4s into FSATs, and 50 more are currently on order. The aircraft from the latest order will be delivered by August 2007.

US Government

The US Federal Aviation Administration (FAA) recently ordered a Bombardier Global 5000, which will be converted for use as an airborne research and development laboratory. Dyncorp, Global Helicopter Technology and Pratt & Whitney Canada recently delivered the first UH-1H helicopter, re-engined with a PT6C-67D turboshaft engine, to the US Customs and Border Protection's San Diego Border Patrol Sector at Brown Field in California. Border Patrol is expected to modify 10 additional Hueys. As well as the new engine, the aircraft is equipped with a tail rotor enhancement kit.

Venezuela

US down-plays MiG-29 plans

Press reports late last year that Venezuelan President Hugo Chávez was planning to acquire 50 Russian MiG-29s, risking a new arms race in Latin America, were played down in Washington. Roger Noriega, senior State Department official for Latin American affairs, said that there were other countries in the region with similar weapons, apparently referring to previous Soviet arms purchases by Cuba and Peru, which led him to believe that Venezuela's plans would not necessarily represent a military escalation.

FAV requirements were quoted as 40 single-seat MiG-29SMTs and 10 two-seat MiG-29UBTs in a comprehensive $5 billion package, which included Vympel R-73 (AA-11 'Archer') close-combat agile IR-guided AAMs and R-27 (AA-10 'Alamo') medium-range radar-homing AAMs, plus air-to-

surface weapons. Venezuelan defence budget restrictions would prevent outright purchase of the proposed Russian arms, for which an initial payment of around $500 million would be followed by trade agreements for supply of commodities and raw materials, such as aluminium, for the balance.

The wisdom of these proposals has also been queried in Washington, where it was pointed out that the FAV funding problems mean that it can barely keep its 22 F-16s airworthy. Its most pressing needs are to replace its dozen or so remaining Rockwell OV-10A Bronco light ground-attack aircraft, supported by EMBRAER EMB-312 AT-27 Tucanos and the first of a dozen Brazilian-built two-seat AMX-Ts now being ordered, for close-support and drug-interdiction roles. Plans were announced in February 2005 to order 12 EMBRAER EMB-314 Super Tucanos from an initial $110 million contract, plus 12 options, to meet this requirement.

AIR ARM REVIEW

Brazil

F-X fighter tenders cancelled

Bidders for the FAB's $700 million F-X combat aircraft requirement, including Dassault Mirage 2000, JAS 39 Gripen, Lockheed Martin F-16, MiG-29 and Sukhoi Su-35 submissions, were notified in February that all tenders were cancelled. Procurement is now planned from alternative approaches.

China

Surveillance aircraft revealed

The Chinese government recently announced that it had begun testing an indigenously developed airborne warning and control system that has been installed on a Russian-built Il-76 airlifter. China plans to use the AWACS aircraft to patrol the 100-mile (161-km) wide Taiwan Strait. China also has been testing an Erieye-style side-looking radar in a fairing mounted above a Shaanxi Y-8 (An-12).

Ivory Coast

Air force to be resuscitated?

Having reportedly destroyed the entire Ivory Coast air force on the ground at Yamoussoukro Airport in Abidjan on 6 November, after two of its Sukhoi Su-25UBKs had attacked and killed nine French military personnel and injured 23 other people near Bouaké, the French and UN peacekeeping missions authorised the repair in January of four Su-25s and several of the helicopters which were apparently not too badly damaged. Considering that French marines fired an Aérospatiale MATRA Roland short-range surface-to-air missile with a 6.5-kg (14.3-lb) pre-fragmented warhead at each parked aircraft, it seems surprising that any are capable of restoration.

Nevertheless, following renewed attacks in the rebel-held north, foreign technicians have been authorised to begin salvaging what they can from the Su-25s, two ex-Botswana BAe Strikemasters, four Mil Mi-24Vs, several Mil Mi-17s and four IAR-330 Pumas. Spares acquisition may also be difficult, since the recent UN Security Council imposition of an Ivory Coast arms embargo.

Japan

Military spending cuts planned

Japan's defence budget for the current 2004-05 Fiscal Year of ¥4.876 trillion ($46.6 billion) is nominally the world's third-largest military expenditure after the US and the UK, although reduced by 1 per cent over FY2003-04. At only 0.974 per cent of Japan's GNP/GDP, it is also well below western estimates

Flight clearance of the Hellenic air force's first production Alenia/Lockheed Martin C-27J Spartan twin-turboprop transport started in mid-December with an initial sortie from Alenia's Turin-Caselle plant. The C-27J's first 83-minute flight by Alenia test-pilots initiated a successful contractual time schedule, to start deliveries of 12 by late January.

of China's real defence spending. Similar 1 per cent reductions are expected in Japan's military budget for the third year in FY2005-06, from 1 April. Medium-term government planning is for further defence economies of 3.7 per cent or ¥24.24 trillion ($231.7 billion), in the five years to March 2010.

Priority funding for two new JMSDF helicopter-carriers and two-tier US-supplied land- and sea-based missile defence systems is planned for 2011 service, to help counter neighbouring North Korea's ballistic missile and nuclear arms threats. Inevitable economies will probably include small reductions in JGSDF troop totals, but will allow increases in missile shield-related contract spending from ¥106.8 billion to ¥119.8 billion in the coming year.

Jordan

Special operations unit formed

The Royal Jordanian AF has announced plans to form a Special Operations Flying Unit (SOFU) that

will include two squadrons and 500 personnel. The SOFU will be tasked with forward reconnaissance, search and rescue, counter-terrorism and transporting special operations troops. Although the organisation will operate under the Jordanian Special Operations Forces, it will remain as a component of the air force and still be reliant on the latter for maintenance and training. It appears likely that the unit will be modelled after the US Army's 160th Special Operations Aviation Regiment and will be equipped with MD500 and UH-60L helicopters. The former are already in service with the RJAF and eight UH-60Ls are due for delivery beginning in September 2006. Jordan is also looking at the possibility of acquiring four CH-47C Chinook helicopters from the Libyan AF.

Qatar

Mirage 2000s withdrawn?

According to press reports, agreement was reached late last year between the Doha and Indian governments for

Air Power Intelligence

The test flight programme of the IDF/AF's Boeing AH-64D Saraf (Serpent) was completed in August 2004 with four Hellfire launches in the US (two RF-guided and two laser-guided). Delivery of three Sarafs on 3 April 2005 allowed the Hornet Squadron at Ramon to begin AH-64D operations. The unit had ceased flying the AH-64A Peten (Python) on 1 February to prepare for the transition.

Replacement of the IDF/AF's SOCATA TB-20 Pashosh (Warbler) fleet is gathering momentum with additional Beech A36 Hofit (Stint, illustrated) deliveries and the presentation of the Pashosh fleet to prospective buyers at Sde Dov in early February.

acquisition by the IAF of nine Qatar Emiri air force Dassault Mirage 2000-5EDAs and two two-seat 2000-5DDAs. They were delivered to the QEAF only in 1997, but their operation was not economically justifiable in the absence of a tangible threat. Some reports claim that Indian interest in the QEAF Mirages was stimulated by earlier negotiations between Qatar and Pakistan for their possible acquisition. The QEAF will now have no combat element, apart from any weapons that may be carried by its half-dozen Alpha Jets.

Russia

Defence budget details
Russian military spending in 2005 will total Rbl 573 billion ($20.5 billion), according to Defence Minister Sergey Ivanov. This figure – about Rbl 70 billion ($2.5 billion), or some 40 per cent over the Rbl 411.473 billion ($14.72 billion) 2004 allocations – represented 2.69 per cent of GDP. Nearly half the new budget, or 41.8 per cent, is for personnel costs, followed by 15.7 per cent for training, and 14.2 per cent for procurement.

R&D receives 13.4 per cent (Rbl 62.8 billion), while 8 per cent (Rbl 11.8 billion) is allocated for repairs and maintenance, and 3.7 per cent for capital construction. About two-thirds of the budget will be spent on the strategic nuclear triad and ground forces, with the remainder divided equally between the RFAF and the navy. Rbl 45 billion ($1.6 billion) is earmarked for equipment upgrades.

According to deputy defence minister for arms procurement General Alexei Moskovsky, planned 2005 acquisitions include four new Topol-M ICBM systems "to maintain genuine nuclear parity with the US", a new Tu-160 'Blackjack' strategic bomber, new fighters, two prototype night-attack Mil Mi-28Ns, and Kamov Ka-60 utility helicopters.

Also announced was procurement of Russia's first non-nuclear air-launched long-range cruise missile, to be launched from RFAF Tupolev Tu-95MS and Tu-160 bombers. Designated Kh-555, the new weapon is based on the original Kh-55MS (NATO AS-15 'Kent') Tomahawk-type air-, surface-, or submarine-launched nuclear missile. With a range of 2-3,000 km (1,079-1,619 nm), this has long been operational in the Russian strategic systems arsenal, and integrated with Tu-95MS and Tu-16 mission systems.

FY2005 defence budget funding includes a new Tu-160, to be completed from the long-static Tupolev/Kazan production line, for 2006 delivery to RFAF's Long-Range Aviation. Lt Gen. Igor Khvorov, the 37th Air Army's special purpose supreme commander, said that the Tu-160 would be upgraded with new digital avionics and equipped for nuclear missile launching, and while another would be modified to carry up to 45 tonnes of bombs. Fourteen Tu-160s would then be in current RFAF service, following the loss of one, with its crew of four, after an in-flight fire following take-off from its Engels air-base in September 2003.

United Kingdom

Nimrod unit plans
In a preliminary move to meet proposed reductions in UK military aircraft and aircrew announced in the recent Defence White Paper, the Air Force Board announced in December that No. 206 Squadron, in the maritime reconnaissance and ASW Wing at RAF Kinloss, would disband from 1 April 2005. This will leave two slightly larger Nimrod MR squadrons, Nos 120 and 201, with establishments increased from 8 to 10 crews; and the Operational Conversion Unit, No. 42(R) Squadron at Kinloss.

The RAF has about 26 Nimrod MR.Mk 2s on establishment, although not all are in current service. Some have long been scheduled for virtually entire structural rebuilds, new Rolls Royce BR710 turbofans, and completely new digital mission systems by BAE Systems as MRA.Mk 4 maritime, reconnaissance and attack aircraft. Under a £2 billion ($3.76 billion) plus December 1996 contract, amended in 2002, BAE Systems was contracted to build 18 Nimrod MRA.Mk 4s.

The contract was further renegotiated in 2003, to comprise three design and development phase aircraft, and an option for 15 production MRA.4s. More recently, on 21 July 2004, Defence Secretary Hoon indicated that Nimrod MRA.Mk 4 requirements would be met by about 12 aircraft. These would be true multi-role types, however, equipped for delivery of precision-guided strategic and tactical attack missiles, in addition to surveillance, ASW, ASuW and SAR missions.

United States

Knighthawks deploy
USS *Bonhomme Richard* (LHD 6) became the first Pacific Fleet amphibious assault ship to embark the MH-60S Knighthawk when the ship deployed in late December 2004. The helicopters are assigned to HC-11 Detachment 4 and are normally stationed at NAS North Island, California. When deployed aboard the amphibious ships, the Knighthawks conduct search and rescue (SAR), logistics support and maritime naval special warfare training support duties in support of Expeditionary Strike Group Six.

KC-130J deploys
The USMC marked a milestone when the first new KC-130J tankers, assigned to VMGR-252, arrived at Al Asad, Iraq, on 13 February 2005. The squadron replaced VMGR-452, which had been supporting operations with its KC-130Ts since arriving in Iraq in August 2004. The 'Yankees', a Marine Corps Reserve squadron, subsequently returned to Stewart ANGB, New York.

StarLifter ends Antarctic ops
The USAF marked the end of another era in aviation history when the last C-141C StarLifter used in support of the Antarctic 'Ice Flights' left Christchurch, New Zealand, on 6 February 2005. Although ski-equipped LC-130s fly missions to remote sites throughout the continent, the StarLifters have shuttled personnel and equipment between McMurdo Station and the Support Forces Antarctica headquarters in Christchurch since landing on the ice for the first time on 13 November 1966. C-141C serial 66-0152, which is assigned to the 452nd Air Mobility Wing at March AFB, California, was the last aircraft to leave New Zealand. The final StarLifter mission departed from McMurdo's Pegasus ice runway on 4 February. In October 1999 a C-17A assigned to the 62nd Airlift Wing at McChord AFB, Washington, became the first Globemaster III to land in Antarctica. With the departure of the StarLifter, the C-17A assumes the responsibility for the air bridge to New Zealand.

Navy changes
One of two Naval Reserve airborne early warning squadron that operated the E-2C was deactivated on 31 March 2005. Before standing down, however, VAW-78 'Fighting Escargots' transferred its complement of four Hawkeyes from Naval Station Norfolk to VAW-77 at NAS Atlanta, Georgia. The 'Night Wolves', which are primarily engaged in counter-narcotics operations, will later gain two additional E-2Cs to bring their complement to six.

The Naval Research Laboratory's (NRL) Flight Support Detachment at NAS Patuxent River, Maryland, was redesignated as Scientific Development Squadron One (VXS-1) on 13 December 2004. Originally established in November 1962, as a field site of the NRL, the squadron operates four NP-3D Orions.

Army aviation
The Minnesota Army National Guard has announced plans to build a new aviation facility at St Clair Regional Airport that will support six UH-60As assigned to the 1256th Medical Company and six CH-47Ds that will be operated by a new unit. The Minnesota guard recently held ceremonies at its base on St Paul's Holman Field that marked the retirement of its last UH-1H helicopters.

The last six UH-1Hs in service with the Colorado Army National Guard's 2-135th AVN at Buckley AFB were also retired recently and flown to Temple, Texas.

Three attack helicopter companies assigned to 1st Armored Division's (1AD) 1st Battalion, 501st Aviation Regiment have relocated to Fort Hood, Texas, and begun their transition from the AH-64A to the AH-64D. The 'Flying Dragons' will spend one year training with the 21st Aviation Brigade before returning to Hanau.

Both the 1st Infantry (1ID) and 1AD in Germany will begin their reorganisations as part of the US Army's ongoing aviation transformation this spring. The 1AD's 1-1st Cavalry, which is stationed at Armstrong AHP in Büdingen, will ship its OH-58Ds stateside. The plan for the 1ID's 1-4th Cavalry, which is normally based at Schweinfurt Army Heliport (AHP), differs slightly and the unit will ship its Kiowa Warriors stateside directly from Iraq when the unit concludes it deployment. After being reconditioned, the OH-58Ds will be redistrib-

All change at Oceana

As part of the accelerated phase-out of the F-14, the FRS (VF-101 'Grim Reapers') delivered the last new Tomcat crews to the Fleet in the spring of 2005. Its final DACT detachment to NAS Key West, Florida, ended on 5 February, and its last carrier qualification currency landings were flown in the early summer. Above is an F-14 landing on *Roosevelt* in March. The squadron will be disestablished on 30 September, around the time the last ever Tomcat cruise gets under way (VF-31 and VF-213 aboard *Roosevelt*). As Tomcat squadrons return from cruises they are standing down to convert to the Super Hornet. Recent additions to the F/A-18F roster are VFA-103 (left, also on *Roosevelt*) and VFA-11 (below left). Super Hornet training at Oceana is handled by VFA-106 (below), which acts as the East Coast FRS for both F/A-18C/D and F/A-18E/F.

uted to the 10th and 25th Infantry Divisions at Fort Drum, New York, and Wheeler AAF, Hawaii. An additional attack helicopter battalion, equipped with AH-64Ds, will replace the cavalry units in each of the divisions. Although the ground assets within the cavalry units will continue to operate in support of the respective divisions, the aviation troops will be deactivated on 15 June 2005.

The US Army's 542nd Medical Company (Air Ambulance) will relocate to Fort Campbell, Kentucky, from Camp Page, Chunchon, Republic of Korea, by April 2005. The unit's relocation is the result of an earlier decision to reduce the number of troops stationed in Korea. The 542nd operates a fleet of 15 UH-60A Blackhawk helicopters.

The closure of Camp Page has also resulted in the decision to relocate the 4-2nd Aviation to Camp Eagle in Wonju. The 'Gunfighters' operate the AH-64D as part of the 2nd Infantry Division (Mechanized).

USAF news

The 1st Fighter Wing welcomed its first flyable F/A-22A on 18 January 2005 when serial 02-4029 arrived at Langley AFB, Virginia, from Tyndall AFB, Florida. The aircraft is on loan from the 325th Fighter Wing. Langley's first Raptor, serial 03-4041, was scheduled to arrive in May 2005.

On 10 February 2005 Air Mobility Command grounded 30 C-130E aircraft when a series of inspections of the centre wing box structure revealed cracks. The 43rd AW at Pope AFB, North Carolina, is responsible for 14 of the 30 grounded aircraft. A further 50 aircraft, comprising C-130E, C-130H, C-130H1 and HC-130N/P models, were placed on restricted flight status to minimise wing stress and increase the safety margin

The Illinois Air National Guard's 182nd Airlift Wing took delivery of a pair of C-130H3 airlifters on 20 January 2004, marking the beginning of the unit's transition from the C-130E. The Peoria unit will receive six further examples by March 2006.

The 943rd Rescue Group (RQG) was activated at Davis Monthan AFB, Arizona, on 12 February 2005. Assigned to the 920th Rescue Wing (RQW) at Patrick AFB, Florida, the unit is responsible for the HH-60G-equipped 305th Rescue Squadron (RQS) and the newly established 306th RQS. With no aircraft assigned, the latter is made up of pararescue jumpers (PJ). The new group has also assumed responsible for the PJs assigned to the 304th RQS at Portland IAP, Oregon. The latter squadron had previously been assigned directly to the 920th RQW.

The 388th Electronic Combat Squadron (ECS) was activated at NAS Whidbey Island, Washington, on 15 December 2004. The unit, which is assigned to the 366th Operations Group (OG) at Mountain Home AFB, Idaho, is responsible for the USAF personnel assigned to the joint/expeditionary EA-6B squadrons at Whidbey. Personnel were previously assigned to Det 1, 366 OG.

USAF initiatives announced

On 1 December 2004 the USAF announced several initiatives that will assign new missions to its reserve forces and combine Air Force Reserve Command (AFRC) and Air National Guard (ANG) units with active compo-

On display at the Ecuadorian air force festival at Guayaquil in late 2004 was this upgraded Kfir CE. Of note is the Rafael Python IV carried on the outer wing panel. This IR-guided short-range AAM has high off-boresight capability. Ecuador's Kfirs are operated by Escuadrón de Caza 2113 'Leones' at Taura air base.

nent USAF organisations under the Future Total Force concept.

During 2005 the AFRC's 419th Fighter Wing will relinquish its F-16Cs but will continue to fly the fighters as an associate unit with the co-located USAF's 388th FW at Hill AFB, Utah.

The Virginia ANG's 192nd Fighter Wing (FW) at Richmond IAP will 'partner' with the 1st FW at Langley AFB, Virginia, as it transitions to the F/A-22A Raptor. Little information has been released regarding how this will affect the 192nd's operations, however it is believed that the unit will relocate to Langley and exchange its F-16Cs for Raptors, taking on associate status.

Air National Guard units in Arizona and Texas will be tasked with operating the RQ/MQ-1L Predator UAVs. It is still unclear whether the organisations will receive their own UAVs. In fact they could support global operations by operating deployed UAVs and their payloads via satellite from ground control stations (GCS) located at their respective home bases.

Active-duty personnel will be assigned to the Vermont ANG's 158th FW at Burlington IAP, where the less-experienced USAF aircraft maintainers will work alongside the more experienced members of the guard.

Air Force Reserve Command personnel will also be integrated into all of the missions assigned to the USAF Air Warfare Center at Nellis AFB, Nevada.

Carrier news

On 1 February 2005 the USS *Carl Vinson* (CVN 70) Carrier Strike Group departed San Diego, California, on a deployment that will take the carrier around the world. The cruise will conclude in Norfolk, Virginia, in approximately six months, where the ship will enter the yards at Newport News Shipbuilding for a three-year refuelling and complex overhaul (RCOH). Carrier Air Wing Nine (CVW-9) is embarked aboard the ship. The deployment marks the final cruise for Sea Control Squadron VS-33 and the last for the S-3B Viking with the Pacific fleet. The squadron will deactivate upon return.

USS *John C. Stennis* (CVN 74) officially changed its homeport from San Diego, California, to Bremerton, Washington, on 1 January 2005. Shortly after arriving in Bremerton, the ship was scheduled to commence a Docking Planned Incremental Availability (DPIA) period.

USS *John F. Kennedy* (CV 67) will change its homeport from Naval Station Mayport, Florida, to Naval Station Norfolk, Virginia, in May 2005. The Navy is however also looking at the possibility of retiring the carrier within the year.

NEWS REPORT

IRAQ

Airpower in the counter-insurgency war

Since the official end of 'major combat operations' in Iraq on 1 May 2003, a ferocious war has been waged between insurgents and forces from the US, UK and their allies as they attempt to secure control of Iraq. The efforts to restore security, rebuild infrastructure and install a democratically elected government have been hampered by fierce opposition from elements within Iraq, and airpower has played a large part in countering these insurgents.

In early April 2003, the arrival of US troops in the Iraqi capital seemed to herald the end of resistance from the forces loyal to President Saddam Hussein. Iraq's uniformed armed forces had put up half-hearted resistance and melted away as US tanks powered north from Kuwait. However, the guerrilla attacks and suicide bombings against US troops, encountered in several Iraqi towns during the brief land war, were a portent of things to come.

When President George W. Bush made his dramatic deck landing on the USS *Abraham Lincoln* off California in a Lockheed S-3B Viking on 1 May 2003 to declare "the end of major combat operations" in Iraq, his troops in the Middle Eastern country were already beginning to detect the first signs of the insurgency that by the end of 2004 would cost the lives of more than 1,000 US troops and result in 10,000 being wounded, on top of tens of thousands of Iraqis dead. The first shots of the insurgency were fired on 28 April in the town of Fallujah, when US paratroopers of the 82nd Airborne

An F/A-18D Hornet from VMFA(AW)-242 turns for take off in support of Operation Iraqi Freedom II at Al Asad. The aircraft carries a single GBU-12 LGB and an AGM-65 Maverick. On the centreline is a Litening targeting pod.

With M1 tanks in the background, an OH-58D lifts off for another scouting mission. The use of helicopters to sweep ahead of ground forces is vital, although there has been considerable attrition among the Kiowa Warrior fleet.

Division were drawn into a confrontation with local people that resulted in 18 civilians being killed and 78 being injured. Within weeks, combat would be raging in several Iraqi towns and cities.

In the first 18 months of the insurgency, airpower played a key role supporting US and coalition troops in action against rebel fighters, providing vital logistic support and assisting the delivery of humanitarian aid. While the air forces were now in a support role to the land forces, the desert geography and extreme climate of Iraq meant it offered US commanders unique capabilities that they have been exploiting to their maximum potential.

Occupation duty

In the wake of the defeat of President Saddam Hussein's government, the US, British and Australian forces dramatically scaled back their airpower in the region from some 1,800 fixed-wing aircraft of all types to less than a quarter of that number. The focus of air operations became support for the 150,000 coalition troops and the re-building of the devastated country. Not surprisingly, the air transport of troops and supplies, and the tactical movement of troops by helicopter, came to the fore.

Combat aircraft were rapidly pulled back from bases in countries surrounding Iraq and major effort was put into opening airports inside the nation to support and sustain the occupation forces. Perhaps the most significant example of this was the withdrawal of all US airpower from Saudi Arabia and the re-location of the US Combined Air Operations Center (CAOC) to Al Udeid Air Base in Qatar. The hi-tech CAOC is packed with computers and communications equipment to give air commanders real-time control of all US and coalition aircraft operating over Iraq, Afghanistan and other parts of the Middle East, and the Horn of Africa.

Al Udeid air base was built in late 2001 to support air operations in Afghanistan and has been steadily expanded since then. In the run-

Whenever the US Army has engaged rebel fighters, the Apache has been heavily involved. Both AH-64A (illustrated) and AH-64Ds have been deployed. They have been particularly useful for pinpoint strikes against particular buildings although, like the OH-58D, they have been a regular target for Iraqi insurgents and have suffered correspondingly high attrition.

Above: Troops of the 101st Airborne take a breather before boarding their waiting UH-60s for a security operation. The Black Hawk is the US Army's principal aerial truck in Iraq.

Right: A gunner watches from the ramp of his CH-47 as it lifts off from an Iraqi base. Hostile fire can come from anywhere, and without warning. Door/ramp gunners maintain vigilance to return and suppress any groundfire they might encounter.

up to the March 2003 attack on Iraq, a sub-CAOC at Al Udeid took over the running of air operations over Afghanistan and the Horn of Africa as part of the US-led 'global war on terrorism'. The CAOC at Prince Sultan Air Base in Saudi Arabia was the nerve centre of the Operation Iraqi Freedom air campaign. During the early summer of 2003 some 400 airlift and 100 close air support sorties were being co-ordinated each day by the CAOC at Al Udeid.

Al Udeid also become the only air base in the Gulf region to host a USAF combat wing, with a mix of F-16 multi-role jets, Boeing F-15C air supremacy aircraft and F-15E Strike Eagles in residence. At any one time a squadron of each type of aircraft was deployed to the base, along with a dozen or so Boeing KC-135 Stratotanker and KC-10A Extender air refuelling aircraft. The Al Udeid wing was also ready to support ongoing US operations in Afghanistan.

A northern transport and tanker base was established at Incirlik Air Base in Turkey to support fighter operations over northern Iraq, and high-priority transport aircraft flying direct from the continental USA heading for Iraq that needed to be refuelled in flight over the eastern Mediterranean.

Within Iraq itself, two squadrons of Fairchild A-10A Warthog close air support jets were deployed to forward operating bases at Tallil in the south of the country and Kirkuk in the north. This was subsequently reduced to a squadron of F-16s at Kirkuk. These aircraft were to assist US Army troops that might need close air support. Additional air support could

News Report

US Army Aviation Order of Battle, Iraq, 1 May 2003

244th Theater Aviation Brigade
1-147th Aviation (Echelon Above Corps Command Aviation)	UH-60
G/149th Aviation	CH-47
Det/A/1-159th Aviation	UH-60
1/A/5-159th Aviation	CH-47
B/5-159th Aviation (Attached)	CH-47D
Det/A/1-214th Aviation	UC-35
Det/C/1-214th Aviation	UH-60
E/111th Aviation (Air Traffic Service)	
Flight Detachment/Operational Support Airlift Command (Kuwait)	C-12

513th Military Intelligence Brigade
B/224th Military Intelligence Bn (OPCON)	RQ-5 Hunter, RQ-7 Shadow

V Corps

4th Brigade, 3rd Infantry Division (Baghdad)
1-3rd Aviation	AH-64D
Team A/2-3rd Aviation (OPCON)	UH-60

4th Infantry Division (Taji/Tikrit)
A/15th Military Intelligence Bn	RQ-5 Hunter
Tactical Unmanned Aerial Vehicle Contract Log Support Team	

4th Brigade, 4th Infantry Division
1-4th Aviation	AH-64D
2-4th Aviation (General Support Aviation)	UH-60
1-10th Cavalry	OH-58D

DREAR/101st Abn Div (Kuwait)
C/3-101st Aviation	AH-64D

2nd Brigade/101st Abn Div (Mosul)
2-101st Aviation	AH-64D

101st Aviation Brigade (Mosul)
1-101st Aviation	AH-64D
3-101st Aviation	AH-64D
6-101st Aviation	UH-60

159th Aviation Brigade (Mosul/Tikrit)
4-101st Aviation	UH-60
5-101st Aviation	UH-60
7-101st Aviation	CH-47
9-101st Aviation	UH-60

2nd Brigade/82nd Abn Div (Fallujah)
TF 1-82nd Aviation
A/1-82nd Aviation (+)	OH-58D
A/1-17th Cavalry	OH-58D

11th Attack Helicopter Regiment (Balad)
2-6th Cavalry	AH-64D
6-6th Cavalry	AH-64D
1-227th Aviation, 1st Cavalry Division (Attached)	AH-64D
5-158th Aviation	UH-60

12th Aviation Brigade
B/158th Aviation	UH-60
B/159th Aviation	CH-47
F/159th Aviation	CH-47
A/5-159th Aviation	CH-47
2 x Aircraft/1-214th Aviation (Attached)	CH-47
F/106th Aviation	CH-47

2nd Armored Cavalry Regiment (Light) (east of Baghdad)
- 1st Squadron, 2nd Armored Cavalry Regiment (Kuwait)
- 2nd Squadron, 2nd Armored Cavalry Regiment
- 3rd Squadron, 2nd Armored Cavalry Regiment (Kuwait)

3rd Armored Cavalry Regiment (western Iraq)
2th Aviation Brigade	4 x CH-47
Medical Evacuation Aircraft/30th Medical Brigade	6 x UH-60

I Marine Expeditionary Force (Reinforced)
south of Baghdad, withdrawn from Iraq by end June 2003

Marine Expeditionary Force HQ Group

15th Marine Expeditionary Unit (Special Operations Capable)
Marine Medium Helicopter Squadron 161 (Reinforced)	CH-46E
Detachment/Marine Heavy Helicopter Squadron 361	CH-53E

24th Marine Expeditionary Unit (Special Operations Capable)
Detachment/Marine Attack Squadron 231	AV-8B
Marine Medium Helicopter Squadron 263 (Reinforced)	CH-46
Detachment/Marine Light Attack Helicopter Squadron 269	AH-1W/UH-1N
Detachment/Marine Heavy Helicopter Squadron 772	CH-53E
Detachment/Marine Heavy Helicopter Squadron 461	CH-53E

3rd Marine Air Wing (Reinforced)

Marine Air Group 16 (Reinforced)
Marine Medium Helicopter Squadron 263	CH-46E
Marine Heavy Helicopter Squadron 462	CH-53E
Marine Heavy Helicopter Squadron 465	CH-53E

Marine Air Group 29 (Reinforced)
Marine Medium Helicopter Squadron 162	CH-46E
Marine Medium Helicopter Squadron 365	CH-46E
Marine Heavy Helicopter Squadron 464	CH-53E
Marine Light Attack Helicopter Squadron 269	AH-1W/UH-1N

Marine Air Control Group 38 (Reinforced)
Marine Unmanned Aerial Vehicle Squadron 1	RQ-2 Pioneer
Marine Unmanned Aerial Vehicle Squadron 2	RQ-2 Pioneer

Marine Air Group 39 (Reinforced)
Marine Light Attack Helicopter Squadron 169	AH-1W/UH-1N
Marine Light Attack Helicopter Squadron 267	AH-1W/UH-1N
Marine Medium Helicopter Squadron 268	CH-46E
Marine Medium Helicopter Squadron 364	CH-46E
Marine Light Attack Helicopter Squadron 369	AH-1W/UH-1N

be provided by US Navy jets from the aircraft-carrier that was kept on station in the Arabian Gulf.

Humanitarian aid

The three week-long American *blitzkrieg* assault on Iraq came to a climax on 4th April when US Army tanks stormed Baghdad airport and were filmed driving past the burnt-out remains of Iraqi Airways aircraft. Within days the government of Saddam Hussein had collapsed. Even before the final fall of the Iraqi government, US Army engineers had repaired the bomb craters in the runway and USAF Lockheed C-130 Hercules transports were landing to help the military build-up for the final assault on the capital. Further south, British Army Challenger tanks were crashing through the perimeter fence at Basra airport and had soon turned the largely undamaged terminal building into the British headquarters in southern Iraq.

In the immediate aftermath of the fall of the old government, US, British and Australian forces moved to establish control of Iraq's main airports to allow them to support their occupation effort, and to allow the delivery of military supplies and humanitarian aid.

Immediately behind the combat troops came units specially trained and equipped to get 'austere' airfields up and running. The USAF's 447th Air Expeditionary Wing took over the now renamed Baghdad International Airport to turn it into a major transport hub to support the thousands of US troops trying to bring law and order to the Iraqi capital. Construction teams from the 1st Expeditionary Red Horse Group and the Civil Engineer Maintenance, Inspection and Repair Team made good repairs to the runway and established an airport lighting system. A constant stream of Hercules, Boeing C-17 Globemaster III and Lockheed C-5 Galaxy aircraft were now landing at the airport, along with the first flights of aircraft chartered by non-governmental organisations and aid agencies. Royal Australian Air Force air controllers took over running the air traffic control for the airport.

Other USAF units opened up airfields at Talli in the south, Mosul and Kirkuk in the north and several remote sites in the western desert. These later airfields were predominately used by military aircraft. At Basra a 500-strong team

While the OH-58D is primarily a scout, it has the ability to shoot back using either Hellfire missiles or 2.75-in (70-mm) rocket pods. Both weapons are carried by this aircraft lifting off for an observation mission.

of RAF personnel set up a Deployed Operating Base (DOB) to run the airport as a transport hub for British forces in the Middle East. Soon, between 20 to 50 fixed-wing and 50 helicopter flights a day were using the airport. The majority of these were military but they also included a large number of charter flights carrying troops and humanitarian aid. This later category included a high-profile visit by flamboyant Virgin Atlantic boss Sir Richard Branson and one of his 747s.

The US and British military quickly moved to set up a system of navigation aids at the airports they controlled, and installed ground radar at them as well to provide limited approach coverage. There was no country-wide radar coverage and all air traffic was controlled procedurally by the Regional Air Movements Control Center (RAMCC) inside the CAOC at Al Udeid Air Base in Qatar. The RAMCC staff issue 'time slots' and transponder codes for aircraft

Above: Deployed on rotation to Al Udeid in Qatar, a squadron of F-15Es provides PGM 'muscle' over Iraq. This aircraft is from the 335th FS 'Chiefs, of the 4th Fighter Wing at Seymour Johnson AFB.

Right: A-10s were deployed to Kirkuk AB in Iraq to provide close air support. The deployment was subsequently taken over by F-16s. This Warthog carries a Maverick, a GBU-12, a Mk 82 and a rocket pod.

aiming to land at Iraqi airfields. The RAMCC negotiated overflight procedures for entering Iraqi airspace from Kuwait, Jordan, Syria and Turkey.

While the main terminal and other facilities at both Baghdad and Basra airports were largely undamaged in the war, the collapse of Iraq's power grid, telephone and sewage systems meant that the military had to bring in almost every resource they needed to run flight operations. US and British officers who inspected the airports after the end of the war

commented that their main problems stemmed from more than a decade without maintenance due to UN sanctions, rather than war damage. The collapse of law and order resulted in hundreds of troops having to be diverted to guard these airports to prevent them being looted by desperate civilians trying to scrape an existence. Drastic security measures, including ground surveillance radars and machine-gun positions, were established around the airports.

Civil control

To fill the power vacuum in Iraq, early in the summer the US and British government eventually formed the Coalition Provisional Authority (CPA), headed by Ambassador Paul Bremer, to run the country. Using his almost unlimited powers, Bremer set up teams of foreign experts to run each Iraqi ministry and government agency until Iraq regained its sovereignty in June 2004, ahead of the election of a democra-

F-16s routinely patrol over Iraq from two bases – Al Udeid in Qatar where a squadron of active-duty SEAD aircraft is based, and Kirkuk, from where the Air National Guard mans a close air support detachment. For the latter role the Litening pod is routinely carried along with GBU-12 laser-guided bombs, as carried by the pair of Virginia ANG aircraft above. JDAMs are also employed, carried at right by a Montana ANG 'Red Tail'.

News Report

Rebuilding the Iraqi air force

Iraq's airspace remains firmly under the control of US-led forces and it will be many years before Baghdad enters the market for new fighter aircraft to replace the hundreds of combat jets purchased by Saddam Hussein's regime. All of these are now rotting in desert 'aircraft graveyards' around Iraq, and the new government has had to start from scratch to rebuild its air force. With the formal ending of the occupation at the end of June 2004, the new Iraqi air force was reformed with surplus US and Jordanian equipment.

The first of eight new Seabird Aviation SB7L-30 Seekers unarmed observation aircraft were delivered in July 2004 to the 70th Squadron, and air patrols of key infra-structure such as oil pipelines and road routes began in September 2004 from Basra International Airport. Eight more Seekers are held on option. In January 2005 the US handed over the first of three refurbished ex-USAF C-130Es. Eight new Sama CH2000 light aircraft have been ordered via a contract with the US Army Aviation and Missile Command. The United Arab Emirates has also offered to provide four Bell JetRanger helicopter and seven Aerocomp Comp Air 7SL observation aircraft by December 2004, with only four of the 7SLs being delivered to Basra airport to date. The JetRangers are to be formed into No. 3 Squadron to operate from the Iraqi Air Force's new base at Al Muthana AB, at Baghdad International Airport.

Members of Iraq's former air force were recruited

A MiG-25R is dug out of its sandy 'grave' at Al Taqaddum (above), while an HMH-465 CH-53E lifts a Chengdu F-7 (right) at Al Asad. Soviet-supplied aircraft like these have no place in the reborn Iraqi Air Force.

for the new air force, and over 100 trainees are currently receiving conversion-to-type instruction from the Royal Jordanian Air Force in Amman, Jordan. A new military air base at Baghdad International Airport is currently being constructed.

Given Iraq's pre-occupation with counter-insurgency operations it is very likely that its requirements for helicopters and surveillance aircraft will grow over the coming year. Attack helicopters would provide much needed fire support and there have also been suggestions that UAVs may be purchased to improve intelligence-gathering.

Aircraft from the Al Udeid wing (379th AEW) parade for the camera. Led by an MPRS-equipped KC-135T, the group includes a 366th Wing F-15E, 31st FW F-16s and two Tornado GR.Mk 4s from the RAF's detachment in Qatar.

As usual, the main burden of intra-theatre transport is borne by the broad shoulders of the Hercules. The USAF's principal transport hub is at Balad, to where C-5s and C-17s bring materiel for onward delivery by C-130.

tic government some time in the future. The CPA now found itself responsible for all of Iraq's aviation infra-structure, its aviation administration and its state-owned airline.

To get Iraq's airports open, the CPA was able to draw on $150 million allocated to the Ministry of Transport to fund emergency repair work. As part of its multi-billion infrastructure contract, the US construction giant Bechtel was contracted, via the US Agency for International Development, to begin emergency projects at both Baghdad and Basra airports. The work at Baghdad included installation of a 5-MW emergency power generator, refurbishment of Terminal C, restoration of sewage and water supplies, installation of security checkpoints, passenger handling facilities, air conditioning and refurbishment of the air traffic control tower and fire station.

A similar programme of works was contracted for Basra, including renewing the power supply, heating, air conditioning, sewage and painting new runway markings. In May 2003, the CPA contracted the airport facilities management companies Skylink Air and Logistic Support Incorporated to assess and manage the running of civil aviation at five airports in Iraq.

Offensive operations

Insurgent attacks began to escalate during the summer of 2003 and US C-130s operating out of Baghdad airport were soon fired at by guerrilla fighters armed with shoulder-launched man-portable surface-to-air missiles, or Manpads. Commercial operators mostly got cold feet about opening routes into Baghdad after the US military issued security warnings. DHL and Jordanian carrier Royal Wings were the only airlines to continue to offer commercial services into Baghdad but they suspended

USAF pararescuemen ('PJs') practise fast-roping from HH-60Gs of the 64th Expeditionary Rescue Squadron at Balad. From this base the unit provides rescue cover for northern Iraq, while another squadron is positioned at Tallil to cover the south.

RAF operations

The UK's contribution to ongoing occupation operations in Iraq is predominantly an Army responsibility, but the RAF is deployed in strength to provide air support in a variety of ways. An Air Commodore based at the CAOC in Al Udeid is responsible for ensuring RAF operations are closely integrated into US air activity in the Middle East.

A detachment of six Tornado GR.Mk 4 bomber/reconnaissance aircraft (drawn in rotation from RAF Marham and RAF Lossiemouth) is maintained in the Gulf region should British troops require close air support. To date this has not been required, with the aircraft being predominantly involved in photographic reconnaissance missions with the Joint Reconnaissance Pod. The detachment was based at Ali Al Salem until it moved to Al Udeid in Qatar in 2004 during major renovation and construction work at the Kuwaiti base. A pair of VC10 tanker/transports from No. 101 Squadron is based at Bahrain's Muharraq airport to support RAF and US Navy fighters.

Two Nimrod maritime patrol aircraft drawn from Nos 120, 201 and 206 Squadrons are forward-based at Seeb in Oman for maritime reconnaissance, and they have also been used to patrol overland oil pipelines in Iraq to prevent sabotage.

Five Puma helicopters from No. 33 Squadron and five Chinook helicopters from Nos 18 and 27 Squadrons were based at Basra International Airport to provide air mobility for British troops during 2004, along with eight Westland Lynx AH.Mk 7/9s and Gazelle AH.Mk 1s of the Army Air Corps. Royal Navy Sea King HC.Mk 4s replaced the Pumas at the end of 2004. The airport is guarded by an RAF Regiment field squadron. A number of No. 7 Squadron Chinooks equipped for special forces tasks are also in Iraq and are known to have operated around Baghdad.

An RAF Hercules C.Mk 4 fires flares as it approaches Baghdad International Airport on 1 July 2003, the first military transport to land at the reopened airfield.

An HS 125 transport aircraft from No. 32 (The Royal) Squadron) is based in the region to move senior commanders around. Two Hercules C.Mk 4/5 transport aircraft from Nos 24 and 30 Squadrons are based at Basra and Al Udeid for intra-theatre transport missions, including flying into Baghdad to support the British mission at Coalition headquarters in the city. Older Hercules C.Mk 1/3 aircraft of Nos 47 and 70 Squadrons are also tasked for operations in Iraq from bases in the UK, Cyprus and Al Udeid.

VC10 tanker/transport aircraft from No. 10 Squadron and C-17 transport aircraft from No. 99 Squadron provide strategic airlift of troops and supplies into Iraq from bases in the UK, via a regional air transport hub at Al Udeid.

their services in November after a DHL Airbus A300 cargo aircraft was hit by a Manpad missile and only just managed to turn around and land at Baghdad. The crew survived by an amazing feat of airmanship in which the stricken freighter was landed with just engine thrust available for control. Nevertheless, the attack highlighted the threat to aircraft operating into the Iraqi capital's airport.

During the first six months of the occupation, the first port of call for air support for US Army troops was their division's organic aviation brigades. These had around 140 Boeing AH-64D Longbow Apache attack helicopters, 90 Bell OH-58D Kiowa Warrior armed scout helicopters, 300 Sikorsky UH-60 Black Hawk transport helicopters and 80 Boeing CH-47D Chinook heavylift helicopters forward-deployed with the occupying troops, and could be overhead trouble spots in a few minutes. During the summer and autumn of 2003, the US Army mounted a series of helicopter-borne search and destroy operations into the heart of the infamous 'Sunni Triangle' in a bid to track down supporters of the growing insurgency against the occupation.

The first of these operations was mounted in June 2003 around Tikrit and Mosul, into the heartland of the old regime. Dubbed Operation Peninsular Strike, it involved Chinook heavylift helicopters of the 159th Aviation Regiment backed up by Apache gunships and other support helicopters. The first combat loss was a 101st Airborne Division AH-64D shot down on 12th June 2003 in western Iraq, the first known incident linking Iraqi insurgents to the loss of a US helicopter. British troops were also in action in June in Maysan province, south of Al Amara, when a patrol of six Royal Military Policemen was attacked and six soldiers killed. A Royal Air Force Chinook came under fire during a rescue mission to pick up wounded British soldiers.

Perhaps the high point for US Army helicopter crews in July 2003 was the raid in the northern city of Mosul that resulted in the deaths of Saddam Hussein's two sons, Uday and Qusay. OH-58Ds of the 101st Airborne Division's aviation brigade participated in the assault on the villa where the two men were hiding, blasting the building with 2.75-in (70-mm) rockets.

In turn, the insurgents began targeting the US Army helicopters with small-arms fire, rocket-propelled grenades and Manpads. In a 21-day

A detachment of Tornado GR.Mk 4s is at Al Udeid for attack and recce. The aircraft below, seen at Basra, carries a Joint Reconnaissance Pod under its belly. A single HS 125 from No. 32 (TR) Squadron (right, at Mosul) provides British forces with intra-theatre liaison.

News Report

The full RAF support machine is involved in Iraq, including regular supply flights to Basra by C-17s (left), TriStars (above), VC10s and Hercules. VC10 tankers are based in Bahrain to support the RAF's Tornados and other coalition aircraft. Canberra PR.Mk 9s (below) have also been deployed to Basra in the post-war period for mapping and other reconnaissance duties.

period beginning on 25 October 2003, Iraqi insurgents are believed to have shot down six US Army helicopters, including four UH-60s, one Boeing CH-47D Chinook and one AH-64. Five of the attacks involved rocket-propelled grenades (RPG). On 2 November, a US Army National Guard Chinook carrying 42 troops was shot down near Fallujah by insurgents with a Manpads, perhaps an SA-7 or SA-14, killing 16 people and injuring a further 26. Five more US Army helicopters were brought down in a month, with the loss of 41 US personnel killed and more than two dozen wounded. In response, the US Army launched a major effort to re-equip its helicopters in Iraq with new defensive systems.

US Army sources were very critical of the predictable flight profiles flown by its helicopter pilots, which made it easy for insurgents to position themselves near US bases and wait for helicopters to fly by. In December 2003 changes in tactics were introduced that considerably inhibited the ability of the insurgents to attack US Army helicopters.

In response to the escalating insurgency, the US commander in Iraq, Lieutenant General Ricardo Sanchez, ordered a series of integrated offensive operations to be launched around Baghdad, dubbed Operation Iron Hammer, which involved the use of fixed-wing aircraft dropping bombs for the first time since the end of the 'war' in April.

On 12 November 2003 the 1st Armoured Division's 3rd Brigade began its assault on the city of Baghdad, targeting Saddam loyalists and other insurgents. Many of the targeted buildings, including an abandoned dye factory, were suspected of being used as staging grounds by insurgents and were deemed areas of interest. Advanced munitions such as 2,000-lb (907-kg) JDAM GPS-guided bombs were dropped on suspected improvised camps and 1,000-lb (454-kg) bombs were dropped on targets in Kirkuk. Coordinated US strikes included the use of Lockheed AC-130 gunships. Targets in Tikrit, Baqouba and Fallujah were attacked by heavy artillery, battle tanks, attack helicopters, F-16 fighter-bombers and gunships. This was followed up by Operation Ivy Cyclone II, which saw air support being provided by aircraft of Carrier Air Wing 1 from the USS *Enterprise* in the Arabian Gulf.

Even the capture of Saddam Hussein in December 2003 did not seem to put a lid on the insurgency that was now spreading throughout the country, with bombings and gun attacks on coalition troops now a daily occurrence.

Protection for the Lynx fleet

Royal Navy Lynx HMA.Mk 8s embarked on its Type 23 frigates are now deploying to the Middle East with enhanced force protection capabilities to counter the maritime terrorist threat. This equipment package is an evolution of the one rushed into service in 2002 as an Urgent Operational Requirement (UOR) for operations in Iraq, dubbed Operation Telic by the UK Ministry of Defence. It includes full night vision goggle-compatible cockpit, third-generation night vision goggles (NVG) and 0.50-in (12.7-mm) calibre heavy machine-guns with Raufoss-supplied armour-piercing high explosive incendiary ammunition. Lynxes in the Middle East routinely fly with door gunners to free up the flight observer to operate the helicopter's sensors and electronic warfare systems.

The man-portable surface-to-air missile (Manpad) missile threat to helicopters operating in Iraq continues to concern Royal Navy Lynx crews, and naval sources suggest that a defensive aid suite is being looked at for the helicopter during its next major upgrade.

The RAF's contribution to the Support Helicopter force in Iraq comprises five-aircraft detachments of Chinooks (above) and Pumas (below). Additional Chinooks are also believed to be deployed to support Special Forces. The Puma is marked with the badge of 1563 Flight, which was established at Basra and operated until the aircraft returned to Benson in late 2004. The aircraft supported the Black Watch's deployment to Camp Dogwood during the USMC's Fallujah operation.

The Ship's Flight Lynx HMA.Mk 8 from HMS Norfolk is seen at the port of Umm Qasr. An addition for Royal Navy Lynxes operating in the Iraqi theatre is the 0.5-in (12.7-mm) machine-gun in the starboard cabin door.

Sea King operations

846 Naval Air Squadron deployed in the early summer 2003 for six months to provide helicopter support to British troops trying to pacify Iraq's second city. On a daily basis three Sea Kings were available for tasking to fly troops and cargo around the British zone in southern Iraq. These include flying soldiers on 'Eagle' vehicle check point (VCP) patrols to intercept suspicious vehicles.

"The Incident Response Team (IRT) aircraft is on 24-hour alert, at 15 minutes notice to move (MNM) in daytime and 30 MNM at night, to respond to medical emergencies," said Royal Marines Captain George Hunt, who was attached to 846 Squadron. "We've had 40 call-outs or 'shouts'. Ten of these guys survived who would not have without the IRT aircraft. The IRT aircraft carries a Royal Army Medical Corps medical team, a two- or three-man explosive ordnance disposal (EOD) section and an RAF or Defence Fire Service firemen to cut people out of road traffic accidents. The package is adjusted as needed. We always carry medics but mix and match between the firemen and EOD specialists."

The Sea King HC.Mk 4 crews operating in Iraq have praised the enhanced performance of the improved Gnome engines that were fitted earlier in 2003 to optimise the helicopter for operations in hot climates. "We have new engines that can run at higher temperatures, which gives us a 3,000-lb [1360-kg] payload increase in temperatures up to 45°C [113°F]", Lieutenant Commander Tim Watkins, 846 Naval Air Squadron's detachment commander at Basra International Airport, said. "In the summer we were still lifting 15 troops and had over an hour's endurance. That is something a Puma can't do and something we could not do before. We can deliver a reasonable load anywhere in theatre."

"The threat here is worrying but we have not been targeted and we have our defensive aid suite (DAS)," said Lieutenant Alastair Jenkins, a Sea King pilot. "No

Four Sea King HC.Mk 4s from 846 NAS line the ramp at Basra. As well as providing transport for the policing forces, the Sea Kings also undertake emergency medical extraction missions. The detachment alternates with the RAF's Puma force.

one has fired at our helicopters since the war. We have seen firing in built-up areas particularly at night but it was not aimed at us, it was just celebratory. We try to minimise the risk from rocket-propelled grenades. We work hard to make it difficult for them to hit us, using different routes and unpredictable patterns. At the Palace [in downtown Basra] and other high-risk zones there are guard towers and we land inside the [army] perimeter. We do not land in towns because of the risk and sometimes we break-off pursuit of cars when we see them go into built-up areas. I've had a few 'shouts' on the IRT side – last week I dealt with a soldier injured in a road traffic accident with spinal injuries and we put in a spinal brace and saved him a very uncomfortable Land Rover ride to Shaibah hospital – in under the 'golden hour'. In those cases you see that you make a difference."

The shift from war-fighting to post-war stabilisation tasks meant a major change took place in the control of air support in May 2003. During the war, senior USAF officers in the CAOC were allocated large areas of Iraq where they could strike at will. From May onwards almost every air strike would have to be conducted under the positive control of a forward air controller (FAC), operating with US or coalition troops on the ground inside Iraq or airborne in an army helicopter. An air support operations centre inside the US ground forces headquarters in Baghdad made requests for air support to the CAOC, and then handed off inbound strike aircraft to FACs near the scene of any action to conclude the final phase of the attack. Only when FACs were happy that the attacking aircraft had correctly identified the targets would they be allowed to clear the pilots to drop ordnance. Each US Army division was assigned USAF tactical air control parties from Air Support Operations Squadrons manned by fully-qualified FACs, who worked closely with their army counterparts and were usually in the thick of any action.

Direct support for British Army units is provided by Army Air Corps Lynx and Gazelle helicopters. Both in-service variants of the Lynx are deployed: the skid-equipped AH.Mk 7 (below) and the wheeled AH.Mk 9 (right). The helicopters fly a variety of missions, including scouting/top cover for ground operations, insertion of security teams, and emergency reinforcement response. The Army's helicopters were deployed forward along with the Black Watch during the Army's Camp Dogwood deployment in late 2004.

The use of fixed-wing close air support was still a rare occurrence until April 2004 when the fighting rose to a new intensity. In the early part of 2004 the USAF and US Navy were flying about 30-40 fixed combat sorties each, but these were almost all combat air patrols that were not usually called into action. The usual tempo of air activity at this time was between 135 and 140 airlift and tanker sorties each day, supported by 10 to 15 reconnaissance sorties flown mainly by General Atomic RQ/MQ-1 Predators unmanned aerial vehicles based inside Iraq.

Airlift

The insurgents continued to mount Manpads missile attacks on military and civil aircraft using Baghdad International airport. To counter these threats, increasing use was made of the cover of night to protect flights but attacks continued. In December 2003, guerrillas hit a

News Report

US Navy airpower in the Gulf

The US Navy has maintained a carrier strike group in the Gulf region continuously since the start of operations in Afghanistan in October 2001. On a daily basis the Gulf carrier commits aircraft to both Iraq and Afghanistan, flying offensive and combat support missions. In a typical six-month Gulf cruise a carrier's air wing will launch 650 combat sorties 'over the beach'. The most 'in-demand' US Navy assets are Northrop Grumman E-2C Hawkeyes and EA-6B Prowlers to monitor airspace and intercept suspicious communications broadcasts, respectively. The E-2Cs are particularly valued because the USAF has pulled all its Boeing E-3 Sentry AWACS force back to its home base.

As well as its shipborne aviation assets, the US Navy maintains considerable shore-based capabilities in the region. Lockheed P-3C Orion aircraft are forward-based at Muharraq International Airport in Bahrain, close to the US Navy's Middle East headquarters. The Orions regularly patrol Middle East waters looking for suspicious maritime activity and they occasionally detach aircraft to Masirah in Oman and Djibouti to extend their endurance. Pakistani Navy Dassault Atlantic crews recently visited Bahrain to co-ordinate joint patrols with their US maritime patrol colleagues. Overland surveillance missions are also flown over Iraq and Afghanistan from Kandahar and Bagram. Muharraq is also home to logistic support squadrons of Sikorsky UH-3H Sea King and CH-53E Sea Stallion helicopters that supply US Navy ships in Middle East waters.

US Middle East carrier deployments 2003-2005

April-October 2003	USS *Nimitz*
October 2003- February 2004	USS *Enterprise*
February 2004 – July 2004	USS *George Washington*
July 2004 – November 2004	USS *John F Kennedy*
November 2004 – April 2005	USS *Harry S Truman*

Right: The aircraft of CVW-17 are prepared for operations aboard John F. Kennedy during November 2004. At the time the air wing was heavily involved in Operation Al Fajr, the assault on Fallujah.

US Air Force C-17 transport with a surface-to-air missile shortly after it took off from Baghdad, causing the engine to explode. The aircraft returned to the airport and landed safely, with only one of the 16 people aboard slightly injured. On 8 January 2004, a USAF C-5 transport with 63 passengers and crew limped safely back to Baghdad's airport after being hit by fire from insurgents.

In early 2004 the USAF began a major airlift effort to replace the 130,000 US troops in Iraq with 110,000 fresh soldiers, requiring more than 3,000 flights into Ali Al Salem Air Base and Kuwait International Airport. The troops were then moved into Iraq by truck or C-130 flights. Airlift efforts intensified over the summer as insurgent ambushes on US supply convoys made the US military rely more heavily on airlift to get troops and supplies into Iraq. By late 2004 the CAOC's air mobility division was typically organising more than 140 intra-theatre C-130 sorties and more than 30 KC-135 and KC-10 air-refuelling sorties a day – the latter offloading close to 1.3 million lb (589680 kg) of fuel daily to receivers. US air mobility assets also moved on average more than 1,700 passengers a day.

Marines arrive

As part of the rotation of troops, the US Marine Corps was drafted in to replace US Army troops based around the 'hot-spot' towns of Fallujah and Ramadi, as well as taking over

Above: An F-14B from VF-103 launches from 'JFK' during the Fallujah operations. The Tomcat is almost certainly seeing its last combat action over Iraq.

Left: Another Navy veteran for which OIF II is its last action is the Sea King. Used to support Navy vessels in the theatre, the UH-3H is flown by HC-2 Det 2 'Desert Ducks' from Muharraq in Bahrain.

Below: Hornets have borne the brunt of carrier attack operations over Iraq. Here a VFA-34 F/A-18C launches during Operation Al Fajr in November 2004.

Land-based patrol aircraft have played a significant part in the post-war policing operation, and US Navy P-3s and RAF Nimrods are in-theatre to support operations in both Iraq and Afghanistan. Right is a VQ-1 'World Watchers' EP-3E Aries II Sigint-gatherer, reversing into its parking space, while below crew members head towards their P-3C from VP-1.

responsibility for patrolling the western desert out to the Jordanian and Syrian borders where foreign fighters were suspected of having infiltration routes.

The 25,000-strong I Marine Expeditionary Force (I MEF) was configured for its Iraqi mission as a Marine Air Ground Task Force (MAGTF), with the 3rd Marine Air Wing (3 MAW) providing the air support element to the combat troops of the 1st Marine Division. Many of these troops were veterans of the 2003 invasion of Iraq.

3 MAW's main air element was provided by the 16th Marine Air Group, which had two squadrons of Sikorsky CH-53E Sea Stallions for heavylift, two squadrons of Boeing CH-46E Sea Knights for troop transport and a light attack squadron with AH-1Ws and Bell UH-1N Hueys. Although the I MEF relied at first on the USAF to provide fixed-wing combat support, it provided its own FAC teams and co-ordination elements at major ground headquarters to ensure close air support arrived on time and in the right place.

The bulk of 3 MAW bedded down at Al Asad Air Base in northwest Iraq, and a major effort began to improve the infra-structure at Al Taqaddum Air Base, including repairing the runway to allow Lockheed KC-130 Hercules to use it. Two squadrons of KC-130Ts were deployed with 3 MAW to provide a link to I MEF's main logistic bases in Kuwait, as well as to provide air-to-air refuelling support. Al Taqaddum was a few miles from Fallujah and allowed detachments of AH-1Ws to be forward-deployed to provide close air support for Marines patrolling the dangerous town.

A major job assigned to I MEF was to prevent infiltration of foreign fighters into Iraq, so air patrol of key roads and border crossing points became a top priority for 3 MAW. Cobra and Huey patrols were augmented by IAI/AAI RQ-2 Pioneers from Marine Unmanned Aerial Vehicle Squadron 2, which relayed back real-time imagery of locations of interest to I MEF headquarters.

When the battle for Fallujah was at its height in April and early May 2004, AH-1Ws were airborne around the clock to support front-line Marines, using 20-mm cannon fire and AGM-114 Hellfire missiles. Marine FACs were also up with the front-line troops to direct USAF F-16s, F-15Es, AC-130H/U gunships and

US Air Order of Battle, Fallujah battle, April/June 2004

IRAQ
447th Air Expeditionary Group, Baghdad Intl Airport/Camp Sather
304th/101st Expeditionary Rescue Squadron	HH-60G
20th Special Operations Squadron	MH-53M

506th Air Expeditionary Group, Kirkuk AB
107th Fighter Squadron	F-16

332nd Air Expeditionary Wing, Balad AB
64th Expeditionary Rescue Squadron	HH-60G
46th Expeditionary Reconnaissance Squadron	RQ/MQ-1 Predator
USAF air transport hub	

3rd Marine Air Wing, 16th Marine Air Group, Al Asad AB
Marine Medium Helicopter Squadrons 161/261/764	36 x CH-46E
Marine Heavy Helicopter Squadrons 466, 465	CH-53E
Marine Attack Squadrons 542 and 214	10 x AV-8B

Al Taqaddum Air Base forward operating base
Marine Light/Attack Helicopter Squadrons 167/775	36 x AH-1W, 18 x UH-1N
Marine Unmanned Aerial Vehicle Squadron 2	RQ-2 Pioneer
507th Medical Company (US Army)	UH-60

407th Air Expeditionary Group, Tallil AB
332nd Expeditionary Rescue Squadron	HH-60G

KUWAIT
Ali Al Salem AB
USAF air transport hub and C-130 operating base
Special Operations Command operating base	AC-130, MC-130, MH-53M
Marine Aerial Refueler Transport Squadrons 352/234	KC-130
71st Rescue Squadron	MC/HC-130P

Kuwait International Airport
USAF air transport hub

QATAR
Al Udeid AB
Combined Air Operations Center (CAOC)
379th Air Expeditionary Wing
555th Fighter Squadron	F-16CJ
Detachment 4th Fighter Wing	F-15E
bomber detachment	4/6 x B-1B
363rd Expeditionary Fighter Squadron	F-15C
340th Expeditionary Refueling Squadron	KC-135
379th Expeditionary Refueling Squadron	KC-10
745th Expeditionary Airlift Squadron	C-130

AT SEA
Carrier Air Wing 7, USS George Washington Carrier Strike Group
(replaced by John F. Kennedy July 2004, Harry S. Truman November 2004)
Fighter Squadron 11	10 x F-14B
Fighter Squadron 143	10 x F-14B
Strike Fighter Squadron 82	12 x F/A-18
Strike Fighter Squadron 131	12 x F/A-18
Strike Fighter Squadron 136	12 x F/A-18
Carrier Airborne Early Warning Squadron 121	4 x E-2C
Electronic Attack Squadron 140	4 x EA-6B
Sea Control Squadron 31	8 x S-3B
Fleet Logistics Support Squadron 40 Detachment	2 x C-2A
Helicopter Anti-Submarine Squadron 5	4 x SH-60F, 2 x HH-60H

11th Marine Expeditionary Unit (Special Ops Capable), USS Peleliu
Marine Medium Helicopter Squadron 166 (reinforced)	AH-1W, CH-46E, CH-53E, AV-8B, UH-1N

24th Marine Expeditionary Unit (Special Ops Capable), USS Saipan
Marine Medium Helicopter Squadron 263 (reinforced)	AH-1W, CH-46E, CH-53E, AV-8B, UH-1N

US ARMY HELICOPTER UNITS
1st Armoured Division (south Baghdad, Karbala, Najaf)
4th Aviation Brigade
1st Squadron, 1st Cavalry Regiment	OH-58D
1st Battalion, 501st Aviation Regiment	AH-64D
2nd Battalion, 501st Aviation Regiment	UH-60

2nd Armoured Cavalry Regiment (south Baghdad, Karbala, Najaf)
4th Squadron	OH-58D/UH-60

1st Cavalry Division (Baghdad region)
4th Aviation Brigade
1st Battalion, 7th Cavalry Regiment	OH-58D
1st Battalion, 227th Aviation Regiment	AH-64D
2nd Battalion, 227th Aviation Regiment	UH-60
1st Battalion, 25th Aviation Regiment	OH-58D

A squadron of two-seat F/A-18Ds is assigned to the 3rd MAW on rotation. Bringing a splash of colour to an otherwise drab scene is VMFA(AW)-224's 'boss-bird' (above), seen at Al Asad in January 2005. The 'Bengals' took over the Iraq deployment from VMFA(AW)-242 'Bats' (left). This aircraft carries a typical mixed load, including Litening pod, AGM-65 Maverick, GBU-12 LGBs and a GBU-38 JDAM.

US Navy F/A-18C Hornets on to targets. The shooting-down of a USAF Special Operations Command MH-53M Pave Low helicopter outside Fallujah during the battle lifted a brief veil on the role of special forces operations in central Iraq by the mysterious Task Force 21.

On 8 April, one F/A-18 Hornet from Strike Fighter Squadron (VFA) 131 flying from the nuclear-powered aircraft-carrier USS *George Washington* (CVN 73) in the Arabian Gulf even conducted a strafing run against a rebel position with its 20-mm cannon.

At noon on 9 April, Marines and Coalition forces unilaterally suspended combat in Fallujah in order to hold meetings between members of the Governing Council, the Fallujah leadership and the leadership of the anti-coalition forces, to allow the delivery of additional supplies by the relevant departments of the Iraqi government, and to allow the residents of Fallujah to tend to their wounded and dead. This, however, proved to be only a temporary truce and the US Marines were soon trading fire with rebels inside the city.

The first battle of Fallujah resulted in the 3 MAW calling up additional reinforcements that arrived in early May, including an additional squadron of CH-46Es and AH-1Ws. They were later followed by two squadrons of AV-8B Harriers. At the same time, two Marine Expeditionary Units (MEUs) were ordered to Iraq to boost the US garrison in the country.

Sadr's rebellion

As Fallujah was going critical in mid-April, Shia militia fighters loyal to the radical cleric Moqtada al-Sadr began uprisings against the US-led occupation in several cities and in Shia districts of Baghdad. When the Sadr revolt broke out, the US Army's 1st Armoured Division was in the process of handing over responsibility for security of the greater Baghdad region to troops of the 1st Cavalry Division. Helicopter units of both divisions' aviation brigades were involved in the subsequent fighting, with Apaches being used to launch missile attacks on Sadr militia positions in eastern Baghdad.

Sadr fighters seized several police stations and government buildings in Baghdad, prompting the US military command to order troops to take them back. AH-64Ds were used to support these efforts and they launched several Hellfire missile attacks on buildings occupied by Sadr supporters. One AH-64D of the 1st Cavalry was lost on 11 April during fighting near Baghdad airport, with the loss of two crew.

To the south of Baghdad, the Polish-led Multi-National Division (Central South) found itself under attack by Sadr militia troops in Al Kut, Karbala and Najaf. The 1st Armoured Division, 1st Infantry Division and 2nd Armoured Cavalry Regiment were sent to restore the situation and help the Poles. Aviation elements of these divisions were closely involved in these operations flying attack, reconnaissance, transport and medical evacuation missions.

By early May US troops were positioned on the outskirts of the holy Shia cities of Najaf and Karbala, and they soon began operations to undermine Sadr's control of them. Armoured raids supported by AH-64Ds and OH-58Ds hit police stations and other positions controlled by Sadr's fighters. The OH-58Ds of the 2nd Cavalry were in the thick of this action, trying to locate Sadr fighters hiding near the holy shrines, which were very sensitive and off-limits to US forces. To try to undermine support among the population for Sadr, leaflet drops were conducted by US UH-60s and Polish PZL-Swidnik W-3 Sokol helicopters.

Above: Partnering the two-seat Hornets at Al Asad is a squadron-sized deployment of AV-8Bs, this pair being from VMA-542.

Left: Marine KC-130s provide refuelling support for USMC aircraft and helicopters, and intra-theatre transport. This Hercules was the first aircraft to land on the repaired runway at Al Taqaddum, which is used as a forward-operating base by USMC helicopters and UAVs.

Shia militiamen were targeted by gunships in the air strikes on Al Amara on 8 May, during the largest air support operation for British troops since the occupation of Basra in April 2003 during the US-led invasion of Iraq. AC-130s were used to support the initial amphibious assault on the Al Faw peninsula by 3 Commando Brigade in March 2003, but this new operation was the first time the British Army had made such extensive use of the AC-130.

Operation Waterloo involved the use of the AC-130 gunships to strike at targets inside the city of Al Amara. In media interviews, Major General Andrew Stewart, the commander of British forces in southeast Iraq, said the AC-130s were employed because of their ability to hit targets with great accuracy to minimise the risk of civilian casualties. The AC-130 strikes involved the use of "thousands of rounds" of 40-mm and 105-mm cannon fire, according to the general.

RAF involvement in Operation Waterloo included BAE Systems Nimrod MR.Mk 2 maritime patrol aircraft, from the RAF Kinloss wing, and Panavia Tornado GR.Mk 4 aircraft. The Nimrods used Wescam electro-optical sensor pods to detect targets for the AC-130s and co-ordinate low passes by Tornados to disperse concentrations of rebels.

New Iraq

At the end of June 2004 the US-led coalition officially handed over sovereignty to an interim Iraqi government, but this act did little to quell the violence and it was now clear that international military troops would have to remain in the war-torn country for many years to counter instability. Even the departure of Ambassador Bremer from Baghdad airport in a USAF aircraft did not stem the violence. A US Air Force C-130 transport plane was hit by small-arms fire after taking off from Baghdad International Airport the day before the handover of sovereignty, resulting in an unknown number of wounded, US officials said. The aircraft was hit west of the

AH-1W Cobras were heavily involved in the Al Fajr operation in late 2004. The HMLA-169 aircraft above are seen at the Al Taqaddum FOB from where most missions into Fallujah were launched. At right is an image from a Cobra's sight as a Maverick hits a target in the town. The Cobra crew had designated the target for the missile, which had been launched by an F/A-18D.

airport. It returned to Baghdad airport and landed safely.

In May the Pentagon said that it planned to keep about 138,000 US troops in Iraq until the end of 2005 at least. The USAF was to retain control of four main air bases at Kirkuk, Tallil, Balad and Al Asad. Balad Airfield was to be the primary air transport hub because of the need to eventually hand Baghdad International Airport over to civilian control.

By July the US command in Iraq had developed a co-ordinated plan to neutralise Sunni and Shia rebels in Iraq. The first phase would concentrate on neutralising Sadr's militia to free up forces for a major show-down with the Sunni insurgents in the autumn.

It was to be the job of the 11th Marine Expeditionary Unit to spearhead the effort to wrest control of Najaf from Sadr's forces. A gradual process of chipping away at territory

Marines run to board a CH-53E at the Al Qaim FOB during the Fallujah operation. CH-53Es can rapidly transport large numbers of troops – and supplies and light vehicles – and are key to the USMC's mobility. In February 2005 one was tragically lost in a crash that killed 31 Marines.

USMC 3rd MAW Air Order of Battle, Iraq region, November 2004

3rd Marine Air Wing, 16th Marine Air Group, Al Asad AB	
Marine Light/Attack Helicopter Squadrons 169, 367	36 x AH-1W, 18 x UH-1N
Marine Medium Helicopter Squadrons 365, 268, 774	36 x CH-46E
Marine Heavy Helicopter Squadron 361	CH-53E
Marine Attack Squadrons 542 and 214	10 x AV-8B
Marine Attack Squadron (All Weather) 242	12 x F/A-18D
Marine Aerial Refueler Transport Squadron 452	KC-130
Al Taqaddum Air Base forward operating base	
Marine Unmanned Aerial Vehicle Squadron 1	RQ-2 UAV
Marine Light/Attack Helicopter Squadrons 169, 367	36 x AH-1W, 18 x UH-1N
507th Medical Company (US Army)	UH-60
Ali Al Salem AB, Kuwait	
Marine Aerial Refueler Transport Squadrons 352	KC-130

Partnering the AH-1W in the HMLAs is the UH-1N. The tranquil landscape may hark from Iraq's historic past, but the trees and buildings could easily hide rebels with guns, Manpads and RPGs.

controlled by the rebel militia was begun, backed up by heavy air support from the unit's own AH-1W Cobras, and a battalion of AH-64Ds and OH-58Ds from the 1st Cavalry Division. Air support from USAF fixed-wing aircraft was also called upon, including laser- and GPS-guided bomb strikes from F-15Es. One strike involved the Marines calling down bombs from a F-15E within a few metres of the Imam Ali Shrine, without causing any damage. There were also nightly AC-130 attacks.

It took until the end of August to finally wear down Sadr's men and their morale was broken after a political deal was brokered with a rival Shia leader to take over control of Najaf's holy shrine.

Further east in Al Amara, British troops were engaged in a series of running battles with Sadr militia troops throughout August. These nightly fire-fights often escalated into major battles, and on one occasion an RAF Regiment forward air controller in Al Amara called in USAF F-16s to bomb rebel positions with laser-guided bombs.

Crushing Fallujah

With the Shia rebellion in the south and Baghdad's Sadr City neutralised by early September, the US command in Iraq rolled out its plans to deal once and for all with the rebel stronghold of Fallujah and other rebel position in the 'Sunni Triangle'. USAF Predator, USM Pioneer, US Army Hunter and Shadow UAVs le a major reconnaissance effort to pinpoint rebe bases and head-off ambushes against U troops. AC-130s and fixed-wing jets conducte nightly strikes into the heart of Fallujah in a attempt to 'decapitate' the leadership of th insurgency by killing rebel commander detected by the UAVs and other intelligenc sources.

For most of the summer and into th autumn, Fallujah was the focus of 3 MAW a operations, as Cobras and Hueys forwarc deployed to Al Taqaddum air base to provid on-call close air support for Marines fighting i the streets of the city. A squadron of F/A-18D was dispatched to Al Asad to reinforce 3 MAV for the coming offensive.

Harriers and Hornets were in demand t drop laser- and GPS-guided weapons. The also flew as top cover for road convoys, usin their forward-looking infra-red targeting pod to detect insurgent ambushes. By the end of th year the 3rd MAW had flown some 200,00 sorties out of Al Asad.

When the final battle for Fallujah unfolded i the middle of November with a major divisiona level assault involving up to 20,000 US, Britis and Iraqi troops, USMC Cobras and US Arm Apaches of the 1st Cavalry were on call t support the attack, with two AH-1Ws being los to hostile fire. During the height of operation: Carrier Air Wing 17 aircraft from USS *John I Kennedy* flew an average of 38 missions a da in support of ground troops.

Coalition aircraft losses and damage in Iraq theatre – 1 May 2003 to 31 January 2005

Date	Location	Type	Unit	Serial	Casualties	Cause	Level of damage/result
2/5/03	south Baghdad airport	Chinook HC.Mk 2	No. 27 Squadron	–	nil	accident after ran out of fuel	badly damaged
6/5/03	Iraq?	OH-58D	US Army	94-0163	1 crew injured	accident	–
9/5/03	Samara	UH-60A	571st Med Coy	86-24507	3 killed and 1 injured	accident, flew into power cables	aircraft lost
19/5/03	Hillah	CH-46E	HMM-364	156424/PF-03	4 crew and 1 rescuer killed	accident	total loss
27/5/03	Fallujah	UH-60A	3rd Infantry Division?	–	–	–	badly damaged
12/6/03	north western Iraq	AH-64D	101st Division	–	–	rebel fire	badly damaged
12/6/03	south western Iraq	F-16CG	421st EFS/388th FW	88-0424 'HL'	nil	mechanical problem	total loss
15/6/03	Al Udeid AB, Qatar	F-16CG	421st EFS/388th FW	88-0421 'HL'	nil	hit parked aircraft	aircraft damaged but repairable
24/6/03	Majar-al-Kabir	Chinook HC.Mk 2	No. 27 Squadron	ZA670	7 injured	Iraqi fire	returned to UK for repair
26/6/03	Baghdad	AH-64D	B/2-101 Avn	97-5039	–	accident	–
27/7/03	Arabian Gulf	EA-6B	USN VAQ-135	158800/NH-503	4 crew rescued	accident	aircraft lost
14/8/03	central Iraq	AH-64D	4th Infantry Division, 1-4th Avn	01-5241	2 crew injured and rescued	accident	–
30/8/03	Iraq	CH-47D	F/159th Avn	88-0098	–	accident	aircraft lost
2/9/03	Camp Dogwood south Baghdad?	UH-60L	2/501st Avn	97-26764	1 killed and 1 injured	accident	badly damaged
7/10/03	Habbaniyah	OH-58D	82nd Airborne 1-17th Cavalry	92-0578	2 crew injured	accident	–
10/10/03	–	AH-64D	US Army	–	nil	brown-out while landing in sandstorm	–
13/10/03	–	OH-58D	US Army	93-0991	nil	brown-out while landing in sandstorm	–
15/10/03	Iraq-Syrian border	OH-58D	82nd Airborne?	–	–	rebel small arms fire	aircraft damaged
17/10/03	–	OH-58D	US Army	–	–	bird strike	aircraft damaged
23/10/03	Kirkuk AB	AH-64D	101st Division	00-5219	nil	mechanical problems	aircraft burned
25/10/03	Tikrit	UH-60L	3-158th Avn	–	5 crew injured and rescued	RPG and small arms fire	aircraft lost
30/10/03	nr Balad	AH-64D	6-6th Cavalry	00-5211	–	fire	aircraft lost
2/11/03	Habbaniyah	CH-47D	F Co, 106th Avn Illinois ANG	–	15 killed and 25 injured	rear rotor pylon hit by Manpad SAM	aircraft lost
7/11/03	Tikrit	UH-60L	B/5-101st Avn	92-26431	6 killed	RPG fire	aircraft lost
15/11/03	Mosul	2 x UH-60L	101st Division, 4/101st Avn and 9/101st Avn	–	17 killed and 5 wounded	RPG fire caused one helo to crash into other aircraft	aircraft lost
21/11/03	–	OH-58D	US Army	92-0605	–	–	aircraft recovered for repair
22/11/03	Baghdad Intl Airport	Airbus A300	DHL	–	–	Manpad SAM	safely returned to airport
25/11/03	–	OH-58D	US Army	96-0040	nil	–	aircraft lost
9/12/03	Fallujah	OH-58D	1-82nd Avn	92-0529	nil	RPG and small arms fire	aircraft lost
9/12/03	Baghdad Intl Airport	C-17A	62nd AW	98-0057	1 injured	Manpad SAM	safely returned to airport
10/12/03	Mosul	AH-64D	101st Division	–	nil	rebel fire?	aircraft lost
11/12/03	Iraq	RQ-1L	15th ERS	97-3036	nil	crashed on runway	aircraft lost
1/1/04	Baiji	UH-60L	101st Division	–	10 injured	accident	aircraft recovered for repair
2/1/04	Fallujah	OH-58D	1-82nd Avn	90-0370	1 killed and 1 injured	shoot-down	aircraft lost
7/1/04	Fallujah	UH-60	US Army	–	9 killed	rebel fire	–
8/1/04	Baghdad Intl Airport	C-5B	22nd AS/60th AMW	85-0010	nil	Manpad SAM	safely returned to airport
8/1/04	Fallujah	UH-60L	571st Med Coy	–	9 killed	Manpad SAM	aircraft lost
13/1/04	Habbaniyah	AH-64D	4th Sqn, 3rd ACR	–	nil	rebel fire	aircraft recovered for repair
23/1/04	Qayyarah	OH-58D	B/3-17th Cavalry	93-0950	2 killed	accident	–
25/1/04	Mosul	OH-58D	3-17th Cavalry	93-0957	2 missing	accident, wire strike	aircraft lost
25/2/04	Haditha	OH-58D	4th Sqn, 3rd ACR	97-00124	2 killed	accident, wire strike	–
27/2/04	Al Udeid AB, Qatar	B-1B	34th EBS/28th BW	86-0139	nil	skidded off runway on landing	aircraft repaired
18/3/04	Fallujah	OH-58D	1-25th Avn	–	nil	mechanical problem	aircraft recovered

I MEF first mounted a major operation to establish a tight cordon around Fallujah to stop any rebels escaping. This involved the British Army re-deploying an armoured infantry battle-group to Camp Dogwood, southeast of Fallujah, to work with US Marines of the 24th MEU. A detachment of RAF Westland Puma HC.Mk 1s of 1563 Flight and Army Air Corps Lynx AH.Mk 7 helicopters deployed to support the soldiers of the Black Watch regiment. They flew liaison missions and inserted patrols near suspected rebel bases. One Lynx was hit by ground fire on 10 November and a crewman was badly injured.

Within two weeks Fallujah had been swept clear of rebel fighters, with the US Marines claiming that an estimated 1,200 insurgents had been killed. Coalition losses comprised 38 US, five British and six Iraqi soldiers killed. Three of the US fatalities were non-battle related injuries. Around 275 US troops were wounded in the battle as well.

Above: Devoid of any markings apart from the 'USA' on the tailboom, this Mi-17 at Basra is thought to be operated by the CIA. The Agency also operates Predator UAVs in-theatre.

Right: Other elements of the multi-national forces bring their own support aircraft with them. This Italian navy SH-3D/H was visiting Basra.

While US and coalition ground troops have borne the greatest burden in the increasing guerrilla war, their aviator comrades have also been heavily involved in almost every major combat operation. The most highly praised aviators are the medical evacuation helicopter crews who, time and time again, have flown into the heart of danger to lift out wounded soldiers and civilians.

Elections were held at the end of January 2005, with a good turnout among the Shia and Kurdish elements, but Iraq remained locked in conflict. Insurgents continued to kill US and Iraqi forces on a daily basis. The Pentagon was reportedly preparing plans to maintain a strong garrison in Iraq until the end of the decade, and it remains hard to see an end to the conflict.

Tim Ripley

Date	Location	Type	Unit	Serial	Casualties	Cause	Level of damage/result
30/3/04	Al Taqaddum FARP	2 x AH-1W	HMLA-775	163947/164595	2 injured	accidental collision	aircraft lost
7/4/04	Baqouba	OH-58D	1-25th Avn	92-0583	nil	rebel fire	aircraft recovered for repair
11/4/04	west Baghdad	AH-64D	1-227th Avn	–	2 killed	rebel fire	aircraft lost
13/4/04	Fallujah	MH-53M	20th SOS	–	3 injured	RPG and small arms fire	destroyed to prevent capture
16/4/04	–	CH-47D	C-193rd Avn Hawaii ANG	90-0183	nil	accident	destroyed to prevent capture
21/4/04	Fallujah	OH-58D	–	–	–	–	aircraft recovered for repair
26/4/04	Kut	OH-58D	US Army	–	nil	mechanical problem	aircraft lost
12/6/04	Taji AB	OH-58D	A/1-25th Avn	94-0162	2 injured	accident	aircraft lost
24/6/04	Fallujah	AH-1W	USMC	163939	nil	rebel fire	–
27/6/04	Baghdad Intl Airport	C-130	RAAF	–	1 killed	small arms fire	safely returned to airport
5/7/04	Fallujah	CH-46E	HMM-161	–	3 injured	RPG and small arms fire	pilot able to land aircraft safely
16/7/04	Balad	AH-64A	1-1st Avn	–	nil	–	aircraft recovered for repair
19/7/04	Basra	Puma HC.Mk 1	1563 Flight	XW221	1 killed and 2 injured	accident	aircraft recovered for repair
22/7/04	Arabian Gulf	F-14 and F/A-18	US Navy	–	nil	accident on *Kennedy* during landing	aircraft repaired
28/7/04	Ramadi	2 x AH-1W	HMLA-775	–	1 killed	rebel fire	both helicopters landed safely
5/8/04	Najaf	UH-1N	HMM-166	160439	? injured	rebel fire	aircraft landed safely
8/8/04	Najaf	OH-58D	1-25th Avn	96-0015	nil	rebel fire	aircraft recovered for repair
12/8/04	western Iraq	CH-53E	HMH-365/HMM-166	157742/YH-06	2 killed and 3 injured	accident during night mission	aircraft lost
17/8/04	Balad	MQ-1L	46th ERS	–	–	–	aircraft lost
31/8/04	northern Arabian Gulf	UH-3H	HC-2	–	3 injured	accident landing on USS *Prebble*	aircraft recovered for repair
4/9/04	Tal Afer, nr Mosul	OH-58D	US Army	–	2 injured	rebel fire	aircraft recovered for repair
6/9/04	Fallujah	RQ-2A UAV?	VMU-1	–	–	rebel fire	wreckage captured by rebels
8/9/04	Al-Buaisa, nr Fallujah	CH-46E?	USMC	–	4 injured	rebel fire	–
11/9/04	Mosul? northern Iraq	C-130H	95th AS/386th EAS	–	nil	rebel fire	safely returned to airport
12/9/04	Fallujah	A-15 UAV	USMC	–	–	rebel fire	wreckage captured by rebels
13/9/04	Al Asad AB	CH-53E	HMH-361	–	nil	accident	aircraft recovered for repair
21/9/04	Tallil	UH-60A	1-224th Avn Louisiana ANG	–	3 injured	accident	aircraft recovered
7/10/04	Camp Areifjan, Kuwait	HH-60H	HS-15	–	7 injured	take-off accident	–
16/10/04	Baghdad	2 x OH-58D	1-25th Avn	94-0172/97-0130	2 killed and 2 injured	collision during night mission	aircraft lost
10/11/04	south Baghdad	Lynx AH.Mk 7	UK AAC	–	1 injured	rebel small arms fire	safely returned to Camp Dogwood
10/11/04	Camp Dogwood	Lynx AH.Mk 7	UK AAC	–	nil	rebel mortar fire while on ground	aircraft recovered for repair
10/11/04	Fallujah	AH-1W	USMC	161021	1 injured	RPG and small arms fire	captured by rebels and destroyed
10/11/04	Fallujah	AH-1W	USMC	165363	nil	RPG and small arms fire	–
12/11/04	north Baghdad	UH-60	US Army	–	nil	rebel fire	aircraft recovered for repair
12/11/04	Fallujah	AH-64D	US Army	–	nil	rebel fire	safely returned to airport
12/11/04	Fallujah	AH-64D	US Army	–	nil	rebel fire	safely returned to airport
12/11/04	Fallujah	2 x OH-58D	US Army	–	nil	rebel fire	safely returned to Taji
9/12/04	Mosul	AH-64A	US Army, 1-151st Avn SC ANG	–	2 killed	accidental collision at night	
9/12/04	Mosul	UH-60A	US Army, 4-278th TN ANG	–	4 wounded	accidental collision at night	
15/12/04	Karbala	W-3 Sokol	Polish Army, 25 BKP	–	3 killed and 4 injured	engine failure	
29/12/04	Mosul?	MC-130H	USAF, 15th SOS/16th SOW	–	some passengers injured	hit building work on runway during landing	
13/1/05	Hit	UAV	US Army	–	–	rebel small arms fire	
15/1/05	Mosul	OH-58D	US Army	97-1322	nil	accident	aircraft recovered
21/1/05	Kuwait training area	AH-64D	US Army, 3-3rd Avn	–	1 killed and 1 injured	accident	aircraft written off
26/1/05	Ar Rutbah	CH-53E	USMC, HMM-361	–	31 killed	cause unknown	aircraft lost
28/1/05	Baghdad	OH-58D	US Army, 1-7th Cav	–	2 killed	casue unknown	aircraft lost
30/1/05	near Al Taji	Hercules C.Mk 1P	RAF, No. 47 Squadron	XV179	10 killed	cause unknown	aircraft lost

DEBRIEF

EPAF M3 update
Lockheed Martin F-16A/B MLU

With the recent ending of the M3 OT&E (operational test and evaluation), the EPAF (European Participating Air Forces, consisting of Belgium, Denmark, the Netherlands and Norway) are ready to introduce new capabilities to their F-16 fleets that will maintain the now-ageing aircraft's viability until replacements enter service.

The M3 upgrade consists of several hardware and software changes. Hardware upgrades cover many of the avionics systems used in the F-16 MLU (Mid Life Update) of the late 1990s. The MLDEEU (mux loadable data entry electronic unit) is replaced by the CDEEU (common data entry electronic unit), while the EUPDG (enhanced upgraded programmable display generator) is replaced by a CPDG (colour programmable display generator). The MMC (modular mission computer) has been expanded with an additional 64 MB muxbus card and almost all the other systems have undergone a software or a so-called OFP (operational flight programme) upgrade.

As part of M3, the MIDS (Multifunctional Information Distribution System) has been installed to provide the aircraft with Link-16 capability, as well as for the use of TACAN. The addition of Link-16 in the F-16 is a giant step forward as this system will be one of the major requirements for participation in future operations. Owing to the way the MIDS handles signals for the HSI (Horizontal Situation Indicator, a new EHSI (electronic HSI) had to be installed. Provisions for JHMCS (Joint Helmet-Mounted Cueing System) will be installed in all aircraft, and in the cases of Denmark and Norway a complete system is fitted. Other new capabilities will be the option to carry IAMs (inertial aided munitions) such as JDAM (Joint Direct Attack Munition), WCMD (Wind-Corrected Munitions Dispenser) and JSOW (Joint Stand-Off Weapon). To cater for this, the correct cabling is installed in the wings to allow the use of the Mil Std 1760 weapons interface protocol.

In 2002 the EPAF – together with the USAF – began testing the M3 upgrade. The DT&E (Development Testing and Evaluation) phase was conducted at Edwards AFB, but also included an EOA (Early Operational Assessment) in Europe at Leeuwarden AFB. This EOA was a useful opportunity to pinpoint potential shortcomings in an operational environment at an early stage. DT&E at Edwards involved a USAF M3 aircraft along with an example from Norway.

For the EOA the Europe-built M3 Lead the Fleet (LTF) aircraft were used. This set a challenge to the modification depots as they were on a tight schedule to deliver the LTF aircraft in time. Link-16 was one of the items which received particular focus during the EOA, and was the first time that EPAF F-16s had operated with the system. The Danish LTF aircraft was equipped with a JHMCS kit, representing an opportunity for other EPAF air forces to also test this system. JHMCS kits (by VSI) were later provided to the rest of the LTF aircraft just before the beginning of the OT&E.

With the conclusion of the EOA the DT&E phase was ended. OT&E Part 1 that followed began in March 2004 at Ørland, Norway.

J-210 is one of the first M3-upgraded aircraft. The EOA phase was conducted at Leeuwarden in the Netherlands, and involved participation of a NATO E-3A, a Patriot air defence unit and a ground-based monitoring unit with JTIDS capability.

M3 aircraft have no external differences, but there are visible changes in the cockpit, such as the new EHSI (lower screen in the centre) and the disappearance of the Mode Select Switch which was next to it.

During four weeks of operational testing the team was able to test thoroughly the new and updated systems. Also, some of the new weapons in the inventory could be tested for real. One of the big successes was the JDAM GPS-aided bomb.

In August 2004 the team was ready for the second phase of the OT&E. Part 2 took place at Nellis AFB, Nevada, during Red Flag 3-04. The Red Flag exercise was chosen as it represents the ultimate testing opportunity in a near-real situation. During this joint EPAF operation it was possible to use Link-16 with other Red Flag participants, suitably equipped 'players' including USAF F-15s, F-16s and E-3s, and E-3s from NATO. The EPAF also managed to drop JDAMs for the first time in such a scenario.

The M3 upgrade is not only intended for EPAF F-16s but also for USAF F-16 Block 40 and Block 50 CCIP (Common Configuration Implementation Program) fighters (with a total of around 650).

While M3 is finding its way to the operational fleet, M4 testing has begun and fielding will begin while M3 upgrades are still being performed. M4 will be a software upgrade only (even numbers are software upgrades and odd numbers for hard- and software changes) although a replacement for the MMC is imminent. The M5 upgrade is being discussed, although there are no firm plans as yet. This is likely to be the last major upgrade for the US/EPAF F-16 fleet.

Danny van der Molen

The six EPAF M3 aircraft are seen during their deployment to Nellis for participation in Red Flag 3-04, and the M3 OT&E Part 2. The group comprised two aircraft each from the Netherlands and Norway, plus single aircraft from Belgium and Denmark. Refuelling and transport support for the deployment was provided by a KLu KDC-10. The formation routed through Gander, Newfoundland.

ARH hand-over

AAAC receives Eurocopter Tigers

Above: The first of Australia's 22 Tiger ARHs (A38-001) taxis across the ramp at Oakey during the hand-over ceremony, followed by the two types it is replacing in the armed reconnaissance role. The Tiger force will be concentrated at Darwin in the north.

In a ceremony at the Oakey Army Aviation Centre on 15 December 2004, the first of 22 Tiger Armed Reconnaissance Helicopters (ARH) were formally handed over to the Australian Army Aviation Corps. Five days later the first Australian-assembled Tiger flew for the first time from Australian Aerospace's Brisbane facility.

The first four Tigers were constructed by Eurocopter, and two remain at Marignane until certification and training is complete. The pair handed over at Oakey (ARH-1 and -2) had arrived in Brisbane aboard an Antonov An-124 three weeks earlier, and were reassembled at the Australian Aerospace facility.

The first Tiger will go to the Aerospace Research and Development Unit (ARDU), for the final integration and live-firing trials of Lockheed Martin's AGM-114K Hellfire II. These trials will take place at Woomera between March and May 2005. The second helicopter will be used by the School of Army Aviation at Oakey to conduct crew training development, and will be joined by others from the Australian assembly line throughout 2005.

Australian Tigers are based upon the French Hélicoptère d'Appui Protection (HAP) version, and are armed with a GIAT 30-mm cannon, 70-mm rockets and Hellfire (utilising the digital M299 launcher). A laser designator is incorporated into the Strix sighting system.

No. 161 Reconnaissance Squadron of the 1st Aviation Regiment, at Robertson Barracks in Darwin, will be the first operational unit to receive Tigers in 2006. Sister unit No. 162 Squadron will relocate from Townsville to Darwin and begin conversion in 2007, relegating the elderly Kiowa to the training role.

Nigel Pittaway

Above: The AAAC is in a period of transition. As well as the Tiger ARHs, 21 NH90s are being procured (as the MRH-90) for the transport role, in which they will serve alongside Sikorsky S-70s. The elderly Iroquois is expected to serve until 2007 at least.

Right: When the ARH procurement decision was taken in August 2001 a choice had yet to be made concerning its missile armament. The AGM-114K Hellfire (here represented by loading drill rounds) was subsequently chosen over the Euromissile HOT-3. Unguided weaponry comprises the 70-mm rocket and the GIAT AM-30781 cannon under the nose.

The Tiger ARH will directly replace the Kiowa (such as the FLIR-equipped example above) in the reconnaissance role with Nos 161 and 162 Squadrons. It also restores an attack capability lost briefly with the retirement of the Bell Iroquois (right) from the armed Bushranger role, although the UH-1Hs will serve as transports for some time yet.

Debrief

'Team 60'

Swedish ambassadors

Sweden's 'Team 60' is well-known and well-respected, despite the team having a modest international profile. While the team is busy within Sweden (it made 21 'home' appearances during 2004) it only appeared outside the country twice last year. 'Team 60' has not had an easy time of late. Swedish defence cuts and base closures have put it under extreme pressure, forcing the team to move home and robbing it of several qualified pilots. For a while the six regular display aircraft were reduced to a four-ship, but there has been consistent support for the team at the highest levels within the air force command. This has assured the survival of 'Team 60' when some thought it might perish.

'Team 60's' job is not simply to 'do' air shows. Instead, it acts as a public face of the Swedish military at important national events, such as festivals or sporting fixtures. Like all air force aerobatic teams, the mission is to serve as a recruiting tool. Another softly-spoken objective is to represent Sweden – crucially Swedish industry – abroad. Typically, these visits are to countries with particular links to Sweden, such as Gripen customers or potential Gripen customers. Among the ranks of the latter is Switzerland, which is preparing to find a replacement for its Northrop F-5Es. With an eye on that requirement, and in support of Gripen International, the Swedish Air Force sent a sizeable contingent of Gripens and 'Team 60' to participate in the 2004 Swiss Air Force show, at Payerne. For 'Team 60' the September event was doubly important as it had been specially selected to participate in the 40th anniversary salute for the 'Patrouille Suisse'.

The decision to go to Payerne was made in early January and was supported very firmly by Sweden's air force chief General Jan Andersson. Although 'Team 60' had returned to its six-aircraft complement, two of its pilots were new and two "almost new", said team leader Major Christian Ziese. 2004 was Ziese's fifth year with the team and his second as 'No. 1'. He is all too familiar with the hard work and effort behind every 'simple' air show appearance. "An 18-minute flying display in Switzerland actually equals about 50 hours in the air to get eight aircraft there and back," he noted. "Preparing for that is inevitably going to affect our work here at base, because the team's pilots are the core of the pilot training instructor group. Having us away makes a huge difference – and the same is true for the Gripen guys coming from Ronneby."

The September 2004 visit by 'Team 60' to Switzerland was an epic journey for the little SK 60s and certainly not routine. One of the team's aircraft is seen here at altitude, in the shadow of the Swiss Alps, shortly before letting down for Payerne.

'Team 60' has not had it easy. At the end of 2002 the team's home at F10 Wing, Ängelholm, was shut down and the Swedish Air Force flying training school moved to F16 Wing, at Uppsala. During its stay there, the team could only field a four-ship and, because that base was also marked for closure, operations would have to move again. In September 2003 the SK 60s moved to Malmen, just outside Linköping. Malmen is a congested base with busy airspace. It hosts most of Sweden's military trials flying, while the Saab factory – with its own intensive test and development flying

'Team 60' made its first appearance as a six-ship in 1976, but the Swedish Air Force has a tradition of aerobatic flying that dates back to the 1930s. The team's pilots are all Flygvapen instructors and its aircraft are part of the central pool of SK 60 trainers, based at Malmen.

Debrief

During 2004 'Team 60' trialled this new grey scheme (above) on one of its jets as the team considered changes to its traditional 'half-camouflage' colours (right). The underwing smoke pods carried by the Saab SK 60Ws are a relatively new innovation. Their introduction, in 2001, was forced by the aircraft's re-engining with Williams-Rolls FJ44 turbofans. The SK 60's previous Turboméca Aubisque (RM9) turbofans had been rigged with smoke generators welded to the tailpipes, but this was not possible with the more modern FJ44s (known as the RM15 in Swedish service).

programme – is seconds away in the air. While living at Malmen, 'Team 60' now uses the old Brouvalla air base (closed in 1992) as its rehearsal location.

The team has seven SK 60 jets that wear its distinctive blue-and-yellow scheme, applied over the aircrafts' regular camouflage. All of Sweden's SK 60s have been upgraded with Williams-Rolls FJ44-1C turbofans (replacing the original Turboméca Aubisques), to become SK 60Ws. The re-engining programme, launched in the mid-1990s, was applied to 106 aircraft in all. It delivered improved thrust response, better fuel efficiency and much greater reliability – but not a significant increase in overall thrust. There are 35 SK 60s allocated to the two elements of the Flygvapen flying school (the GTU and the GFU), so 'Team 60' absorbs some 20 per cent of available assets there. The seven 'Team 60' jets are flown as regular aircraft within the school syllabus until the display season starts – at that point fitting them with their underwing smoke pods removes them from the training pool.

The team's operation is very self-contained. The Sk 60s need no external support equipment, not even a tow-bar. Their self-deployment skills puts the caravan that accompanies some other air force teams to shame. For the Swiss visit, 'Team 60' took its seven two-seat SK 60s plus one four-seat transport variant. The FJ44s give the jets about two and a half hours endurance, so the hop to Payerne was made directly, after a stop-off at Ljungbyhed (the former F5 Wing) in southern Sweden. A month's planning and preparation was thrown into disarray 24 hours before departure when a problem at Eurocontrol forced the entire journey to be replanned overnight. For the pilots the most important issue was to 'think imperial'. All the SK 60's instruments are calibrated in metric, so each team member had a conversion chart to calculate his speed in knots and the appropriate flight levels within airways (all expressed in feet). In a tightly controlled process, one jet handling all ATC comms for a group of others (the transit was flown as a three-ship and a four-ship formation). The new reduced vertical separation minima enforced in European airspace – and a lack of TCAS – meant that close attention had to be paid to flight levels at all time.

The highlight of the Payerne show was a mass formation flypast of five teams – the 'Frecce Tricolori', 'Team 60', 'Patrouille Suisse', 'Team Aguilla' and the 'Red Arrows' – saluting the Swiss team on its 40th birthday. Either by co-incidence or brilliant foresight the formation contained exactly 40 aircraft. The teams formed up and flew past twice, in an arrow formation led by the 'Patrouille Suisse', at a height of 100 m and a separation of what was supposed to be 50 m but was more like 30 m. There was no way to rehearse – the teams briefed it in the morning and then flew it. Christian Ziese said, "I couldn't stop smiling. In my mirrors I could see the sky full of aircraft. Below, there were 140,000 people watching us. We will never be in anything quite like that again."

Training for the 2005 display season – the team's 29th year – started in March. Although 'Team 60' expects to fly fewer displays in Sweden this year, it will be making two special appearances at European shows – the Austrian Air Force's 50th anniversary celebration at Zeltweg (June) and the 45th anniversary of Italy's 'Frecce Tricolori', at Rivolto (September).

Robert Hewson

PHOTO REPORT
CRUZEX 2004

Between 7 and 20 November 2004 the Força Aérea Brasileira staged one of the largest exercises to have been held in South America – Exercise Cruzeiro do Sul (Southern Cross, or Cruzex) 2004. The event focused on combined air operations in a low-intensity scenario, using NATO-style procedures.

Below: Flying from Fortaleza, five F-5Es from 1º Grupo de Aviação de Caça 'Senta a Pua' flew as Red Force adversaries. These aircraft carry the locally developed Mectron MAA-1 Piranha missile, although standard Sidewinders can also be carried on the wingtips. Rafael Python 3s can be carried on underwing pylons.

Above: A Brazilian F-103E (Mirage IIIEBR) leaps into the air from the runway at Natal, which hosted the majority of the Blue Force aircraft. Brazil's single Mirage unit is the 1º Grupo de Defesa Aérea, normally based at Anapolis. The F-103 is due to be retired by the end of 2005.

Below: Representative aircraft from the four nations that participated in Cruzex 2004 pose at Natal. In the previous exercise, in 2002, Venezuela attended only as an observer. The chance to participate was gratefully accepted to provide the FAV with vital COMAO experience.

Above and left: Grey-painted F-5Es from 1º/14º Grupo de Aviação 'Pampa', based at Canoas, flew from Natal during the exercise as part of the Blue Force. These ex-USAF aircraft differ from the F-5Es of 1º GAvCa not only in their paint scheme, but also by the lack of dorsal fin and refuelling probe. However, as the Brazilian F-5 fleet goes through the F-5EM/FM upgrade process (also known as F-5BR), the 'Pampa' aircraft will be repainted in the jungle scheme and will have dorsal fins and probes installed. It is expected that the upgraded aircraft will be supplied with Rafael Python 4 missiles. Upgrading the two F-5 units will partially bridge the gap between the impending retirement of the Mirages and the introduction of a new fighter. In early 2005 the long-running F-X BR programme for new-build aircraft was terminated, and second-hand aircraft are now being studied.

Right: Better known for its training role, the EMBRAER Tucano also has armed roles in the Brazilian air force. These were practised during the exercise by 10 AT-27s (variously marked as 'T-27' and 'A-27') deployed to Recife. They were used to provide armed escort to Blue Force CSAR helicopters, and also to cover extraction operations by C-95 Bandeirantes. The aircraft came from 1º and 2º Esquadrão of 3º Grupo de Aviação at, respectively, Boa Vista and Porto Velho.

Above: Six EMBRAER AT-26 Xavantes of 1º/4º GAv 'Pacau' deployed away from their home at Natal to join the Red Force, flying from Mossoro. The Xavantes put on a good show during the exercise, even managing to penetrate the Blue Force air defence to attack Natal.

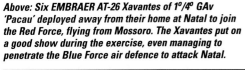

Below: Among the six F-103Es of 1º GDA deployed to Natal was this aircraft, known as 'Papagaio' (parrot), which still carries the special scheme applied in 2002 to celebrate 30 years of Brazilian Mirage operations. In the right hands and with good tactics the Mirage is still a viable air defence tool, but it is increasingly difficult and expensive to support, and is to be withdrawn without immediate replacement.

Above: Co-produced with Italy, the A-1 (AMX) is a useful light multi-role combat aircraft. Three units sent A-1s to Cruzex, including 1º/16º Grupo de Aviação 'Adelfi' at Santa Cruz, which flew fighter-bomber missions from Natal as part of the Blue Force. As A-1s go through overhaul or upgrade, the original two-tone grey scheme is replaced by a green/grey 'Amazon' scheme, which is considered much more effective over Brazil's rain forest.

Above: Blue Force's second AMX unit was 1º/10º Grupo de Aviação 'Poker' – reconnaissance specialists from Santa Maria. The unit is equipped with RA-1s, in both RA-1A single-seat and RA-1B two-seat forms, the latter being illustrated. Reconnaissance capability is provided by a centreline camera pod made locally by Gespi, which contains traditional wet-film sensors. A real-time capability is to be introduced shortly thanks to the Rafael Recce Lite system, based on the Litening designation pod.

Right: Red Force was assigned three A-1As in the attack role. Here one of the 3º/10º GAv aircraft prepares to take off from Fortaleza, where it was based for the duration of the exercise. Other Red Force aircraft included two EMBRAER P-95A Bandeirulha patrol aircraft.

Surveillance missions of an undisclosed nature were flown for Blue Force by an EMBRAER R-99B of 2º/6º GAv flying from Recife.

2º/6º GAv also provided a pair of R-99A AEW aircraft for controlling Blue Force aircraft. Additional AWACS coverage was provided by a single French E-3F.

As in the previous Cruzex 2002, the Armée de l'Air sent a sizeable contingent. The 'teeth' of the French deployment were three Mirage 2000Ns from EC 3/4 (above), and three Mirage 2000Cs (below) and a single 2000B from EC 1/12. Deployment support was provided by a pair of Boeing C-135FR tankers from ERV 93 (left), of which one stayed for the duration of the exercise. Transall C.160s and a DC-8 ferried in equipment and ground support personnel, while EDCA 36 sent a single E-3F Sentry. The use of AEW aircraft was a very welcome facet of Cruzex 2004, especially for the Venezuelan and Argentine contingents, who had little previous experience of operating under AWACS control.

Although not part of Cruzex, Brazil's newest combat aircraft was much in evidence at Natal during the exercise, including demonstration flights for some visiting Venezuelan officers. The A-29A (single-seat, left) and A-29B (two-seat, above) versions of the EMB-314 Super Tucano have entered FAB service and are undergoing operational work-up with a small unit known as Grupo Alpha. This was scheduled to form the basis of a 'new' 2º/5º GAv 'Joker', which retired its AT-26 Xavantes at Natal in December 2004 in anticipation of becoming the first operational A-29 squadron.

Argentina's Blue Force delegation comprised six A-4AR Fightinghawks from Grupo Aéreo 5 de Caza and a single KC-130H from Grupo 1. Ground personnel arrived by Boeing 707. The Fightinghawks operated mainly in the attack role, although they carried AIM-9 Sidewinders (left). To promote interoperability, the KC-130H mainly refuelled French and Brazilian aircraft, while the A-4s tanked from Brazilian or French aircraft. Venezuela was originally to have sent a 707 tanker, but this did not materialise.

Above: Blue Force established a combat search and rescue force at Campino Grande, equipped with UH-1H, UH-50 (Helibras HB 355 Esquilo) and CH-34 (Super Puma) helicopters from the FAB, plus two AS 532 Cougars from Venezuela (illustrated). The FAV aircraft are flown by Escuadrón 102 'Piaros' of Grupo Aéreo de Operaciones Especiales 10.

Left: Four F-16As and a single F-16B from the FAV's Grupo Aéreo de Caza 16 (161 and 162 Escuadrones) were deployed to Natal. Venezuelan F-16s have been upgraded with assistance from Israel to carry the Litening targeting pod for laser-guided bombs, and Python IV air-to-air missiles.

Three of the five FAV F-16s wore special fin markings applied in 2003 to celebrate 20 years of F-16 operations. The single F-16B (below) taxis with one of the five Litening targeting pods acquired by the Fuerza Aérea Venezolana for its F-16 force. The wingtip missiles are standard AIM-9L Sidewinders.

For Cruzex 2004 Venezuela sent a pair of two-seat Mirage 50DVs (above and below) and one single-seat 50EV (left) from Escuadrón 33 'Halcones'. Ground personnel were ferried in by FAV Boeing 737. Some of Venezuela's mixed Mirage fleet was upgraded by Dassault in the late 1980s/early 1990s to Mirage 50 standard, complemented by additional aircraft from other sources. The upgrade involved fitting Atar 09K-50 engines, Cyrano IVM-3 radar (50EV only) and refuelling probe in a lengthened nose section, Serval radar warning receivers, ULISS 81 INS and canard foreplanes. The parent Grupo Aéreo de Caza No. 11 is known as the 'Diablos' (devils), and the group's badge was painted on a single Mirage 50DV (below) in 2003 to commemorate the 30th anniversary of the Mirage entering Venezuelan service.

HALE/MALE
Unmanned Air Vehicles
Part 2: 21st Century warfighters

The General Atomics Predator has been the most successful of the recent UAVs, having reached 10 years of combat operations in 2005. This RQ-1 carries the General Atomics/Sandia APY-8 Lynx synthetic aperture radar in a large nose bulge, as well as a Wescam 14-in (35.5-cm) EO/IR sensor turret. This view highlights the slots into which the slender undercarriage units retract.

Within just a few years of entering service, UAVs have demonstrated their ability to at least partially replace manned systems in certain roles, including precision attack, and further development will only increase their capabilities. UAVs are unlikely to completely replace manned systems, but they are here to stay and can no longer be ignored.

Operation Allied Force, the campaign to remove Bosnian Serb forces from Kosovo in April 1999, turned out to be the first full-scale conflict where UAVs would play a major role. Predators operated widely over Kosovo, tracking enemy and civilian movements and finding targets. According to USAF Secretary Whit Peters, "for the first time ever, we've integrated UAVs… into our strike forces." The USAF fused Predator imagery with information from the Northrop Grumman E-8C Joint STARS [Joint Surveillance and Target Attack Radar System] radar platform, and with images from reconnaissance satellites. Said Peters: "That allowed us to target off of the Predator video, a capability we have never had up to this day until Kosovo."

One problem, though, was communicating Predator information to the fighter pilot – who was seeing the target scene from a different aspect and altitude and in a very different timeframe, and who, unlike the Predator operator, was at personal risk. Later, in 2003, USAF Secretary Jim Roche vividly described the process: "When they'd see a tank between two red-roofed buildings, the Predator pilot

Illustrating three areas of UAV development is this Northrop Grumman family group, comprising the RQ-8 Fire Scout short-range tactical UAV, RQ-4A Global Hawk high-altitude, long-endurance surveillance UAV, and the X-47A Pegasus, a UCAV technology demonstrator.

CeeBee and Have Lemon – arming the UAV

The idea of using a UAV for weapons delivery had surfaced as long ago as 1953, when studies were made into arming the Ryan BQM-34 Firebee target drone. However, it was not until late 1964 that Ryan was given a contract by the US Army Missile Command to demonstrate this capability. Under Project CeeBee, Ryan modified a Firebee with underwing pylons to carry two 250-lb (113-kg) bombs, and further modified it with extended wings to carry two 518-lb (235-kg) SUU-7 cluster bombs. To cater for the extra weight, the drone was launched using a powerful ASROC rocket booster. Test launches were conducted successfully at White Sands Missile Range in New Mexico.

Studies continued, and were given added impetus in the late summer of 1970 when the Soviets installed a large number of SAM and AAA batteries along the Suez Canal. As well as threatening Israeli operations, the Egyptian air defences displayed what kind of reception NATO might face in central Europe. To counter the threat of layered air defences, the Pentagon instigated a programme named Have Lemon, which covered a number of technologies intended to provide what had become the latest military 'buzzword' – defence suppression. Refined precision-guided stand-off weapons were a key technology, as was the use of unmanned aircraft to attack and degrade the air defence network in advance of strikes by manned systems.

On 4 March 1971 Teledyne Ryan received a contract to modify four Firebees to BGM-34A/Model 234 configuration, with underwing pylons. By 14 December the BGM-34A was ready for the first live missile launch. Carrying an AGM-65

One of the BGM-34As is seen under the wing of a DC-130E, armed with an inert AGM-65 Maverick for captive carry trials. The drone also carried a dummy bomb to port, which counterbalanced the missile and was dropped when the Maverick was launched. Ryan's studies showed that asymmetric flight was possible, which would have allowed an operational BGM-34 to launch two separate missile attacks.

Maverick on the starboard pylon, the drone was dropped from a DC-130 over the range at Edwards AFB, and fired the Maverick from 800 ft (244 m) at 2 miles (3.2 km) out from the target. The ground controller, who had real-time imagery from both a TV camera in the drone's nose and the Maverick's seeker, guided the missile to a direct hit – the first launch of a missile from a UAV. A second demonstration a week later achieved the same results. On 10 and 15 February 1972 two more successful strikes were demonstrated, this time using the 'Stubby Hobo' electro-optically guided glide bomb. AGM-45 Shrike anti-radiation missiles were also tested in 1972.

By this time air operations in Vietnam had escalated dramatically, and the USAF was considering using Model 234s against the Vietnamese SAM sites. However, the SAM sites were well camouflaged and were difficult enough to find for manned aircraft. The armed drones were not deployed, and the concept was quietly put back on the shelf. Over 25 years later, it was dusted off again and became operational reality with the Predator/Hellfire combination.

David Donald

These BGM-34s are armed with AGM-45 Shrike anti-radar missiles. This weapon was tested but was not ideal as it would lose guidance if the target radar was shut down during the missile's fly-out. Note the counterbalancing Mk IV bombs.

or systems operator would try to talk the eyes of the A-10 pilot onto the tank. But the people flying the Predator were not people who were schooled in close air support, or the tactics of forward air control.

"USAF air component commander General John Jumper," said Roche, "described what ensued as 'a dialogue of the deaf'. The Predator operator would say, 'Sir, it's the tank between the two red-roofed buildings'," Roche explained. "Of course, the A-10 sees 40 buildings, all with red roofs. The operator of the Predator is looking through a soda straw, at 10-power magnification. He says, 'Well, if you look over to the left there's a road right beside the two houses. A tree line is right next to that. A river is running nearby'. Forty-five minutes later, the A-10 might be in the same zip code, but certainly hasn't gotten his or her eyes on the target."

To address this problem, the USAF decided to fit Predator with a laser designator. Under a two-week quick-reaction programme, the Wescam 14 sensor ball (Wescam had acquired the original supplier, Versatron) was replaced by the Raytheon AAS-44, originally developed for Navy helicopters, which included a laser designator and ranger. It allowed the Predator operator to designate a target directly, producing a laser spot that could be detected by the AAS-35 Pave Penny laser spot tracker (LST) on the A-10, or it could guide a laser-guided bomb (LGB) released from a fighter. It was known as the 'Kosovo ball' although it was not used operationally during that conflict, but it was tested at Nellis in 1999.

Arming the Predator

After the war, the USAF programme office restored the Predators to their original configuration, because the 'Kosovo ball' had not been fully qualified. However, in 2000, General John Jumper, who had commanded air operations in Kosovo, became the leader of USAF Air Combat

Flying from Indian Springs with a 53rd TEG crew, RQ-1L 97-3034 launched an inert AGM-114C and scored a direct hit on 16 February 2001. Another inert missile was fired five days later using the satellite link, and on the same day a live missile was fired using the line-of-sight link. The RQ-1 was configured with the AAS-44 'Kosovo ball' for the initial tests, but subsequent trials introduced the AAS-52 MTS-A turret and AGM-114K Hellfires, which were first fired at China Lake on 30 August.

Focus Aircraft

Right: An early RQ-1K is seen with the fairing removed from its Lockheed Martin Ku-band satellite communications antenna, which measures 30 in (76 cm) in diameter. The vehicle is fitted with the original Versatron (later Wescam) 14TS Skyball EO/IR turret. The small aperture in the extreme nose houses a TV camera to provide the operator with forward vision for landing. At the time the photo was taken in 1996, the Predator system was being utilised on operations over Bosnia from Taszar in Hungary. In 1999 Predators were heavily involved in operations over Kosovo, and were also deployed to Kuwait for operations over Iraq.

Above right: Based on Gnat technology, General Atomics offers the smaller Rotax 582-powered Prowler II for long-endurance (18 hours) tactical surveillance. It first flew in June 1998.

Below right: On 23 August 2002, as part of GA-ASI's ongoing development of the Predator, 94-3011 demonstrated the RQ-1's ability to launch another UAV – the FINDER (Flight Inserted Detector Expendable for Reconnaissance). The tests were conducted in support of the CCAS (Chemical Combat Assessment System), in which the Predator launches one or two FINDERs equipped with Spider chemical sensors. Each FINDER weighs around 57 lb (25.8 kg) and has a wing which slews into position after launch. For CCAS, the Predator itself can be configured with the PIRANHA (Predator Infrared Narrowband Hyperspectral Combat Assessor) chemical sensor. During the first test the FINDER flew a 25-minute pre-programmed mission after launch, before being recovered by a ground operator at Edwards AFB.

Right: Following the successful integration of Hellfire, the Predator was tested with air-to-air armament in the form of the 23-lb (10.4-kg) Raytheon FIM-92E Stinger. Up to four missiles can be carried in twin-tube launchers. By January 2003 Stinger was deployed operationally, and had been launched in an engagement with an Iraqi MiG-25, in which the Predator was shot down.

Command. Jumper was less than pleased to discover that the lasers had been removed from the Predators. Not only did he direct that a fully developed designation system should be installed on the Predator force, but he asked the USAF's Big Safari office – responsible for Predator and other semi-black systems – to take the next logical step and fit Predator with a weapon – allowing three months and $3 million to do preliminary tests. (The idea had been briefly mooted in 1998 – but then largely forgotten.)

The project started in August 2000. The weapon selected for the tests was the AGM-114 Hellfire, designed as an anti-tank weapon for use from helicopters. The programme could be pursued much more aggressively than would be the case in attaching a weapon to a manned aircraft: as USAF programme manager Lieutenant Colonel Kevin Hoffman put it later, nobody was quite sure what might happen when the Hellfire was fired. "There was some concern that a wing might come off or that the vehicle would become yaw-unstable." Phase I tests included a single ground firing and two low-altitude shots – the last of them was guided – and were completed by February 2001. The next step, including higher-altitude launches, was complete by the summer of 2001.

The principal challenge was achieving consistently accurate shots from extreme range and high altitude despite heat, clutter, smoke and cloud. In Phase 1 tests using the AGM-114C missile and the AAS-44(V) sensor, launch altitudes were restricted to 2,000 ft (610 m). Phase II firing used an upgraded version of the AAS-44 – the AAS-52 – known in the Predator community as the MTS-A (Multi-spectral Targeting System). It added a long-range video camera, a laser designator, an eye-safe laser rangefinder and an illuminator, allowing targets to be pointed out for night vision goggle (NVG)-equipped operators on the ground. Later tests used the AGM-114K version of Hellfire, which has an improved guidance system and is better able to re-acquire the target if it is obscured in mid-flight.

Since that time, all new Predators have been delivered as MQ-1L aircraft, with weapon provisions and the MTS-A, and many older RQ-1s have been retrofitted to the MQ-1L configuration.

By late 2000, GA-ASI had delivered 55 Predators to the USAF, against orders for 70 aircraft; the service, by that time, was planning a total buy of 100 air vehicles, a far larger production run than for any comparable UAV. However, the aircraft had yet to be declared fully operational or commence its formal operational test and evaluation (OT&E), mainly because the system had continuously evolved: System 6, the first baseline system including de-icing and a modified engine, had been delivered in 1999 and had spent most of its time deployed overseas.

Also in late 2000, GA-ASI was building prototypes of its next-generation UAV, the Predator B. It was a new air vehicle, sharing only a few components and its basic shape with the Predator A, but used the same satellite link and ground control station. In its initial form, it was three times heavier than the MQ-1L. It was designed to be much more reliable, with turbine power – the Honeywell 331-10 turboprop, the basic powerplant, has a mean time between failure of 150,000 hours, far better than the Rotax engine – and triplicated avionics. Its much larger payload would allow it to carry SAR and electro-optical sensors at the same time.

There were three Predator B models in the initial plan: the basic turboprop-powered model with a 64-ft (19.5-m) wingspan; a higher-altitude, shorter endurance version with an FJ44 turbofan (the engines were located directly above the rear fuselage and were rapidly interchangeable) and an even longer-endurance model, named Altair, with an 84-ft (25.6-m) wing. Originally envisaged as a platform for a commercial airborne cellular communications system, the Altair was first funded by NASA as part of its Environmental Research Aircraft and Sensor Technology (ERAST) programme. By late 2000, Congress had asked the USAF to develop plans for operating a mixture of standard Predators and enlarged aircraft, and to include those plans in its FY2002 budget request.

Global Hawk development

The Global Hawk programme was also making progress in late 2000. The fleet had completed more than 65 flights covering almost 800 flight hours, and aircraft had reached 66,000 ft (20117 m) and performed missions as long as 31 hours. The aircraft had not demonstrated its early design goal of a 24-hour endurance at 6500-km (3,510-nm) radius, but the USAF did expect it to achieve a 24-hour endurance at 2200 km (1,188 nm) – well in excess of any other aircraft's capability – and it still enjoyed a global unrefuelled range.

Global Hawks had deployed successfully away from Edwards – operating out of Eglin AFB, amid a mix of commercial and military traffic – and had flown long-endurance missions over the US and the Atlantic. In May 2000, a Global Hawk operating in support of a NATO exercise, Linked Seas 00, obtained radar imagery of targets in Portugal and transmitted it to NATO Atlantic headquarters. Days later, the aircraft took part in a Navy/Marine amphibious exercise.

In a very important test that started in April 2001, Global Hawk AV-5 was dispatched to Australia for joint trials. The Australian government had been studying Global Hawk for some years, with particular reference to its potential for providing surveillance over Australia's northern coastline, likely areas of land operations in Indonesia, and the Timor Sea. The test represented the first long-distance overseas deployment of a UAV, and also focused on the system's 'reach-back' – its ability to be controlled effectively by a user at a great distance from the system's location.

AV-5 carried the standard sensor suite, with some modifications: radar software was changed to add a maritime moving-target indication mode, and a Litton LR-100 electronic surveillance measures (ESM) system was added. In 238 flying hours, flying 11 missions as long as 23 hours, the Global Hawk demonstrated its ability to perform both over-land and maritime surveillance. Using its maritime MTI mode and the LR-100, it successfully cued its IR sensor to image the carrier USS *Kitty Hawk* – in the region for a joint exercise – at a range of 65 km (35 nm), in enough detail to see the landing signals officer on the deck.

Replacing the Dragon Lady

In late 2000, just before the Presidential election, USAF Secretary Peters floated a plan that would have accelerated procurement and development of Global Hawk by diverting funds from sustainment of the U-2 and allowing the veteran reconnaissance aircraft to retire early. By early 2001, however, the USAF was backing away from that plan. However, the service did accept that – ultimately, and with no certain date – the Global Hawk would replace the U-2.

Underlining Global Hawk's status as a complement to the U-2 – and its potential to replace it in the long term – the USAF selected Beale AFB in California as the first operational base for the type. According to Jumper of ACC: "Co-locating Global Hawk with ... the U-2 mission will ensure Global Hawk transitions smoothly from initial bed-down to full operational capability. It also ensures that cultural issues associated with transitioning from manned to unmanned reconnaissance are in the hands of our current strategic reconnaissance experts at Beale."

Translated, this meant that Global Hawk would eventually replace the U-2, and that the USAF would not buy more of the Lockheed aircraft; but that the Pentagon had rejected the option of setting a firm retirement date for the U-2. Retirement of the U-2 would not be on the table until the Pentagon and Capitol Hill agreed that the UAV was ready to take over all the U-2's missions with equal or superior effectiveness.

Until that happened, the Global Hawk force would complement the U-2 by taking over missions that not only the U-2 could perform. This would allow the remaining U-2s to continue covering the missions that the Global Hawk cannot perform, even if pilot shortages and attrition continued to be problems.

As a result of the tests and the Kosovo experience, the Pentagon decided to bring the Global Hawk into engineering and manufacturing development (EMD) and production as soon as possible. The first step, taken in late 1999, was to keep the production line warm by ordering two more vehicles, to be delivered in mid- and late 2001; a definitive engineering and manufacturing development (EMD) contract was signed in March 2001.

Although some users wanted to modify the Global Hawk – adding payload and onboard power, for instance, to cater for sensors that were carried by the U-2 – the Pentagon settled on a staged development programme. The first EMD

Predator at war

The Predator was first used operationally over Bosnia in 1995, and then in Kosovo in 1999. The type played a major part in surveillance operations during Operation Enduring Freedom in Afghanistan in 2001 (left) and Iraqi Freedom in 2003 (above). For the latter campaign nine RQ-1s and seven MQ-1s (and possibly more) were deployed to two locations, believed to be Ali al Salim in Kuwait and Azraq in Jordan. On occasion up to five Predators were flown simultaneously to support ground operations.

Since the end of Operation Iraqi Freedom RQ/MQ-1 Predators have been flying in Iraq on reconnaissance patrols. They are particularly useful for scouting ahead of security forces, and for keeping watch over security operations. The main operating locations are the former Iraqi AF bases of Tallil (where this aircraft are seen) and Balad.

Left: An MQ-1L lands at Tallil, clutching an AGM-114K Hellfire on its starboard pylon. At least two Predators have been lost in accidents since the major combat phase of OIF was declared over – one at Tallil on 11 December 2003 and one at Balad on 17 August 2004. During a routine patrol over Iraq on 27 September 2004 the Predator fleet notched up its 100,000th hour, of which almost 70,000 were in combat. In 2005 the Italian air force sent Predators to Tallil for their first operational deployment. The RQ-1s are operated by the 28º Gruppo Velivoli Teleguidati, attached to 32º Stormo at Amendola.

fleet increased its total time by almost one-third in only six months – and according to one source there were as many as 32 Predators flying during the Iraqi operation. The 83rd Predator was delivered to the USAF in April 2003 and production was continuing, against orders for 125 aircraft for the USAF and six for Italy. The USAF was preparing to stand up three new squadrons at Indian Springs.

The US wasted no time in deploying the AGM-114 Hellfire on Predator: "several dozen" attacks had been carried out by the end of 2001, according to DoD officials. In March 2002, fire support from a Predator was credited with saving a team of Army and USAF special operations troops caught in a losing fight with Taliban militia on a 10,000-ft (3048-m) peak in Afghanistan, after their helicopter was knocked out. "Two F-15E Strike Eagles and two F-16s had already strafed the enemy, and the F-16s had already dropped three 500-lb bombs," USAF secretary Jim Roche would say later. Finally, an armed Predator was called in. "After calling for a test shot into the side of the mountain – in fact, at a particular tree – to confirm the accuracy of the weapon, our sceptical combatant commandos allowed the Predator pilot to fire his missile into the enemy position, less than 50 metres from their location. Just as the operator promised over the radio, he hit the target ... turning the battle for survival in favour of the Americans."

The first publicised Predator attack, on 4 November 2002, killed six suspected al-Qaeda operatives in Yemen. The aircraft was controlled from Djibouti and the operation was monitored from Saudi Arabia. During the Iraq campaign in 2003, a Predator was credited with destroying a ZSU-23-4 Shilka mobile anti-aircraft gun.

A Predator-launched Hellfire, in fact, took propaganda minister 'Baghdad Bob' off the air, remarked USAF secretary Roche. "As entertaining as 'Baghdad Bob' was, it was really time for him to go away. So we found the mobile satellite dish that they had set up – they put it on the roof of a building near the Grand Mosque in Baghdad. And, of course, we weren't going to use a 500- or 2,000-lb bomb that close to the Grand Mosque. Also, that satellite dish was parked about 150 ft away from the main Fox News antenna."

An F-15E pilot, call sign 'Ivana', and her system operator were tasked to use a Predator to take out the antenna system. "She flew her aircraft into downtown Baghdad very slowly so as to be very quiet, and hits the antenna and its generator and structure with a Hellfire missile," Roche said later. "They not only took out the dish, but Fox News kept on broadcasting uninterrupted, and the mosque was left unscathed."

A 15th Expeditionary Reconnaissance Squadron MQ-1L rests in its shelter prior to another mission during Iraqi Freedom. It is fitted with the AAS-52 MTS-A turret that provides EO/IR imagery and laser designation for the AGM-114K Hellfire carried under the port wing. In operations over Iraq before and during the war, Predators often patrolled with one Hellfire missile balanced by a pair of Stinger AAMs under the opposite wing. This view highlights the marginal tail clearance on rotation, and the 5° nose-down tilt of the pylons.

The 40-hour endurance GA-ASI I-Gnat ER is a development of the I-Gnat, but uses the Predator's longer wings and Rotax 914 engine. It also has a deeper forward fuselage. A Ku-band satcoms antenna in a bulged nose is an option. The US Army uses it in Iraq.

variant, Block 5, would be very similar to the current aircraft, apart from some improvements to its navigation and control systems.

By mid-2001, therefore, the development efforts that had started with the formation of the DARO in 1994 had produced some important results. The Predator was effectively (if not formally) fully operational, and those bugs that were not ironed out were accepted as the price of doing business. The improved Predator B was under development, even though the USAF was not entirely sure of how it would fit into its operations; and the Global Hawk was showing itself to be highly dependable, considering that it was a new and experimental design. Then came the 9/11 attacks, leading to major changes in every aspect of military operations.

UAVs were among the first weapons deployed in the US war against terrorism – first in Operation Enduring Freedom (OEF) in Afghanistan, and then in Operation Iraqi Freedom. Details of many of those operations have not yet been released, but both Predator and Global Hawk were as active as possible – bearing in mind that Global Hawk was still very much a developmental programme in September 2001.

Predator: from strength to strength

By early 2003 the MQ-1 Predator had logged a total of 65,000 flying hours with the USAF, half of them in combat. The pace of operations was indicated by the fact that the Predator logged its 50,000th hour in late October 2002 – the

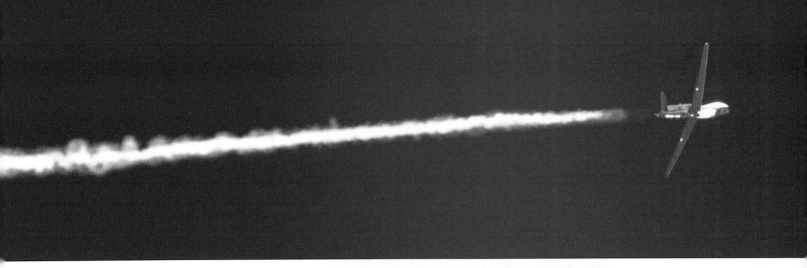

The Hellfire missile itself has been tailored to the kinds of targets encountered by Predators with warhead changes. A thermobaric – high-temperature incendiary and concussion – warhead has been fielded for use in buildings and caves, and steel and tantalum 'sleeves' have been designed to make the warheads more effective against civilian-type vehicles used by paramilitary and terrorist forces.

The 'fighter Predator'

Hellfire was not the only weapon fitted to the Predator. Late in 2002, the Big Safari office launched a programme to fit FIM-92E Stingers – designed as surface-to-air missiles – to the Predator as a defence against hostile aircraft. The programme started with safe carriage tests, continued with four simulated air-to-air engagements against an F-16C, and concluded with four tests with unguided weapon simulators (blast test vehicles, or BTVs) and four live missile firings. The entire project took only 60 days.

At least one fighter-versus-Predator engagement took place over Iraq in January 2003, and video was leaked to US media. Although the Predator lost the fight, the fact that a Predator could be carrying a missile no doubt forces an attacker to be more circumspect – which was probably why the video was leaked in the first place.

In another quick-reaction programme, USAF Special Operations Command established a system which allows Predator imagery to be received as streaming video onboard the AC-130 Spooky/Spectre gunship. Voice communication between the gunship's tactical controller and the UAV operator, both seeing the same picture, allows the gunship to hit the right target. Known as the remote operations video enhanced receiver (Rover), it was deployed operationally at the end of 2001. Another Rover variant has been developed for ground forces.

A more complex system is Scathe Falcon, a palletised version of the Predator GCS that can be installed on a C-130. A control and communications antenna is installed beneath the transport's fuselage. The goal was to extend line-of-sight range – in circumstances where the Ku-band satellite link might not be available – and to carry out Hellfire engagements at maximum range.

Overall, the USAF – in testimony to Congress in 2004 – characterised the Iraqi operation as Predator's first 'networked' operation. "By using both in- and out-of-theatre control stations with beyond line-of-sight aircraft control, we provided the Combined Forces Air Component Commander (CFACC) with additional capability and redundancy to simultaneously control five aircraft over the battlefield, three of which were controlled via reach-back from the United States," assistant USAF secretary Marvin Sambur said. The Predator's satellite link made it possible to operate more Predators in the Iraqi theatre without overloading the satellite links at the CFACC headquarters. "This combined reach-back operation allowed our units to increase their operational flexibility, more efficiently manage manpower, minimise forward footprint, and reduce our high operations tempo. Predator has extended its success this past year by providing tactical imagery directly to ground combat forces, and by providing targeting information to AC-130 gunships."

Also operational with export customers and 'US government agencies' is the I-Gnat, a cousin of the Predator that

differs principally in not having a satellite link. This lightens the aircraft and gives it a longer endurance than Predator – more than 50 hours. In 2003, the US Army acquired a number of I-Gnats to provide overhead surveillance around troops on the ground, avoiding some of the surprises that befell supply columns in the Iraqi operation. They are fitted with the APY-8 Lynx lightweight SAR, developed by General Atomics and Sandia National Laboratories. The 115-lb (52-kg) Ku-band radar is claimed to represent a breakthrough in size and performance, with a 1-ft (0.3-m) resolution at 30 nm (55 km) and a 4-in (0.1-m) resolution at 13.5 nm (25 km). Maximum range is 46 nm (85 km).

By early 2003, as many as 30 USAF or Central Intelligence Agency Predators had been lost due to accidents or hostile action. According to several sources, however, most of these accidents were ascribed to operator error or misjudgments, including loss of situational awareness on landing – Predator is not autonomous, but is flown remotely with the aid of a forward-facing camera – and inadvertent flight into bad weather. One Predator was lost when the operator flew it into 'a hazardous cloud'. By late 2002, the Predator fleet had a cumulative loss rate of 32 class A mishaps – resulting in the loss of the aircraft, but excluding combat losses – per 100,000 hours. That rate is approximately 10 times that of the F-16.

Some older Predators were deliberately sacrificed. Aircraft that were either due for overhaul or in need of modernisation were stripped of equipment and sent to

RQ-4A Global Hawk AV-3 (98-2003) pulls contrails at altitude (above) and sails over Edwards AFB (left), from where most flight testing takes place. This aircraft has seen a lot of active use: it was one of two Global Hawks that participated in Operation Enduring Freedom in late 2001, and by the time of the invasion of Iraq in March 2003 was the only Global Hawk available for operations (AV-4 and AV-5 having both been lost during OIF missions). During both conflicts it operated from Al Dhafra in the UAE, which also played host to a detachment of 9th Reconnaissance Wing U-2Ss. The RQ-4A was flown by the 12th Expeditionary Reconnaissance Squadron, with both USAF and Northrop Grumman personnel. Global Hawk missions over Afghanistan often reached nearly 30 hours in duration and provided theatre commanders with exceptional intelligence. AV-5 was lost in December 2001, the cause being determined as an incorrectly installed nut that caused an actuator rod to be bent out of shape, in turn causing the RQ-4 to spin during its let-down and a subsequent break-up of the vehicle. AV-4 was lost on 10 July 2002 after a loss of power and reported engine vibrations, later traced to a fuel nozzle malfunction. The controllers elected to perform an emergency landing at an airfield in Pakistan, but the aircraft struck a ridgeline as it attempted to glide in. Initial investigations identified problems with the terrain database in the vicinity of the airfield. At the time of its loss it carried 42 OEF mission symbols.

Above: Painted white, the position of the Global Hawk's various antennas is clearly visible on AV-3. The underwing blade antennas serve the UHF line-of-sight datalink, while the two blisters under the rear fuselage serve the X-band Common Datalink and Inmarsat back-up C² system. The large bulge under the fuselage houses the SAR/GMTI radar. Also visible are the 'T' antennas for the UHF satcoms on top of the engine and DGPS on the side.

orbit over suspected hostile forces – who rapidly learned that shooting at a UAV is a very good way of declaring yourself as a legitimate target.

Global Hawk over the battlefield

Global Hawk was also a combat veteran by early 2003. Out of 3,000 flight hours in the programme, almost 1,500 hours had been logged in combat as part of Operations Enduring Freedom and Iraqi Freedom. The aircraft had been deployed to a base in the region – either Al Dhafra in the UAE or Pakistan, according to differing reports – supported by a mixed force of Northrop Grumman and USAF personnel, and has repeatedly self-ferried from the US to the Middle East.

Two Global Hawks – AV-3 and AV-5 – were sent to the Middle East in October 2001, flying one-stop via Australia on routes that avoided densely packed civilian airspace. Early tasks included the monitoring of a riot at a detention camp in north-eastern Afghanistan; AV-5 was operating hundreds of miles away, but was quickly retasked to help map the fires inside the camp. *Ad hoc* tasking, controlled directly from the Combined Air Operations Center (CAOC), became routine, with the help of software and hardware modifications to the MCE.

In the attack on al-Qaeda's redoubt in Tora Bora, Global Hawk's IR sensor proved valuable in detecting campfires and lookouts on ridges, confirming which of the many cave shelters in the region were active. After 17 missions, however, AV-5 crashed at the end of December 2001, due to a technical problem; AV-4 was deployed to replace it, but it in turn was lost in July 2002.

This left AV-3 as the only Global Hawk available for the invasion of Iraq. In this case, the UAV worked mainly with strike assets, checking 'kill boxes' and target areas, a few hours ahead of strike and suppression of enemy air defences (SEAD) operations, and providing updated information on targets that had entered or left the area. The UAV's radar was a key factor in allowing air strikes to continue during a sandstorm in late March, in conditions that blinded daylight video and IR sensors. According to some reports, Global Hawk was used to monitor Iraqi force movements in the path of the coalition forces advancing towards Baghdad and was instrumental in multiple tank kills.

Global Hawk performed well, according to the Pentagon. One statement noted that, "although accounting for only three percent of the total ISR sorties during the 26-day campaign, it provided over half of the time-critical targeting data against Iraq's air defence assets." In the light of this experience, the USAF is pushing ahead with improved and enlarged versions of both the Predator and Global Hawk.

Hunter-killer – the MQ-9

The USAF has adopted the GA-ASI Predator B as a so-called 'hunter-killer' UAV, designated MQ-9A. As noted above, three Predator B prototypes were being built in mid-2001. After 9/11, the USAF took delivery of two Predator B aircraft – the first prototype, flown in early 2001, and the second aircraft, which was to have been built as a jet but was delivered with a turboprop. (The third aircraft with extended wings was completed as NASA's Altair.)

The first of seven pre-production MQ-9As ordered under a risk-reduction contract – distinguished by a deeper fuselage and with a 4500-kg (9,921-lb) gross weight – made its first flight in October 2003. Deliveries of this batch of aircraft will be completed in late 2005. By mid-2004, GA-ASI was under contract to deliver a total of 13 MQ-9As by the end of 2006. The company is expanding its manufacturing capacity with a new 15000-m² (161,464-sq ft) facility at Sabre Springs, near San Diego, which will be dedicated to producing airframes for the Predator B and standard Predator.

The USAF plans to acquire 60 MQ-9s. Service entry will follow initial operational test and evaluation (IOT&E), which is set for 2007, but the type may well be operational on a limited basis before that time: "You can't discount our ability to introduce a weapon system quickly if requirements exist," Lieutenant Colonel Eric Mathewson, chief of the UAV requirements division at Air Combat Command, remarked in mid-2004. (At least one Predator B has been deployed to what is euphemistically referred to as 'an area of interest'.)

Key new features of the MQ-9A include MIL-STD-1760 stores management – allowing the vehicle to carry a variety of smart weapons – integration of the APY-8 Lynx synthetic aperture radar (SAR), and the MTS-B, a larger version of the Raytheon Multi-spectral Targeting System developed for the MQ-1. It packs a massive 56-cm (22-in) diameter folded-optics telescope, to spot and identify targets from high altitude.

AV-6 was the penultimate ACTD Global Hawk, first flying on 23 April 2002. It is seen taxiing at Edwards AFB (right), and with its Northrop Grumman E-8C J-STARS stablemate (above). Both aircraft are important elements in the increasingly networked US ISR (intelligence, surveillance, reconnaissance) architecture.

HALE/MALE UAVs

The MQ-9A carries both radar and the MTS-B full time – they are alternatives on the MQ-1 – and has an external payload of 3,000 lb (1361 kg). Three stores stations are installed under each wing – rated at 680 kg (1,500 lb) inboard, 300 kg (661 lb) on the middle station and 70 kg (154 lb) outboard. The pre-production aircraft are armed with Hellfire – not entirely suitable for this high-altitude platform because of its limited slant range – but tests were under way with the 500-lb (227-kg) GBU-12 laser-guided bomb by mid-2004. The MQ-9A can carry the Joint Direct Attack Munition (JDAM) and other weapons that have to be programmed before release. The primary weapon, according to one USAF source, will be the 500-lb GBU-38 JDAM, and the GBU-39 Small Diameter Bomb will be carried when it is available.

The MQ-9A has an endurance of up to 48 hours, although 24 hours will be more typical with an external payload, and will cruise at up to 48,000 ft (14630 m) – outside the envelope of many air defence systems. It has a low infra-red signature and – with the help of a new low-visibility paint scheme – is hard to detect visually, but its aluminium propeller rules out a low RCS, and it will be mainly used in areas where friendly forces have established air superiority.

A new sensor to be tested on Predator B is General Atomics's Foglite, a laser radar (Lidar) sensor that is claimed to be capable of penetrating moderate cloud cover, smoke, dust, foliage and camouflage. The short-wave IR laser system can be mounted in the laser designator bay of the MTS-B sensor and is intended to identify targets with enough accuracy to permit weapon release.

Air-to-air missiles

Mathewson confirmed in mid-2004 that the USAF was studying the use of air-to-air missiles – both the infra-red guided AIM-9X and the radar-guided AIM-120 – on the MQ-9. Airborne targets could either be detected and tracked by other systems, using the MQ-9 as a forward-located airborne launch platform, or by the UAV's own radar. Although the Predator B has been flying with the APY-8, the USAF is launching a separate competition to select a radar for the production aircraft. Anticipating that the new radar may have a 'power-hog' active electronically scanned array (AESA), GA-ASI is installing a beefy 45-kVA alternator on the MQ-9 and its derivatives, and Honeywell is modifying the engine to handle such loads.

In fact, Predator B was emerging in 2004 as a direct rival to Global Hawk, and was being pitched to a number of potential customers. The first Predator B was used by the Department of Homeland Security in October/November 2003 for a trial named Operation Safeguard, flying missions more than 30 hours long over the southern border of the US. The operation "resulted in numerous arrests and confiscations" according to GA-ASI CEO Tom Cassidy. Undetectable at its cruising altitude, the Predator B could detect people and vehicles crossing the border and vector DHS helicopters and ground teams to intercept them.

The UK's Joint UAV Experimentation Program (JUEP) used a Predator B to test the Goodrich DB-110 Raptor day-night long-range oblique photography (LOROP) reconnaissance pod, which is in UK service aboard the Tornado GR.Mk 4. The Predator B is being studied as a possible replacement for the RAF's long-serving Canberra PR.Mk 9s.

Known as AF-1, RQ-4A 02-2000 was the first production aircraft to be rolled out from Northrop Grumman's Antelope Valley Manufacturing Center at Palmdale. By the time this event occurred in the summer of 2004 the Global Hawk was already a veteran of two wars. Subsequent Air Force deliveries (from AF-4) will be of the enlarged RQ-4B variant.

Above: Global Hawk's interim colour scheme highlighted the location of the aircraft's main antennas, and also made it more visible. A new scheme was introduced in which just the wings are white, displayed here by AF-1 at Edwards.

Left: Airmen check over RQ-4A Global Hawk AF-3 (02-2010) at Beale AFB, California. It was the first air vehicle to be handed over for service with the 12th Reconnaissance Squadron of the 9th Reconnaissance Wing. The aircraft wears the 9th's well-known 'BB' tailcodes and four-cross finstripe. The choice of the 9th RW as operator for the Global Hawk was logical, as the wing has a long history of strategic reconnaissance and is the current operator of the USAF's U-2S fleet.

Focus Aircraft

Above: The first Predator B cruises over southern California on an early test flight. The prototype made its first flight from GA-ASI's flight test facility at El Mirage on 2 February 2001. The Predator B typically carries an 800-lb (363-kg) payload up to an altitude of 50,000 ft (15240 m).

General Atomics MQ-9 Predator B

Buoyed by the success of its Predator A, and in response to demands for ever larger payloads and better performance, GA-ASI began the company-funded development of a larger design, known as Predator B, in 1999. The Honeywell TPE331-10T turboprop was used for greater reliability and to provide more power, both for propulsion and for onboard systems. The two prototypes were rated at 7,500 lb (3402 kg) MTOW, but this was raised in subsequent aircraft to 10,000 lb (4536 kg) and the undercarriage strengthened for – among other things – the carriage of 1,000 lb (454 kg) more fuel.

Wearing the 57th Wing's 'WA' codes, this 11th RS MQ-9A is one of the pre-production batch delivered to the USAF after the two prototypes had been handed over. The first of them flew on 17 October 2003. Noticeable is the deeper fuselage of the later model. At least one MQ-9 is believed to have been deployed to the Middle East for operational evaluation in 2004. Among the many overseas services interested in the Predator B is the RAF, which conducted a series of tests in 2004/05 using the Predator B configured with a DB-110 LOROP sensor (as installed in the Tornado's Raptor pod) in an underwing pod.

Production RQ-4

Global Hawk passed a milestone at the end of October 2004 when the third production aircraft, AF-3, became the first RQ-4A to arrive at Beale AFB to join the newly formed 12th Reconnaissance Squadron – an event which had been long delayed due to operational demands. At the end of 2004, the system had logged more than 5,000 flight hours, including 2,000 hours in operational deployments to the Middle East.

AF-3 is part of the third low-rate initial production (LRIP) batch of Global Hawks. This group is due to comprise six aircraft – four for the USAF and two for the US Navy's Global Hawk Maritime Demonstration (GHMD) programme – and deliveries are due to be completed in October 2005. It will include the first three enlarged RQ-4B vehicles, which will be the standard production vehicle from that point. The RQ-4B features a longer fuselage, a new wing with 4.3-m (14-ft 1-in) greater span, a revised landing gear and a 150 per cent increase in available electrical power, to 25 kVA. Payload increases from 900 kg (1,984 lb) to 1350 kg (2,976 lb), and the new vehicle is 2500 kg (5,511 lb) heavier than the RQ-4A.

Increased payload and power reflect the need to carry multiple payloads, such as electronic surveillance measures, and the power requirements of active electronically scanned array (AESA) radars. The RQ-4B is intended to be the first platform for the Raytheon/Northrop Grumman Multi-Platform Radar Technology Insertion Program (MP-RTIP), with flight tests starting in 2006. The MP-RTIP is a modular radar, designed around a number of basic components that can be reconfigured for different platforms. It is also intended to be used on the E-10A battle management platform, the intended successor to Joint STARS.

Advantages of the AESA include its ability to switch its beam almost instantaneously from one target to another; for example, it will allow Global Hawk to maintain simultaneous coverage of multiple, widely separated moving targets, and will also be able to search for airborne targets such as fighters and cruise missiles while continuing to conduct ground surveillance.

Export prospects

With endurance UAVs as an established part of US forces, and proven in combat, more operators – in the US and worldwide – are planning to acquire them. Global Hawk's range and payload are such that it is tightly controlled under the Missile Technology Control Regime (MTCR), but Australia and Japan are considered priority sales targets. The Australian Defence Force has budgeted $570-$750 million in its current defence plan to acquire an operational squadron of five multi-mission UAVs, the Global Hawk being the leading candidate – a final decision is expected in 2005 and Australia expects the aircraft to enter service in 2009/11. Japan has an outstanding requirement for a high-altitude, long-endurance UAV. Japan is likely to seek a major role for domestic industry, and Northrop Grumman is likely to propose that Japanese companies should concentrate on sensors and processing rather than duplicating US-funded work on the air vehicle.

Northrop Grumman and EADS have been teamed since 2000 on the EuroHawk project, aimed at meeting a German requirement for a signals intelligence (SIGINT) platform to replace a fleet of Breguet Atlantic aircraft. In October 2003, a Global Hawk deployed to Nordholz, Germany, for a three-week, six-sortie demonstration of an EADS-developed SIGINT package, working with its own exploitation system on the ground. In late 2004, Northrop Grumman was expecting a contract for a single EuroHawk, a modified Global Hawk with an EADS-developed signals intelligence payload, and a ground station. Four more SIGINT EuroHawks could be followed by up to five image-intelligence aircraft, the company believes.

Ground crew chock an MQ-9 as it returns from a sortie. The vehicle appears to be fitted with the original Wescam 14 turret rather than the more capable MTS-B intended for operations. Although it is virtually a new aircraft, Predator B uses the same links and ground stations as its predecessor. GA-ASI is studying the use of overwing conformal tanks to increase fuel capacity.

IAI's Heron has also been achieving export sales. Known export customers for the Heron include India, where the UAV was fully operational by 2003. Under an agreement with EADS, IAI has developed a version of Heron, named Eagle 1, with a nose-mounted satcoms antenna for over-the-horizon communications. Eagle 1 has been bought by the French Government under its SIDM (Système Intérimaire de Drone MALE, or Interim Medium Altitude Long Endurance Drone System) programme. In 2004, the French Air Force was taking delivery of two ground control stations and three air vehicles. The systems were due to be delivered to the Centre d'Essais en Vol at Cazaux in mid-2004, and are due to be operational at Cognac by the fall of 2005. France intends to operate the Eagle until the definitive EuroMALE UAV enters service after 2011.

One Eagle has been leased to the UK Ministry of Defence, which has operated it in support of the UK's Joint UAV Experimental Program (JUEP) tests. Last year, the aircraft took part in a joint combat rescue exercise in Nevada and a major British Army exercise at the service's training range in Suffield, Alberta. The aircraft performed well in almost 130 flying hours, according to EADS.

Now in the full-scale development phase is the turboprop-powered Heron TP, broadly equivalent to the Predator B. Powered by a 1,000-hp (746-kW) turboprop – believed to be a Pratt & Whitney PT6A – the Heron TP is a 3600-kg (7,936-lb) vehicle with a 26-m (85-ft 4-in) wingspan, cruises at 240 kt (444 km/h) and reaches an altitude of 42,000 ft (12800 m) with a 600-kg (1,323-lb) payload. The advantage of the Heron TP, according to IAI presentations, is productivity: with higher speed, it spends more of its mission time on station, and it carries a greater payload. Compared with the basic Heron, the larger aircraft has twice the productivity, calculated in terms of time-on-station multiplied by useful load. EADS is interested in a version to be known as Eagle 2.

ERAST – NASA high-altitude research

Aurora Flight Sciences has produced a number of high-altitude UAVs for NASA's ERAST programme, beginning with the Perseus POC (proof-of-concept) aircraft in 1991. This led to the Perseus A, of which two were built. This design had to be tow-launched before the Rotax 914 engine could be started. One crashed and the other was retired. The improved Perseus B had a tricycle undercarriage to alleviate the launch problems, and was used for high-altitude work until grounded after a heavy landing. The follow-on Theseus (below) is larger and is powered by two Rotax 912s. Funded under NASA's 'Mission to Planet Earth', the Theseus was delivered in 1996 but was lost in an accident after six flights. A new Theseus II, with TPE331 turboprop, is now in development.

General Atomics has built two types of UAV for NASA's ERAST/Earth Science programmes, the Altus I/II (left), which was based on the Gnat and has been used for thunderstorm research, and the Altair (below). The latter is the third Predator B prototype fitted with long-span wings.

Focus Aircraft

General Atomics MQ-9A Predator B

Fuel is held in fuselage tanks located fore and aft of the centre section, and in integral tanks in the inter-spar areas of the wings. Underwing tanks can also be carried. The high fuel fraction and excellent economy offered by the TPE331 provide the Predator B in standard configuration with an endurance of over 30 hours.

The TPE331 turboprop mounted to face rearward. The air intake feeds air to compressor by ducts on either side of the engine, with exhaust gases being ejected above the engine. Maximum speed is more than 220 knots (407 km

Above: This Predator B carries the 20-in (50.8-cm) diameter AAS-52 MTS-A turret that is fitted to most MQ-1 Predator As. It has a laser designator to provide both generations of Predator with the ability to launch Hellfire. Note the twin forward-facing nose cameras which provide real-time imagery for landing.

This schematic view shows the main features of the Predator B. The nose section houses the main payload, including the Ku-band satcoms antenna dish and undernose EO/IR turret. SAR radar can also be carried in a fairing under the front section. Additional stores up to 3,000 lb (1361 kg) are carried on six underwing hardpoints, and can comprise podded sensors and weapons, the latter including JDAMs, Small Diameter Bombs and AMRAAMs.

Left: One of the Predator B prototypes is seen fitted with the new Raytheon MTS-B turret that incorporates a powerful folding-optics telescope for high-altitude operations, as well as laser designator/rangefinder. The Predator B can simultaneously carry the APY-8 Lynx synthetic aperture radar: RQ/MQ-1s can only carry either the Lynx or MTS.

Having used piston power for its earlier UAVs, General Atomics switched to a turboprop for the Predator B. The Honeywell (formerly Garrett, then AlliedSignal) TPE331 has been around since 1965, and has been improved considerably in the following 40 years. The TPE331-10T that is used in the MQ-9 is based on an engine whose reliability has been proven in a range of twin transports, including the CASA 212, Dornier Do 228, Fairchild Merlin/Metro and BAe Jetstream 31.

Also under study by IAI is the Heron TJ, with a 35-m (114-ft 10-in) wingspan and powered by two Williams International FJ44-2 engines. It would have a 24-hour endurance and reach altitudes of 70,000 ft (21336 m).

Another Israeli endurance UAV is the Hermes 450, produced by Elbit's Silver Arrow division, which received a boost in the summer of 2004 through the selection of Thales to lead the UK's Watchkeeper UAV programme. The 450 and the smaller 180 are the primary air vehicles in the Thales system, the WK450 version being fitted with a SAR. The 450 has also been used by the US Department of Homeland Security for patrols along the southern border.

Reliability is a key element of the Hermes 450, according to Elbit. Most crucial systems are dual-redundant. For example, there are two flight control computers, using a specialised algorithm to cross-check the two computers and identify the system that is failing. Duplication extends to the actuators, the power supply, the gyros, fuel pumps and filters, and ignition. Some components are triplicated. Another part of Silver Arrow's reliability philosophy – shared to some extent by GA-ASI – is that everything is built by the company, including the composite airframe, flight control hardware and software, ground control station and communications links. "It may seem to make economic sense to buy the autopilot from a commercial vendor," remarked marketing director Yair Ganor in a 2003 interview, "but every time you need to make a slight change you have to tackle the issue of proprietary technology." Uniquely, Silver Arrow owns its own engine company: UK-based UAV Engines Ltd (UEL), which supplies the 60-kW (80-hp) Wankel-type rotary engine.

The 450 can be launched from a catapult and recovered to a short, improvised runway, so it is independent of airfields. For corps, theatre and maritime requirements, Silver Arrow has developed the much larger, twin-engined Hermes 1500.

Twin engines serve two purposes. UAV payloads that weigh more than 200 kg (441 lb), Yanor pointed out, "get expensive and unique – you want full redundancy, including propulsion." They also tend to require more electrical power, and the twin engines provide nearly 10 kW.

The 1500 was designed from the outset to accommodate multiple payloads with a high degree of flexibility. "It's not just a heavy payload – it's electrical power and volume," said Ganor. "Some UAVs, you open the payload bay and it's like the hood of a Japanese car – you can't fit a pin in there." The 1500 is designed with volume to spare and the payload bay is located close to the centre of gravity – not in the nose – so that different weights can be accommodated easily. The Hermes 1500 is not yet in production but has been proposed to a number of customers.

In the US, the Navy and Army were both working on requirements for endurance UAVs in 2004. Both services have recognised that endurance UAVs can be controlled by tactical commanders, and respond to their requests for surveillance and imagery, while being based elsewhere. A long-range UAV can support a fleet without flying off a carrier; and a UAV does not need to fly out of a clearing or a stretch of road in order to fly over a battlefield.

BAMS for the Navy

The US Navy's Broad Area Maritime Surveillance (BAMS) programme has emerged from a long study of the degree to which UAVs can help replace the Navy's large (and costly to operate) fleet of P-3 maritime patrol aircraft. Essentially, the service has concluded that the UAV can take over the routine intelligence, surveillance and reconnaissance (ISR) missions performed by the P-3 today, allowing a numerically smaller fleet of Multi-mission Maritime Aircraft (MMA) to concentrate on anti-submarine warfare and littoral surveillance and strike. The Navy selected Boeing to build MMA in July 2004, with a version of the

HALE/MALE UAVs

EuroHawk – German trials

In late October/early November 2003 the first Global Hawk (AV-1, 95-2001) was deployed to Nordholz in Germany to fly a six-flight series of trials with the aim of demonstrating its suitability as a replacement for the German navy's ageing fleet of Atlantic Sigint aircraft. For the flights the RQ-4A carried an EADS Elint payload under the nose, part of the ISISS (Integrated Sigint Sensor System) planned for the EuroHawk. Another EuroHawk payload is a SAR/MTI system with radar in a ventral canoe fairing. Production EuroHawks are based on the larger RQ-4B version, and further proposals include a maritime surveillance platform with search radar.

Two views show the RQ-4A landing at Nordholz during the EuroHawk trials, the EADS Elint payload clearly visible under the nose. The detachment demonstrated the ability of the RQ-4 to operate safely in civilian-controlled airspace up to FL450, which the RQ-4 reaches in about 22 minutes from take-off. Above FL450 the RQ-4 conducted a variety of Elint trials, including near real-time downlinking and inflight retasking. Germany is to acquire five or six EuroHawks for the Sigint mission.

737, and the planned 108-aircraft fleet reflects the fact that it will work with an adjunct UAV.

In mid-2004, the Navy hoped to get BAMS under way during 2005, leading to an initial operational capability (IOC) in 2010, but this was under review as this article was written, and a three-year delay was possible.

Operationally, the goal of BAMS is to keep at least one vehicle in the air at all times in five regions, covering virtually the entire world and allowing the Navy to maintain a continuous maritime picture – rather than recreating it every day with shorter-endurance manned aircraft. The Navy plans to locate BAMS support and ground-control systems at five bases that already host P-3s and will support the MMA: MCB Kaneohe Bay in Hawaii; NASs at Jacksonville, Sigonella in Italy and Kadena in Japan; and Diego Garcia in the Indian Ocean. BAMS is expected to require about 25 vehicles, including reserves.

BAMS is expected to carry a 360° radar – rather than the side-looking systems carried by ground-surveillance UAVs – with SAR, ISAR and ground moving target indication (GMTI) modes; an infra-red/electro-optical sensor; and signals intelligence (SIGINT) sensors. The Navy would also like to see the vehicle carry a communications relay package for over-the-horizon communication between ships, and ship-to-shore communications via satellite.

The prime candidates for BAMS are the Global Hawk and an overwater version of Predator B known as Mariner. Northrop Grumman argues that the Global Hawk – larger, faster and higher-flying than Mariner – is "impervious to weather" and will in practice provide more time on station at long range because it will be above the level of jetstream winds. Moreover, the bigger aircraft inherently packs more payload. The company has not made any sensor selections, but is looking at a radar derived from its own APY-6 – an advanced high-resolution search and targeting system developed by Northrop Grumman's Norden unit and sponsored by the Office of Naval Research.

The Global Hawk could get a boost in the competition thanks to the Global Hawk Maritime Demonstration (GHMD) project, intended to support development of the BAMS requirement. The first of two GHMD aircraft flew in October 2004, and both UAVs will be delivered to the Navy's Patuxent River flight test centre in late spring 2005. Northrop Grumman is under contract to deliver two air vehicles, two launch and recovery units – permitting the aircraft to be operated away from Patuxent River – and a single mission control element (MCE).

The GHMD aircraft are standard RQ-4As, with the addition of Northrop Grumman's LR-100 electronic surveillance measures (ESM) system. This represents a scaling-back of the original plan, under which the aircraft would have been upgraded with the APY-6 AESA radar, an improved EO sensor – including BAE Systems' Dragonfly electro-optical target recognition system – and Titan's Copperfield SIGINT collection system.

Mariner for BAMS

GA-ASI has teamed with Lockheed Martin to propose Mariner for BAMS, with Lockheed Martin as prime contractor, and in late 2004 the two companies were expanding their agreement to cover all maritime UAV users.

Lockheed Martin argues that the Mariner is smaller and less costly than the Global Hawk, and that it represents no greater development risk; the air vehicle is very similar to the Air Force's MQ-9, which should be operational long before BAMS. Also, the Mariner team asserts, the turboprop

During Iraqi Freedom Global Hawk AV-3 was primarily used to provide surveillance of areas prior to either air or land attack, and was particularly concerned with the elimination of air defence assets, as well as plotting Iraqi ground movements. It remained in-theatre at the end of the conflict to provide surveillance cover for ongoing security operations. Flown by the 12th ERS (note the 9th Wing's 'BB' tailcode), it has amassed a large number of operational missions and has been credited with the destruction of numerous targets, as displayed by the impressive 'kill' tally on the aircraft's nose. It is seen here at its forward-deployed location, in front of the portable shelter that is used by the 12th ERS team for maintenance and preparation.

Focus Aircraft

IAI's Heron UAV is used on surveillance missions for the Israel Defence Force. Some reports suggest they are flown by IAI under contract to the military.

Below: This Heron/Eagle 1 is configured with a very large ventral radome for an Elta EL/M-2022 360° maritime search SAR/ISAR radar. The stalk-mounted dorsal antenna is a satellite link for control of the air vehicle and radar information datalink at BLOS (beyond line of sight) ranges. EADS/IAI demonstrated the Eagle at the 2004 Asian Aerospace show at Changi, Singapore. This was the first time a large UAV had performed at a public air show.

Right: The Indian Air Force has operated a fleet of IAI Herons since 2003 on military surveillance operations. The long-endurance UAVs were used in late 2004/early 2005 to aid humanitarian efforts following the Asian tsunami disaster that devastated parts of the Indian and Sri Lankan coastlines.

is more efficient than the RQ-4B's turbofan at low altitudes, so the Mariner will take a smaller hit on mission endurance when it has to drop below cloud cover to image a target with its EO/IR sensor. The Mariner's lower operating altitude is an advantage, the team contends, because shallower graze angles are desirable for a clear picture on an inverse synthetic aperture radar (ISAR) – a system that uses the movement of a ship target to generate a high-resolution radar picture.

There are three external differences between Mariner and Predator B: Mariner carries a 360° surveillance radar in an under-fuselage pod – GA-ASI had flight-tested both the Raytheon Sea Vue and the Telephonics APS-143B by mid-2004 – it has a saddle tank that increases internal fuel capacity by 900 kg (1,984 lb), and it has the 26.21-m (86-ft) wing of NASA's Altair. Maximum take-off weight is increased by 450 kg (992 lb) to 5000 kg (11,000 lb). With the extra fuel, the UAV has an endurance of 49 hours with a full sensor payload – equivalent to 24 hours on station, 2,000 nm (3700 km) from base.

Internally, the Mariner has a new mission payload processor (MPP), developed by Lockheed Martin. It also uses a different mission control station, also developed by Lockheed Martin and based on the company's Valiant workstation, which is itself a member of the Q-70 family of naval and military workstations. The high-capacity wing stations are wired for sensor pods.

Operationally, Mariner is intended to be more autonomous than the basic Predator B, reflecting the fact that radio-frequency data bandwidth is limited at sea: maritime users usually have fewer, smaller satellite antennas and greater distances preclude the use of line-of-sight communications. The MPP is intended to perform onboard sensor cueing and fusion. For instance, if the Mariner's radar detects an unknown ship – without an automated identification system, for example – the MPP will automatically perform an inverse synthetic aperture radar (ISAR) scan and classify it by size and shape. If the target remains suspicious, the UAV will automatically close in – dropping below cloud level if necessary – to image the target with its colour video on infra-red sensors. Only then does the vehicle actually need to send imagery over its satellite datalink, and it can continue to track the target's movement passively or actively. Mariner incorporates automatic landing and take-off systems, which are optional on Predator B.

GA-ASI has used the NASA-sponsored Altair as a Mariner demonstrator, carrying the radar pod. Following initial flights in April off San Diego, observed by US Navy officials, the demonstrator was deployed in July 2004 to the US Coast Guard base at King Salmon, Alaska, flying several missions to observe wildfires and evaluate the use of endurance UAVs for the USCG's operations. It was also the first test of a UAV carrying the USCG's Automated Identification System (AIS), a transponder system analogous to IFF.

Australia, Italy and Spain were showing strong interest in the Mariner in 2004, according to programme officials. The Global Hawk has the inside track in Australia, but the Predator B/Mariner could play both in this contest and in a separate requirement for a UAV to support the country's civilian CoastWatch operation. Mariner, says Cassidy, is exportable to NATO and other allied countries as a surveillance platform.

US Army requirements

The US Army is preparing to select a contractor for its Extended Range/Multi-Purpose (ER/MP) UAV. ER/MP is intended as a division-level asset, capable of acting as a surveillance platform and communications relay above the battlefield, and is also expected to carry weapons. The requirement called for a long-endurance UAV with a satellite link, based on an existing platform – which has limited the competition to versions of the Predator and Heron. The

HALE/MALE UAVs

Army was looking for a number of complete systems, each with 12-18 air vehicles.

Teamed with AAI and Sparta, GA-ASI is offering a Predator variant known as Warrior. The principal change is the introduction of a 'heavy-fuel' engine – that is, an engine burning diesel or JP-5 rather than gasoline, helping the Army reach its objective of having a single, non-volatile fuel on the battlefield. A modified Predator flew on 25 October 2004 with a Thielert Centurion engine, a German-developed turbocharged diesel that has been developed for light aircraft.

Northrop Grumman has joined forces with IAI and Aurora Flight Sciences to promote the Model 397 Hunter II. It is actually a derivative of the single-engined Heron rather than the RQ-5A Hunter, but the name has been selected to emphasise the fact that it will use the same ground station as the Hunter. Both ER/MP teams were awarded contracts in January 2005 for demonstrations, leading to an SDD decision expected in April.

New platforms

Increasing interest in long-endurance UAVs is also manifested in interest in completely new platforms. Such programmes are underway in Europe and Asia as well as in the United States, where a major unmanned combat air vehicle (UCAV) programme is now stressing endurance. After 2010, some planners believe, the USAF could be sponsoring a UCAV the size of a medium bomber, with the ability to perform 100-hour missions over hostile territory.

Some of these programmes have been developed in the 'black world' of classified programmes, and are likely to remain there. One example is the USAF's work on a stealth endurance UAV. Combining stealth with endurance, relia-

EADS/IAI Malat Eagle 1/2

Using the IAI Malat Heron as a basis, the EADS Eagle 1 is a Rotax 914-powered UAV with a bulged forward fuselage housing a satcoms antenna. The Eagle 2 is a PT6A turboprop-powered derivative, similar to the IAI Heron TP.

bility and affordable technology is not easy, as the Quartz and DarkStar experience showed, but the requirement survived DarkStar's 1999 cancellation. The early 2001 incident in which a Chinese interceptor collided with a Navy EP-3E signals intelligence platform, forcing it down in China and compromising Sigint methods, refocused attention on the need for stealth.

Above: The EADS Eagle 1 carries its main payload in the central fuselage, in this case a 360° search radar. IAI offers the similar Heron with an Elta EL/M-2022 maritime radar or an EL/M-2055 SAR radar. The MOSP EO/IR turret is standard.

Left: The EADS Eagle 1 was first revealed (in mock-up form) at the 1999 Paris air show. This aircraft was displayed at a subsequent show, with a cutaway radome showing the installation of the satcoms antenna. The first aircraft flew in 2002 and operations by the French air force began in 2004.

Below left: This Heron has been modified with the Eagle's bulged satcoms nose and an air data probe for flight trials. It retains the stalk-mounted datalink antenna and is fitted with a belly radome. On the tailboom it carries a plate antenna that is probably part of an ESM system. IAI markets the Heron as a possible Sigint platform, and has produced an impression depicting an aircraft liberally covered with similar antennas. The Heron has a maximum take-off weight of 1100 kg (2,425 lb), of which 250 kg (551 lb) can be payload. Useful range, as quoted by IAI, is 200 km (108 nm) using line-of-sight control, 350 km (189 nm) when using an airborne relay, and 1000 km (540 nm) when the UAV is operated autonomously. Standard endurance is 40 hours and a typical operating altitude is 30,000 ft (9144 m).

Focus Aircraft

Elbit/Silver Arrow Hermes 450S

Elbit Systems' Silver Arrow division produces a range of UAVs about which details are sketchy. The Hermes 450S is the mid-range offering with an endurance of around 20 hours. The original Hermes 450 was a twin-engined UAV, with two 36-hp (27-kW) UEL rotary engines in underwing pods, but the 450S has a single pusher rotary engine. The 450 has been widely used by the IDF on surveillance missions, and has been modified to incorporate beyond-line-of-sight (BLOS) control/data transfer capability. It forms part of the successful Thales bid for the UK's Watchkeeper programme.

Above: The Hermes 450's unusual cylindrical fuselage and parasol layout are claimed to be highly efficient from an aerodynamic standpoint, and facilitate access to onboard systems. Standard payloads for the Hermes 450 are TESAR (SAR/GMTI) radar, the DSP-1 'low-end' EO turret, and the Compass 'high-end' EO/IR turret with laser designator option.

Right: UAVs have obvious applications to border surveillance and anti-trafficking missions. Between June and November 2004 the US Border Patrol leased two Hermes 450Ss for an operational demonstration along the southern Arizona border, resulting in numerous arrests. This initiative continued with a pair of US Army RQ-5 Hunters. A Predator B has also been used in US border security trials, and further evaluations are to be undertaken in the near future.

By mid-2001, Lockheed Martin and others were considering a new UAV called U-X. Indications are that it was a large-payload, long-endurance aircraft, capable of carrying similar sensors to the U-2, but that it was not expected to be quite as stealthy as Quartz: it was intended to be able to stand off from threats without being detected, rather than penetrating the IADS. Lockheed Martin looked at a relatively conventional vehicle known as P-610 Distant Star. It would have been a large vehicle, with a wingspan as great as 60 m (197 ft).

However, 9/11 caused plans to change. While Predators and Global Hawk were valuable, they were operating in an environment free from large or medium-sized SAMs. Such systems could rapidly deny airspace to non-stealthy drones. Consequently, Lockheed Martin was awarded a contract in late 2001 to deliver a small, LO UAV as soon as possible. This black-world vehicle – which reportedly resembled a larger version of DarkStar – was used operationally in 2003 to support the war in Iraq.

Perhaps significantly, Williams International has supplied a batch of 1,505-lb (6.7-kN) FJ33 engines for an unidentified government application. The use of the FJ33, which is smaller than the FJ44 fitted to DarkStar, would suggest either that the Lockheed Martin aircraft is very small (in the 5,500-lb/2500-kg class) or that it is a Predator B-sized twin-jet aircraft. There would be two advantages for a twin: the ability to recover the vehicle in case of an engine failure, and the fact that a smaller-diameter engine can have a shorter low-observable inlet and exhaust system, making it easier to package in a short-body, tailless vehicle.

Lockheed Martin has also been reported working on the design of a larger vehicle, known as the Penetrating High-Altitude Endurance (PHAE) UAV. It would have a 3,750-lb (1700-kg) payload and an endurance of more than 16 hours. Lockheed Martin talked about 36-hour plus missions, indicating that the company was considering the use of inflight refuelling.

J-UCAS for reconnaissance

But this kind of classified mission may end up being performed by a vehicle being developed in a public, acknowledged programme: the Joint Unmanned Combat Air System (J-UCAS). In 2004, Boeing and Northrop Grumman – which is teamed with Lockheed Martin – received contracts from the J-UCAS programme office to build prototypes of their UCAV designs: Boeing's X-45C and Northrop Grumman's X-47B. The vehicles are due to fly in 2007.

The idea of endurance has been a factor in the UCAV programme since the mid-1990s, when the idea was first mooted. Lockheed Martin, for example, proposed a UCAV that would be based on an old F-16 airframe, fitted with a long-span U-2-like wing, which would be able to perform missions longer than a single pilot could tolerate. Initially, the DARPA UAV programme – which led to the Boeing X-45A demonstrator – envisaged a small short-endurance aircraft for strike missions; but this changed under pressure from the users and in the light of experience.

The US Navy was pushing for endurance from the start. It had experimented with the Predator – launched from land bases – in the mid-1990s, and General Atomics studied a carrier-based, jet-powered UAV based on the Predator design. The Navy also started to develop a requirement for a Multi-Role Endurance (MRE) UAV. MRE was a sea-based system – whether it would be based on carriers, amphibious warfare ships or small-deck combatants was an open question – which could be used for reconnaissance, suppression of enemy air defences (SEAD) and strike. In April 2000 the Navy issued MRE risk assessment contracts to Boeing, Lockheed Martin, Northrop Grumman and General Dynamics.

By early 2001 the Navy had rolled the MRE requirements into the UCAV-Navy (UCAV-N) programme, also managed by DARPA. When Northrop Grumman unveiled the X-47A

Elbit/Silver Arrow Hermes 1500

Silver Arrow developed the Hermes 1500 in response to demands for larger payloads. Powered by two Rotax 914s, the Hermes 1500 has an unobstructed fuselage for sensor carriage, offering 2 m³ (70.6 cu ft) of capacity. It can lift 350 kg (771 lb) and has an endurance of more than 24 hours at a typical operating height of 30,000 ft (9144 m). Power available for payload operation is 9.8 kW.

The Hermes 1500 was reportedly developed against an IDF/AF requirement for a UAV with large payload and power capacity. The layout easily allows large radar antennas to be carried beneath the centre fuselage, from where they have an unrestricted view across the lower hemisphere once the landing gear has been retracted.

HALE/MALE UAVs

General Atomics Mariner

The Mariner is a joint General Atomics/Lockheed Martin programme aimed primarily at providing a maritime patrol platform for the US Navy's BAMS programme, as well as possible export. The airframe is based on the Altair/Predator B-ER, with long-span wings for a 49-hour endurance. A 360° search radar is carried in a ventral fairing. The third Predator B prototype was completed with the long wings as the Altair for NASA. In 2004 the Altair was used as a Mariner demonstrator, including a deployment to Canada in August.

Right: The Mariner demonstrator is seen on a typical overwater flight. The proposed production Mariner has an 11,000-lb (4990-kg) take-off weight, with a fuel load of 6,000 lb (2727 kg) and payload capacities of 1,150 lb (522 kg) internally and 2,000 lb (907 kg) externally.

Left: The most obvious external feature of the Mariner is the ventral radome for the search radar. The Raytheon Sea Vue and Telephonics APS-143B have been tested.

This image of a container ship was recorded during a Mariner demonstration. Typically, the search radar is used to cue the EO/IR system, which then transmits imagery back to the ground station for vessel identification.

UCAV-N demonstrator, the company noted that the Navy was looking not only for a strike aircraft, but for an intelligence, surveillance and reconnaissance (ISR) platform with an endurance of at least 12 hours.

Early experience over Afghanistan – where USAF fighters had to operate at the absolute limits of their endurance – pushed the USAF, too, towards a larger UCAV. The service's planners rejected some of the concepts behind DARPA's small UCAV: for example, they wanted a vehicle that could be ferried worldwide like a Global Hawk, not loaded into a transport aircraft, and they wanted the vehicle to be able to carry more weapons. Ultimately, the USAF decided that they wanted 'lethal persistence' – the ability to loiter for several hours deep inside enemy territory.

The result was that the UCAV grew, and grew again, until the X-47B and X-45C emerged as stealthy all-wing vehicles, with a gross weight similar to an F-16 and a greater range and endurance than the Joint Strike Fighter. While a 12-hour reconnaissance mission is the baseline, the aircraft will be designed to accommodate auxiliary weapon-bay tanks for longer missions. Both contractors will demonstrate the ability to refuel the UAVs in flight – so that mission endurance is limited by reliability and weapon load. Moreover, Northrop Grumman intends to test the X-47B with a hose-and-drogue pack in one weapons bay, permitting 'buddy-buddy' refuelling deep inside defended territory.

In late 2004, Northrop Grumman J-UCAS programme director Scott Winship disclosed that the company was looking at production UCAVs with a 4.5- to 7-tonne weapon load and a maximum take-off weight of 45 tonnes or more, designed to fly missions of more than 50 hours.

According to Winship, the bigger UCAV resulted from a perceived need for more range and endurance. Northrop Grumman and the USAF saw the UCAV as a way to fill what the company's analysts have dubbed the 'supergap': the inability of existing fighters and bombers to target increasingly capable mobile systems such as surface-to-air missiles (SAMs) and theatre ballistic missiles (TBMs) deep inside enemy territory. "Beyond the IADS [integrated air defence system] our influence is episodic, with satellite and B-2 fly-bys. We need to be there when the target presents itself."

The USAF also sees the UCAV as an ultimate close-air-support weapon. USAF Secretary Jim Roche remarked in 2003 that he and USAF chief of staff General John Jumper "envision the day when we'll have an ultra-stealthy vehicle that orbits over the battlefield. It will talk to the combat controllers and the people on the ground. The person on the ground will dial up what he needs and the dispenser up in the sky will put the desired weapon exactly where it is directed."

Below: In support of the BAMS requirement, two RQ-4As are allocated to the US Navy's Global Hawk Maritime Demonstration (GHMD) programme, drawn from the Air Force's first six-aircraft production batch. This is the first of them, seen landing at Edwards AFB, California, at the conclusion of its first flight from Palmdale on 6 October 2004. Following the completion of trials at the desert base, the two GHMD aircraft were due to transfer to Patuxent River to begin the Navy's operational assessment trials. BAMS is under review and may be delayed, and it is possible there may also be revisions to the associated Boeing MMA manned patrol aircraft programme.

Below left: This model shows how the RQ-4B Global Hawk could possibly look if it is selected for the BAMS mission. The ventral radome houses a 360° search radar that automatically cues the EO/IR sensor turret under the nose.

Boeing X-45A

A product of Boeing's Phantom Works, part of its Integrated Defense Systems division at St Louis, the X-45A is a sub-scale technology demonstrator for Boeing's J-UCAS proposal. Measuring 26 ft 5 in (8.05 m) in length, spanning 33 ft 8 in (10.26 m) and weighing 10,700 lb (4853 kg), the X-45A is powered by a Honeywell F124-GA-100 engine and can carry a payload of 1,500 lb (680 kg) in its fuselage bays. The first of two X-45As made its maiden flight on 22 May 2002 at Edwards AFB. After trials of the X-45A's autonomous operations and weapons release had been accomplished, the two aircraft were flown together to test co-ordinated operations.

Weapons testing with the X-45A began on 18 April 2004, when a 250-lb (113-kg) inert GPS-guided bomb was dropped successfully. The full-scale X-45C is designed to carry eight 250-lb Small Diameter Bombs in its two internal bays.

This model shows how the full-scale X-45C is expected to look when it undertakes its first flight, scheduled for 2007. The $766.7 million contract awarded by DARPA to Boeing covers the construction of three X-45Cs for the J-UCAS programme, along with two mission control elements. Boeing will also integrate the equipment with the J-UCAS Common Operating System and conduct an operational experiment and assessment. The X-45C is designed to deliver a 4,500-lb (2041-kg) payload over a distance of 1,200 nm (2222 km), with a normal mission radius of 1,300 nm (2407 km). On 10 November 2004 General Electric delivered a pair of F404-GE-102D engines for the X-45Cs. Final dimensions are expected to be around 39 ft (11.89 m) in length and a span of 49 ft (14.93 m). The X-45C will weigh around 36,500 lb (16556 kg) and will have a ceiling of around 40,000 ft (12192 m).

Winship noted that the F-35C Joint Strike Fighter (JSF), the longest-range member of the family, offers 1.1 hours of persistence 900 km (486 nm) from the tanker 'safe-line' – where tankers can operate out of range of the IADS. The B-2 can stay on station for 7.4 hours at 900 km and 3 hours at 1850 km (1,000 nm). A scaled-up UCAV could offer B-2-like range (5400-10800 km; 2,916-5,832 nm), combined with the ability to return repeatedly to the tanker and to fly well beyond the limits of human performance.

Overall, says Winship, a force of 100 UCAVs, each with a 50-hour endurance, could operate out of Hawaii, with tankers off the US West Coast, and maintain continuous orbits over one-third of the United States "on a 24/7 basis".

Northrop Grumman's 'cranked kite' is highly scalable, because the relative size and sweep of the outer wing panels can be changed to adjust the trim and performance of the vehicle. Larger, very-long-endurance UCAVs may have two engines, because the likelihood of an engine failure increases with long missions. In late 2004, Northrop Grumman was talking to General Electric about an engine for future UCAV designs, based on the TF34/CF34 family of engines for the A-10 and commercial regional aircraft.

The UCAV is expected to be highly stealthy, combining a natural LO shape – a flying wing with no vertical surfaces – with the latest in materials. It will be designed with an open systems architecture, making it easy to add new payloads and sensors, and will be large enough to accommodate large apertures for electronic and imaging sensors. It should therefore be a very capable ISR platform, and could supersede plans for a dedicated stealth reconnaissance UAV.

Ongoing development at GA-ASI

Among other new UAV programmes is a jet-powered vehicle under development by GA-ASI, tentatively known as Predator C. CEO Tom Cassidy had no comment about the project in mid-2004, "the next generation after Predator B", except to say that it was "going wonderfully well" and was expected to fly in 2005. Reportedly a stealthy design, the new aircraft is different from the FJ44-powered Predator B, which was completed in 2002 but was converted to a turboprop-powered aircraft before it flew. In a 2002 interview, Cassidy said that the company "was not at liberty to discuss" its jet plans, but that "we will have a customer". He also said that a jet was a pre-requisite for carrier-based operations and that the company fully intended to develop a sea-based version of Predator. It seems likely that the GA-ASI vehicle is sponsored either by the Navy or by the CIA – which was GA-ASI's launch customer for the I-Gnat. 'Predator C' could emerge as a major competitor to the Global Hawk in an era of budget cuts. In response, Northrop Grumman considered the Model 396 scaled-down RQ-4, and is now looking at the Model 395 based on the Scaled Composites Proteus.

In Europe, Dassault, EADS and Thales are now collaborating on EuroMALE, a European medium-altitude, long-endurance UAV. EADS will be prime contractor on EuroMALE. Dassault will be responsible for the air vehicle itself, and Thales will develop the ground control system. Sagem has joined the programme, responsible for the line-of-sight datalink and optronic systems, and other nations – notably Sweden and the Netherlands – are expected to join the programme.

The EADS Eagle 2 – alias the IAI Heron TP – may still be a candidate platform for EuroMALE, but Dassault may have other ideas: all the UAV designs that the company has shown to date have been jet-powered, and Dassault favours twin engines for larger UAVs. So far, the only country in the EuroMALE group to have expressed a firm requirement for the vehicle is France. Its MALE requirement, issued in September 2002, calls for a vehicle that adds Sigint, jamming and communications relay to the surveillance and designation functions performed by the Eagle 1 SIDM, so the vehicle is likely to be larger and more powerful than the Eagle 1. France plans to acquire 24 air

vehicles and 20 ground stations after 2010, replacing the SIDM.

Another active MALE programme is in South Africa, where Denel – having successfully developed a short-endurance UAV in the form of the Seeker – is developing the Bateleur UAV. Powered by a pusher engine (either 100-hp/75-kW Rotax 914 or 160-hp/119-kW) Subaru EA-82T), the Bateleur has a 46-ft (14-m) wingspan and can carry a payload of 440 lb (200 kg) for around 24 hours.

Singaporean programme

A mysterious endurance UAV project is Singapore's LALEE (low-altitude long enduring endurance) system. Its existence was first mentioned in 2001, and reports at the time suggested that Singapore Technologies (ST), the developer, had enlisted the help of Burt Rutan's Scaled Composites company to develop the airframe. Despite its name, LALEE is intended to cruise at 60,000 ft (18288 m) – low-altitude is relative to spacecraft – with an 18-hour endurance, carrying a large surveillance radar for overwater and littoral use. It has also been reported that the system will be optionally piloted, and it has been described as having a 737-like wingspan.

These requirements closely matched Rutan's Proteus, flown in July 1998. Proteus is a high-altitude manned research aircraft, but was designed so that it could be modified into a UAV. It is configured to carry sensors or other payloads in interchangeable pods, and has been used for many development programmes since its first flight.

It may seem surprising for Singapore to develop a system like LALEE, but if the country is blocked from obtaining Global Hawk it needs a long-range, radar-equipped system to cover the sea lanes that are vital to its existence and support its defence strategy – which reflects the fact that by the time the war reaches Singapore's own borders it is too late.

By early 2004, it was reported that Rutan had been persuaded to withdraw from any deal with Singapore by the US government (Scaled Composites is 49 per cent owned by Northrop Grumman), and there are some indications that ST may seek cooperation with EADS or IAI – which would no doubt be glad to find a sponsor for the Heron TJ.

Rotary-wing endurance

One of the most remarkable endurance UAVs under development today is the most recent creation of Abe Karem, the designer of the I-Gnat and Predator family. The A160 Hummingbird was developed under contract to DARPA by Karem's Frontier Systems, which was acquired by Boeing in early 2004. It is intended to achieve a range of over 4500 km (2,430 nm) and an endurance of 40 hours, with a 130-kg (286-lb) payload, at altitudes of more than 25,000 ft (7620 m). And, by the way, it's a helicopter.

The first A160 flew in January 2002, but was damaged in a heavy landing in an early test. DARPA funded Frontier to rebuild it in the same configuration as the second and subsequent vehicles, which represent a heavier-payload vehicle with four rather than three main rotor blades. The second aircraft made its first flight in November 2002 and performed a forward flight in February 2003. Tests were continuing in September 2004, and the flight envelope was being gradually expanded.

The unique feature central to the A160 is that the RPM of its main rotor can be reduced to as little as 40 per cent of its maximum value in forward flight, reducing drag and fuel consumption by half. This cannot be done with a conventional articulated rotor because it would lead to catastrophic vibration, so the A160 has a fully rigid rotor. The blades change in pitch, but the blades and hub are hingeless and stiff in the flapping plane – the hub is a solid forging from AerMet 100 high-strength steel – so that the aircraft is controlled as much like a fixed-wing aircraft as a helicopter. Roll and pitch moments are transferred directly from the rotor to the vehicle. (The same 'rigid rotor' design was explored by Lockheed in the 1960s and used in the AH-56 Cheyenne compound helicopter.)

To demonstrate technologies for its UCAV-N/J-UCAS proposals, Northrop Grumman's Integrated Systems sector funded the subscale X-47A Pegasus A, built by Scaled Composites at Mojave. The X-47A is 27 ft 9 in (8.47 m) long and spans just an inch less. It weighs 5,500 lb (2495 kg) and is powered by a 3,200-lb (14.24-kN) thrust Pratt & Whitney JT15D-5C turbofan. Technologies for the X-47A were developed in Northrop Grumman's new Advanced Systems Development Center at El Segundo.

Northrop Grumman X-47A Pegasus

The X-47A Pegasus made its first flight from the Naval Air Weapons Station at China Lake, California, on 23 February 2003. Whereas the Boeing X-45A began life as a demonstrator for a short-range UAV for the USAF, the X-47A was initially a demonstrator for a much longer-ranged US Navy carrier-capable aircraft (UCAV-N). Now, both programmes come under the J-UCAS banner and are vying for merged USAF/USN requirements. The primary aim of the X-47A is to demonstrate its suitability for carrier operations, including low-speed handling and compatibility with carrier landing systems and arresting equipment.

This montage depicts Northrop Grumman's X-47B UCAV-N design landing on a carrier. Much larger than the X-47A, the intended operational aircraft features a 'cranked kite' wing. The view highlights the enormous internal volume of the vehicle, in turn allowing it to carry large fuel loads for long range/endurance. Northrop Grumman is examining scaling up the basic design to create even larger UCAVs with around 50 hours' endurance. These fit well with both the USAF's and US Navy's future plans for a long-range strike/reconnaissance air vehicle with what they term 'lethal persistence'.

Scaled Composites Model 281 Proteus

Officially rolled out on 23 September 1998, Burt Rutan's reconfigurable Proteus had in fact first flown in late July. Funded in part by NASA through its ERAST programme, the Proteus was designed for HALE operations in either manned or unmanned configuration, and has been proposed for a variety of missions ranging from high-altitude research, through commercial TV/communications relay, to military surveillance. It is believed to have once formed the basis for Singapore's LALEE military programme.

Above: The Proteus is powered by two Williams/Rolls FJ44-2E turbofans. The entire centre fuselage is a modular barrel section that can be easily removed and replaced, in turn allowing different payloads to be installed. Payloads can also be carried externally, as here.

The Proteus performs a low flypast in formation with one of NASA's F/A-18As. Proteus was the Greek god who could change his form at will, and the Scaled Composites mimics this ability in part. As well as having an interchangeable centre fuselage and nose section (for manned or unmanned operation), the Proteus has optional extensions for all four wings to cater for various load/endurance requirements. The standard span of the front wing is 54 ft 9 in (16.7 m) and that of the rear wing 77 ft 10 in (23.72 m). With extensions fitted the rear wing can be 91 ft 10 in (27.99 m) in span. Extra large loads can be carried externally, but offset to provide the necessary ground clearance. In such an instance, the wing extensions can be fitted to one side only, offsetting the effects of carrying an asymmetric load. NASA's Hornet fleet, too, is playing a part in UAV development, as two have been modified to investigate mid-air refuelling for unmanned aircraft.

The other contributing factors to the A160's performance are low empty weight leading to a high fuel fraction, and an efficient propulsion system. The goals for the basic vehicle are a 900-kg (1,984-lb) operating empty weight and a 2270-kg (5,004-lb) maximum take-off weight, so that around 50 per cent of the take-off mass could comprise fuel. This is made possible by using composites for almost the entire airframe, including the landing gear.

The A160 made its first flights with a modified, turbocharged Subaru automotive engine, but the Army and DARPA have sponsored several programmes to develop an engine for a production aircraft. Sonex Research is under contract to develop a modified automotive engine, burning JP-5 or JP-8 aviation fuel; Pratt & Whitney was awarded a contract in late 2002 to develop a low-cost, uncooled ceramic turbine for a 370-kW (496-hp) turboshaft intended for the A160 and other advanced vehicles; and Frontier Systems itself has worked on a flat-six turbocharged diesel engine that is expected to deliver 335 kW (449 hp) from a 205-kg (452-lb) package.

DARPA is supporting the A160 programme because of interest from the US Army, which sees strong potential for the vehicle for missions ranging from surveillance and communications relay to armed reconnaissance, the long-range insertion of unmanned ground vehicles and even the exfiltration of special operations crews. In 2004, DARPA and the Army were committed to supporting a five-year programme to mature the A160 technology, running through FY2007.

Boeing plans to develop the A160 to carry a large payload pod under the fuselage and a satellite antenna above the rotor, mounted on the fixed rotor mast. Other options include missile stations, carried on the inside of bulged landing gear doors. (One advantage of the rigid rotor is that it can accommodate a wider centre-of-gravity range than a conventional helicopter.) The A160 is also designed to be fitted with short wings, off-loading the rotor and increasing its maximum speed to an estimated 190 kt (352 km/h).

Search for alternative power

One even more exotic UAV technology was outlined in a USAF Institute of Technology research paper published in early 2003: a revival of nuclear power for aircraft, a technology that consumed a great deal of time and money in the 1950s but was abandoned in the early 1960s.

One of the strategic problems of UAV development is that any technology that can be applied to a UAV – a more efficient engine, lighter, cheaper electronics, better materials – benefits the UAV's main rival, the manned airplane, equally if not more so. Nuclear propulsion is different, for two reasons. Without a crew, it is possible to exploit the only decisive advantage of nuclear power, which is extremely long flight endurance. Also, the problem of shielding the reactor is much less complicated when there are no humans in full-time proximity to the working system.

The AFIT study was based on a Global Hawk, retrofitted with a small nuclear energy source, and powered by an engine modified so that it could run on either nuclear energy – that is, reactor cooling air – or jet fuel. The reactor would not be activated until the UAV was cruising at 40,000 ft (12192 m), and would be shut down at the same altitude before landing. With the Global Hawk's top-mounted engine, the onboard equipment would not have to be wrapped in a heavy shield – a hemisphere encasing the radiation source would be enough.

The study's conclusions were revolutionary: the shield might weigh 2,700 lb (1225 kg), but that would be only 20 per cent as much as the Global Hawk's conventional fuel load. All in all, the nuclear UAV would weigh no more than the standard aircraft, but could stay airborne for weeks on end.

The drawback to the study, though, was that it was based on a heat source called a triggered isomer heat exchanger (TIHE). In 1998, a group of physicists at the University of Texas reported the results of a unique experiment: they had bombarded a small sample of hafnium-178 – an isomer of hafnium, an energy-containing excited state of the basic element – with a dental X-ray.

An ounce of hafnium-178 stores enough energy to boil 120 tons of water, but it normally releases its energy slowly, with a 31-year half-life. But the Texas researchers reported that the X-ray had accelerated the energy release and that the release had slowed when the X-ray was cut off. The energy released was in the form of gamma rays – harmful but dissipated by distance and less persistent than fission products – and the material itself was relatively safe to handle in its normal state.

The potential of triggered isomers attracted a great deal of attention. Not only could they provide a portable, controllable source of long-duration power, but they could lead to an ultra-small nuclear weapon, with 50,000 times

HALE/MALE UAVs

the explosive power of the same weight of TNT. And, since the process does not technically 'split' the atom, such a weapon might be exempt from arms control agreements.

The snag is this: not only has nobody created a device to extract power from a triggered isomer, but other researchers seem to have had difficulty replicating the Texas work, including a well-funded team at the Argonne national laboratory. For some scientists, triggered isomers are perilously close to repealing the laws of physics, and a raging debate has left some of the UT researchers isolated.

Nonetheless, the Air Force paper sparked ideas elsewhere – ideas that may be based on TIHE or on other, arcane or classified developments in reactor design. "There are concepts out there," one source said, "that the only way they'd cause damage in a crash is if they dropped on you." There are materials, the source goes on, that produce energy and need shielding – but not lead – and are not potential bomb materials.

The promise is not just endurance but energy: the ability to power lasers to defend the UAV, or high-power microwave weapons that could force jamming signals even into the enemy's landline-connected systems. This is a serious problem on conventional UAVs: oversize the engine to deliver megawatts of power to such a device, and you lose the cruise efficiency and endurance.

In 2004, a team of industry and government researchers – including the Boeing Phantom Works – were quietly briefing an idea called Air Breathing Satellite (ABS), an innocent name for a nuclear UAV twice the size of Global Hawk. Powered by two modified jet engines with magnetic bearings (so that it would not run out of liquid lubricants) and packing several hundred kilowatts of onboard power, the ABS would have a flight endurance of three years. Missions would include communications – with its onboard power, the ABS could funnel high-rate, secure communications from any airborne platform in the theatre to rear-area commanders or to the US – reconnaissance and jamming.

But even if such energy sources are physically non-explosive, they are room-temperature nitroglycerine in political terms. USAF planners are aware of ABS but very keen to distance themselves from it. It is probably safe to say that if any work is continuing on ABS, it is doing so behind keypad-locked doors; and, as in the case of the original nuclear-powered airplane programme in the 1950s, inflight refuelling might prove to be a more efficient way to achieve the same goal.

The value of persistence

The lesson from operations over Afghanistan and Iraq is that, if there is one 'killer application' for unmanned air vehicles (UAVs), it is long endurance. 'Consumers' of intelligence on the surface – whether in foxholes, ships or submarines – have come to value the unique benefits of persistence. The UAV, circling for hours above its targets, allows the analyst on the ground time to re-look at a suspicious group of vehicles, or at something that may or may not be a camouflaged target, and assess their importance

not just according to their appearance but according to their activity, or the activity around it. For example, lines of civilian vehicles don't all stop at the same moment; military convoys do.

Meanwhile, pilots who are more than willing to face the toughest air defences are less keen to spend 12 hours or more in a fighter-sized cockpit, and there is little argument that human performance declines after such a long period. The USAF calls this UAV advantage 'digital acuity' – the ability to be as sharp after 12 hours as at the start of the mission. But an aircraft with more than 24 hours endurance can be built using off-the-shelf airframe and propulsion technology, and its productivity in missions such as surveillance is hard to beat.

Bill Sweetman

Above: A number of collaborative projects are being discussed in Europe, including the EuroHALE. This impression shows a possible configuration for this vehicle, wearing the markings of one of the interested nations – Sweden.

Left: Several countries have instigated UAV research programmes. Saab flew its SHARC (Swedish Highly Advanced Research Configuration, illustrated) demonstrator in 2002, intended to investigate controls and other systems for a follow-on UCAV demonstrator known as Filur. In 2004 the aerodynamic shape of the UCAV was tested with a small-scale model called 'Baby Filur'. Saab has now joined the Dassault-led Neuron project for a low-observable UCAV.

Boeing/Frontier Systems A160 Hummingbird

Developed by UAV pioneer Abe Karem, the A160 is a rotary-wing UAV that promises to offer the same kind of performance levels as fixed-wing aircraft, thanks to an innovative propulsion system in which the rigid rotor is slowed dramatically when the vehicle is cruising at altitude. Small wings are also being reviewed as a means of offloading the rotor, in turn allowing it to sustain the aircraft on even less power, or enabling it to lift more and fly faster.

An A160 sits outside the Frontier Systems hangar at Victorville, California (formerly George AFB). Frontier was bought by Boeing in 2004 and Hummingbird development now falls under the aegis of the Phantom Works.

Left: Boeing has a DARPA/US Army contract to develop the Hummingbird, which offers unprecedented range and endurance for a rotary-wing aircraft. It is being developed for a number of roles that include unmanned resupply and sensor placement tasks, as well as the more obvious aerial surveillance and attack applications.

SPECIAL FEATURE

Sea Harrier Farewell

Thanks to its exploits in the 1982 Falklands War, the Sea Harrier acquired a reputation out of all proportion to its diminutive size. In updated form it further underlined its usefulness over the Balkans in the 1990s. However, budgetary cuts are now forcing its premature retirement, leaving the Royal Navy without a dedicated carrierborne air defence capability until the JCA arrives in service some time in the next decade. This review looks at the 'Sea Jet' community in its final months of existence.

Sea Harrier FA.Mk 2s from both front-line squadrons are represented by this pair. The lead aircraft wears 800 NAS markings. This unit was the first to be disbanded, ending its 'Shar' days on 31 March 2004. It will be reformed at Cottesmore as a Harrier GR.Mk 9 operator.

On 1 April 2000, Joint Force Harrier was formed to jointly operate the Royal Air Force Harriers and the Royal Navy Sea Harriers. Today, this joint command is still operational and is considered as an overall success: the concept of joint Harrier/Sea Harrier operations has progressed to the point that embarked carrier air groups usually deploy with both types of aircraft.

Soon, all the Sea Harriers will have been withdrawn from operational service, leaving only the RAF Harrier GR.Mk 7/GR.Mk 9 fighters in use, with T.Mk 10/12 trainers in support. With 800 NAS already decommissioned and 899 NAS disbanded on 23 March 2005, only one Sea Harrier squadron remains. However, 801 NAS will also relinquish its last Sea Harriers on 31 March 2006, instead of 2012 as initially planned. "For the Royal Navy, 31 March 2006 will mark the end of the Sea Harrier," explains Commander Tim Eastaugh, Commander of the Sea Harrier Force at Yeovilton. "At that time, the last eight Sea Harrier FA.Mk 2s and two Sea Harrier T.Mk 8s will be withdrawn from use.

"However, two Navy units will be recreated at Cottesmore while an Air Force squadron numberplate will go. The process is quite simple: on 31 March 2006, one RAF squadron [No. 3] will decommission and change numberplate, becoming 800 NAS. In October 2006, 801 NAS will also recommission. Eventually, we will have four squadrons each equipped with nine Harrier GR.Mk 9s flown by 12 pilots stationed at RAF Cottesmore instead of the current structure of three squadrons each with 12 aircraft and 16 pilots. By April 2007, we are aiming to have 24 FAA pilots and 24 RAF pilots serving with those four units, plus FAA and RAF instructors with No. 20(R) Squadron, the Harrier Operational Conversion Unit at RAF Wittering. Royal Navy squadrons will have direct responsibility to embark in aircraft-carriers and ensure that the vessel is capable of mounting strike operations. RAF squadrons will embark too, but for shorter periods, just to retain their own ability to mount embarked operations, and not specifically to train with the ship and task force. In terms of morale, we are collectively confident that the future is promising because of our involvement in the Harrier GR.Mk 7A/GR.Mk 9 and Joint Strike Fighter programmes. At the moment, 15 Royal Navy pilots are already flying RAF Harriers and we are still aggressively transferring pilots to the two RAF stations."

JSF is the future

In the longer term, Joint Force Harrier will convert to the Lockheed Martin F-35B Joint Strike Fighter (called Joint Combat Aircraft, or JCA, in the UK). "For us, operating the JCA will offer many advantages," stresses Commander Lawler. "It will be a 21st Century stealthy airframe with an advanced avionics system. It will be fully supersonic and will offer a more robust bring-back profile. Range will be boosted by 50 per cent at least. In a high-threat environment, the Sea Harrier would have to go in at low-level, whereas the stealthy JCA can fly at high level where fuel burn rates are much lower, and that will make a huge difference. The JCA has been designed from the outset for ease of maintenance, and that will be a welcome bonus. Moreover, thanks to the 3,000-plus aircraft programme, it will be easier and cheaper to upgrade the airframe and avionics systems." The new type will operate from the decks of two new 55,000-ton carriers, HMS *Queen Elizabeth* and HMS *Prince of Wales*, which under current plans will be commissioned in 2012 and 2015, respectively.

"The Joint Force Harrier structure is the path of the future," underlines Commander Tim Eastaugh. "We are using it as a stepping stone to allow the JCA to be brought into service with minimum disturbance. From 2010 onwards, the

Left: *In its last year of operation 899 NAS painted up one of its Sea Harriers in this blue and white 'Admiral's Barge' scheme. The squadron flew Sea Harriers for 25 years before it disbanded in March 2005.*

Right: *This Sea Harrier is armed with two live AMRAAMs for duties over Iraq in the late 1990s. On the outboard pylon it carries a BOL Sidewinder launch rail with integral chaff dispenser, denoted by the bulbous front end of the rail.*

Special Feature

Maintenance-intensive airframe

It is widely acknowledged that the Sea Harrier is a complex aircraft which requires a lot of effort and time for maintenance. No fewer than 120 maintainers took care of the 10 Sea Harrier FA.Mk 2s that served with 899 NAS, the Sea Harrier T.Mk 8s being maintained by SERCO engineers. "The Sea Harrier has an old airframe that requires a lot of maintenance from skilled people," admitted Warrant Officer Neil Alexander, 899 NAS Senior Maintenance Rating. "The aircraft was not built for maintenance. It is manpower-intensive. You must not forget that the Harrier is nothing but a derivative of a demonstrator which was not really intended to end up in front-line squadrons. A lot of things require the wing to be taken off, a process which takes five to six hours. In order to have access to or remove the engine, the wing has to go. In a training squadron environment, it takes a week or a week and a half to perform an engine change. Working round the clock on a ship in shifts, you can't do it in less than a day anyway. The engine weighs about 3,500 lb [1590 kg], and we must use a crane and gantry to remove it."

Since the disbandment of 801 NAS, a lot of spare parts have been reclaimed from the withdrawn aircraft and, at the moment, the supply of spares is not bad. "Because the aircraft is progressively withdrawn from use, no extra spares are being procured," explained Commander Tim Eastaugh, Commander of the Sea Harrier Force. "Therefore, we need to maximise our usage of spares within our existing frame." Thankfully, the Sea Harrier does not require huge quantities of spares because its reliability is rather good. "Sea Harrier reliability is no worse than that of any other fast jet type," stressed Flight Lieutenant Alexandra Hyatt, an RAF maintenance officer on exchange with 899 NAS. "Its only real problem is the amount of man maintenance hours per flying hour because of the engine, wing and fuselage configuration. The large amount of mechanical and electrical connections is a major problem for the engineers too, hence a low overall maintainability."

Operating at sea also creates its fair share of problems. "The ship always has good supplies of spares, but this is a finite stock and we occasionally end up having aircraft sitting in the hangar," confirmed Neil Alexander. "Everything onboard is harder to achieve because of the weather, the ship's motions, the carrier's manoeuvres, the difficulty of moving stores around... We do not have any engine test cell onboard the carriers. This is why each ship carries two spare engines. However, the majority of engine problems occur because of foreign object damage. There aren't any stones on a carrier. It is a much cleaner environment than land bases and we have less engine problems. Every time we detach to another air base, we always send a team to inspect the runways and taxiways because of the high cost and high number of working hours required to perform an engine change. The engine is very robust though, and aircraft have safely come back from flight with massive damage on the compressor blades because of FODs or bird strikes."

A Sea Harrier is seen in the workshop at Yeovilton with its wing removed (above). The wing is lifted off as a complete assembly, providing access to the Pegasus engine, which can then be craned out for maintenance. Compared with the ease with which engines can be slid in and out of modern conventional fighters, this is a difficult and time-consuming process. The thrust from the Pegasus is split into two lateral exhausts (below) to feed the rear nozzles. The front nozzles use cold fan air.

FAA and RAF personnel will be busy preparing the introduction of the type and the first unit should be fully operational by 2014. At the moment, it is anticipated that initial conversion training will be conducted in America."

Sea Harrier training

At the end of March 2005, 899 Naval Air Squadron, the Sea Harrier Operational Conversion Unit, disbanded, marking the end of Sea Harrier conversion training for the Royal Navy.

Based at Yeovilton, Somerset, 899 Naval Air Squadron was the Sea Harrier operational conversion unit for students who had completed basic fast jet training at RAF Valley, Wales. "The syllabus here was divided into two phases," explains Commander Chips Lawler, the last 'Boss' of 899 NAS. "The first one was a conversion to type which comprised 30 sorties. The second one was the conversion to role, divided into 60 to 70 reconnaissance, air-combat manoeuvring, radar training and air-to-ground attack sorties. In a year, we aimed to get six students through. Additionally, we conducted air warfare courses for instructors and refresher courses for pilots who had just completed a ground tour."

To carry out its missions, 899 NAS was equipped with a fleet of 10 single-seat Sea Harrier FA.Mk 2s and four twin-stick Sea Harrier T.Mk 8 trainers. Every year, the fleet logged about 2,400 flying hours. Each instructor within the unit typically flew anything between 200 and 250 hours per annum. Although it was dedicated to Royal Navy fast jet training, the squadron very seldom deployed to sea, the only occasion being the carrier qualifications to give students carrier experience, typically twice a year for a day or two.

Since entering service in the late 1970s, the Sea Harrier has undoubtedly proved its worth in the very capable hands of Royal Navy pilots. "The FA.Mk 2, the latest variant of the Sea Harrier, punches above its weight," stresses Chips Lawler. "Although its airframe dates back to the 1970s, its avionics suite is up to the 90s standard. Its Blue Vixen radar was specifically designed around the AMRAAM missile and we understand a lot about it. The aircraft is not supersonic and it is hardly agile, but, thanks to our stringent training standards, we consider ourselves to be among the top fighters in Europe and probably in the world. Even as a swing-role fighter, the Sea Harrier is reasonably capable. We can deliver laser-guided bombs and we boast a significant self-escort capability. The aircraft is fitted with an internal optical camera that can be optimised for low- or medium-level reconnaissance. Our main restriction in the air-to-surface role is the limited number of hardpoints."

The Sea Harrier can land on many platforms, but combat operations could only be carried out from dedicated carriers. "The Sea Harrier is very flexible, and we have got a general clearance to land on any flat-top carrier," continues Commander Lawler. "Our main limitations are wind direction and the nature of the anti-skid paint on the deck. For instance, when landing on the French *Charles de Gaulle* carrier, we had to send an advanced party to have a look at its flight deck. We did not find any problem and we eventually landed on her.

"The Sea Harrier does not have a big wing and it does not generate much lift before we reach 100 kt [185 km/h]. This is why we either require a ramp to ski jump or a long take-off distance. As a result, operating from the French carrier is not an option as we would suffer from payload restrictions. Nevertheless, we could use the vessel as a diversion or for pre-positioning and logistics. We cross-deck regularly with the Spanish and the Italians, however, as their ships are both equipped with ski jumps.

"The Sea Harrier has a good pedigree to operate from carriers: it is an all-weather aircraft with an excellent instrument system. In peacetime, we can operate down to a 200-ft [61-m] cloudbase and the Blue Vixen radar supports the ability to come into the hover. In fact, we do not have a clearance to do that, but we could in an emergency. We lack an organic inflight refuelling capability, but we can come back onboard with very limited fuel as we do not face any foul deck problem."

Blue Vixen radar

At the core of the Sea Harrier FA.Mk 2's capabilities is the Blue Vixen radar designed by Ferranti/GEC-Marconi (now BAE Systems). This lightweight radar was adopted to supplant the less-capable Blue Fox radar fitted to the earlier Sea Harrier FRS.Mk 1 variant. "The multimode pulse-Doppler Blue Vixen radar has been optimised for a single-seat aircraft and, as such, is largely automated to reduce the pilot's workload," says Chips Lawler. "The radar is operated via the hands on throttle and stick controls, and its track-while-scan mode allows the pilot to track multiple targets at very long range."

Although the pilots at Yeovilton will not confirm it, the radar is said to be able to update navigation data for four AMRAAM missiles simultaneously thanks to a multiple fighter-to-missile datalink capability. It should be noted that the Blue Vixen also offers air-to-surface modes which improve accuracy during air-to-ground firing over mountainous terrain. The radar is unusual in that some elements of the

Until March 2006 801 NAS survives as the last Sea Harrier squadron. The badge comprises a winged trident, and black/white checkers usually adorn the rudder. Aircraft are coded in the '00x'.

system are remotely sited in the rear fuselage and connected by a fibre-optic link to the antenna in the nose.

The Sea Harrier FA.Mk 2 could deliver Sea Eagle anti-ship missiles but the weapon was withdrawn from use a few years ago because of a change of emphasis in the force's role. It was felt that there was no viable threat anymore for the Sea Harrier to be used in anti-ship strikes.

Missile firing

Live missile firings are sometimes carried out in order to check combat efficiency. "We do regularly fire live missiles, two AMRAAMs and two Sidewinders per year on average," reveals Commander Chips Lawler. "For the AMRAAMs, we deploy to Point Mugu, in California. We could fire them in Scotland, but we like to do an in-depth analysis of each firing because you need to know what happened, whether it is a success or a failure. That process is more cost-effective in the USA."

The latest firing campaign in California involved two Sea Harriers and five pilots. "I got very lucky to be selected," declares Lieutenant Ian Sloan. "I was the next senior front-line pilot who had never shot before. Five aircrews deployed to the USA but only two actually fired. For us, California is great because the weather is rarely an issue there. In the past, multiple firings have been conducted, but this time we fired only two single shots. The firings were real trials to expand the envelope of our missile knowledge. We shot at Phantom radio-controlled drones guided by US Navy operators. As a precaution, we had two aircraft ready to fire each time, and both of them got airborne, one acting as a primary shooter and the other as a spare.

"The AMRAAM I fired was rail-launched. It was great to watch it streak away from the jet with a characteristic 'whoosh' sound and a strong impression of speed. There was absolutely no noticeable aircraft movement as it went. I could see the flame of the rocket motor, but not a lot of smoke. It's an image I am going to remember for a while.

"For us, the difficult point was reaching the planned firing parameters. We had practised it only once in the air before but we had 'flown' the profile in the simulator numerous times. However, it is not that technical, so that any average front-line pilot could do it. The test was under the umbrella of the Operational Evaluation Unit which is part of the squadron."

Advanced tactics

Royal Navy fighter pilots are trained to a very high standard: they frequently practise dissimilar air combat tactics against other types and regularly deploy abroad and to various ranges to hone their skills. "Obviously, the Sea Harriers mainly train in the southwest of England but, during conversion courses, 899 NAS also regularly detached to RAF Waddington to take advantage of the North Sea Air Combat Manoeuvring Instrumentation range," says Commander Lawler. "Additionally, we always tried to go three times per course to other bases, generally one in the UK and two abroad, to broaden our students' experience. Usually, we liked to work with somebody who can simulate a good Fox 1 capability, that is semi-active radar-guided missiles, and even a robust Fox 3 capability, that is active radar missiles such as the AMRAAM or the MICA. This is the reason why our main adversaries were the RAF Tornado F.Mk 3s, the USAFE F-15s, the French

The key to the Sea Harrier FA.Mk 2's success is the Blue Vixen radar, which not only supports AMRAAM missiles, but allows the Sea Harrier to function in a 'mini-AWACS' role. The latter was particularly useful over the Balkans in the 1990s, when 'Shars' were often called upon to plug gaps in NATO AWACS coverage.

In its final year of operations 800 NAS also produced a special-scheme aircraft. Dubbed 'Red 22', the aircraft had the squadron badge painted on its spine, and carried the legend '800 NAS. 1980-2004' on the nose.

Four 801 NAS Sea Harriers are parked on deck after returning from a training sortie. Each has a tow truck attached for rapid manoeuvring around the deck. In recent times Sea Harriers have often operated in mixed air groups with RAF Harriers.

Mirage 2000-5s, and the Dutch and Belgian F-16 MLUs. For electronic warfare training, the RAF range at Spadeadam was a most useful destination."

Compared with that of the earlier Sea Harrier FRS.Mk 1, the Sea Harrier FA.Mk 2's cockpit is dominated by two multifunction displays. However, it looks outdated in comparison with the latest fighters being introduced, such as the Typhoon, the Super Hornet or the Rafale. "With only two screens and a head-up display, the FA.Mk 2's cockpit is not particularly ergonomic," underlines Lieutenant Ian Sloan. "That said, the navigation kit with its INS and its GPS and the radar are particularly good. The radar is user-friendly and very automatic. The left screen is generally used as a multipurpose display for navigation data or as a tactical display in the air-to-air attack mode. In that case, we have the elevation data on the left page and the god's eye view on the right one. The small additional radar warning receiver gives us the bearing and type of threat. We were getting the ASRAAM missile and the JTIDS datalink but, unfortunately, we lost the funding for that."

For a defensive counter-air mission from a carrier, the Sea Harrier pilots like to launch as a four-ship formation to set up a combat air patrol (CAP) pattern with two aircraft looking down-threat and two up-threat. "Once airborne, we always operate with our radar switched on and we usually get a picture from one of our surface ships," says Ian Sloan. "The Blue Vixen automatically selects the pulse-repetition frequency it thinks best to maximise detection range. If a threat shows up, we immediately commit on that. Our tactics are based around the capabilities of our jet. For instance, we do not like to go into the visual arena. We much prefer engaging our targets from long range and that is why our main configuration comprises four AMRAAMs. For target designation, we use our left thumb to move a target-marker on the radar screen. When faced with multiple enemies, we can ripple-fire our AMRAAMs in quick succession."

If possible, Sea Harrier pilots will always try to avoid being drawn into a close-range combat. However, if required, Sea Harrier aircrews will honour the threat thanks to the formidable capabilities offered by the Sea Harrier's unique airframe/engine combination. "The Sea Harrier is very responsive, especially in roll. It turns well for a high-wing loading aeroplane, but evidently not as well as an F-16 or a Mirage 2000. We can use the nozzles to generate a lot of nose-pointing. The well-known 'Viffing' manoeuvre, that is vectoring in forward flight, is a great defensive manoeuvre, but is, in fact, more useful offensively to get into position for an attack."

Deck operations

While at sea, deck alert status varies from Alert 60 – routine – to Alert 5, with the pilot sitting in the cockpit and ready to go in less than five minutes. "We either launch individually or as a pair, with two aircraft lined up one behind the other," says Ian Sloan. "The pair's take-off is the most common, and we either launch as an operational pair in a five-second stream, or as a semi-operational pair with a fighter launching and the other waiting behind. In the first case, the first aircraft has just gone airborne when the second is already rolling. The ski jump effectively makes the runway three to four times longer, depending on the configuration. For launches, the carrier will systematically be steaming into wind and the jet would have a 400- to 500-ft [122- to 152-m] deck run with a full load of four missiles and two drop tanks." Sometimes, in very hot weather, 'wet launches' with water injection are carried out. This cools the turbine to allow more fuel into the combustion chamber so that the engine can be run at a higher rpm, hence

India's Sea Harriers

Time may be fast running out for the Royal Navy's 'Shar' fleet, but the type still figures prominently in the plans of the Indian Navy, and is scheduled to do so until at least the end of the decade.

India selected the Sea Harrier to replace its long-serving Sea Hawks in October 1978, having maintained interest in the Harrier throughout its development. The Harrier underwent sea trials aboard India's carrier – INS *Vikrant* – in July 1972, but it was to be another 12 years before the type became an operational reality for the Indian Navy.

India's initial order covered six Sea Harrier FRS.Mk 51 single-seaters, supported by two Harrier T.Mk 60 trainers. The first were delivered on 21 December 1982 to RNAS Yeovilton, where a 'new' Indian Naval Air Squadron (INAS) 300 began to form. On 16 December 1983 the first three aircraft arrived at INS *Hansa*, the shore base at Goa-Dabolim. Five days later Indian Sea Harriers went to sea aboard *Vikrant* for the first time. In February 1984 the first T.Mk 60 was delivered, and in July 1984 INAS 300 was declared operational.

Initially Sea Harriers operated from the conventional deck of *Vikrant*, which needed to retain its catapults and arrester gear to accommodate the Breguet Alizé. The final Alizé launch in May 1987 allowed the carrier to enter refit for the last time, from which it emerged with a 9° 45' ski jump. A Sea Harrier made the carrier's first ski-jump launch on 14 March 1990. However, the elderly 'Majestic'-class carrier was nearing the end of its useful life. It undertook its last cruise in November 1994 and was officially decommissioned on 31 January 1997.

By this time, of course, the Indian Navy was operating a 'new' carrier in the form of INS *Viraat* (ex-HMS *Hermes* of Falklands fame). The vessel, complete with 12° ski ramp, had been acquired in 1986 and was commissioned into the Indian Navy on 20 May 1987 to take over the

Above and right: In October 2004 INAS 300 painted two of its Sea Harriers (603 and 614) in a new light grey scheme with toned-down markings. The national insignia consists of a pale saffron and green roundel, with white deleted, while the squadron's leaping tiger badge is presented in outline (contrast with the original high-visibility markings above). This aircraft carries CBLS practice bomb carriers.

carrier commitment while *Vikrant* entered refit. To extend the IN's Sea Harrier capability, and provide sufficient aircraft to equip both carriers, additional aircraft orders were already in place. The first, placed in November 1985, covered 10 FRS.Mk 51s and one T.Mk 60. In October 1986 a third order was placed, for seven FRS.Mk 51s and another trainer. In 1999 two modified ex-RAF T.Mk 4s were delivered to bring total acquisitions to 23 FRS.Mk 51s and six two-seaters. That year *Viraat* entered a two-year refit from which it returned to service in July 2001. The vessel is expected to be decommissioned in 2010.

Until 1990 Indian Navy Harrier pilots trained in the UK, undergoing basic Harrier conversion with the RAF's 233 Operational Conversion Unit at Wittering. To cut costs, training was moved to INS *Hansa*, where the Sea Harrier Operational Flying Training Unit (SHOFTU) was stood up on 16 April 1990. SHOFTU was merged with the Kiran-operating INAS 551 on 28 May 1995, adopting the new identity of INAS 551B 'Braves' on that date (INAS 551A continued the work of the original INAS 551, providing fleet requirements and basic fighter training using Kirans). The 'Braves' operate the two-seat Harriers, and draw single-seaters from the pool as required. All Sea Harriers wear the leaping tiger badge of INAS 300 'White Tigers'.

Prospective Sea Harrier pilots first go through the Indian Air Force's fighter training programme, undertaking phase 2A at the Fighter Training Wing at Hakimpet (Iskras and Kirans) and phase 3 on the MiG-21 at Chabua or Tezpur. Once back with the Navy, the students undergo simulator and basic helicopter training before converting to the Harrier. The students work up through instrument, night and formation flying before learning air combat manoeuvring. At least 50 vertical landings are required at the shore base before the student goes out to the carrier for two trips in a two-seater. Four solo deck landings are then undertaken to fulfil carrier qualification requirements.

Above and below: The two 'new scheme' Sea Harriers are parked aft on Viraat along with an original scheme aircraft. Also on deck are three Sea King Mk 42Bs from INAS 330 'Harpoons' and a single HAL Chetak from INAS 321 Hansa Flight 'Angels'. The latter is based at Goa as a detachment from INAS 321 at Mumbai. Viraat also plays host to the Ka-28s and recently delivered AEW Ka-31s of INAS 339 'Falcons'.

#	Label
123	Ammunition magazine, maximum capacity 150 rounds
124	Ammunition feed chute
125	Gun gas vent
126	Cartridge case ejector
127	30-mm ADEN cannon
128	Fuselage centreline pylon
129	Lateral missile pylon
130	Drain mast
131	Hydraulic system ground connections, port and starboard
132	Engine fan duct bleed-air connection to reaction control system
133	Forward fuselage lateral fuel tank, port and starboard
134	Digital engine control unit
135	Accessory equipment gearbox
136	Twin alternators
137	Engine bay access panels
138	Oil tank
139	No. 1 UHF antenna
140	Cockpit conditioning system ram air intakes
141	Engine intake front fan
142	Intake suction relief doors, free-floating
143	Engine bay zone 1 venting air intake
144	Forward-retracting nose wheel
145	Trailing axle nose wheel suspension, leg shortens on retraction
146	Taxiing light
147	Starboard air intake
148	Extending boarding step
149	Nose wheel doors
150	Nose undercarriage wheel bay
151	Intake boundary layer bleed air duct
152	Hydraulic accumulator and nose wheel retraction jack
153	Cockpit air conditioning pack
154	Intake boundary layer spill duct
155	Ejection seat headrest
156	Pilot's Martin-Baker Mk 10H ejection seat
157	Canopy emergency release
158	Cockpit sloping rear pressure bulkhead
159	Kick-in boarding step
160	Starboard side console panel
161	Lower UHF antenna
162	Doppler antenna
163	Radar power supply unit
164	Rudder pedals
165	Control column
166	Instrument panel shroud, twin multi-function CRT displays
167	Canopy miniature detonating cord
168	Port air intake
169	Sliding cockpit canopy
170	Demountable inflight refuelling probe
171	Port side console-mounted engine throttle and nozzle angle control levers, HOTAS controls
172	Pilot's head-up display
173	Windscreen panel
174	Windscreen wiper
175	Cockpit front pressure bulkhead
176	Junction box
177	Attitude reference platform
178	Oblique camera
179	I-band transponder antenna
180	Pitot head, port and starboard
181	Nose reaction air control valve
182	Blue Vixen radar equipment module
183	Fresh air ram intake
184	Windscreen washer reservoir
185	Yaw vane
186	MADGE/IFF antenna
187	Radome, hinged position
188	Scanner tracking mechanism
189	Planar radar scanner
190	Glassfibre radome

producing more thrust while minimising engine life consumption.

For every launch and recovery, there is a Sea Harrier pilot in 'Flyco' (Flight Control), acting as a Landing Signal Officer. "On the deck, we will always have live nosewheel steering," explains Lieutenant Sloan, "but, once you are lined up and straight, you do not need it very much during the launch. When ready, you hold 55 per cent power against the brakes and, when you think the timing is good, that is when the ship starts to pitch down, you slam the power, check 100 per cent, and then release the brakes. Normally, you should hit the ramp as it is going up. The front oleo does compress but you don't really feel it. Initially, the exhausts are straight aft during the deck run but, when you reach the ramp exit, you bring the nozzles down 35° relative to the horizontal. This angle offers the best compromise for lift and acceleration. You instantly feel a good upwards momentum and you leave the ramp at anything between 90 and 100 kt [167 and 185 km/h]. Once you are airborne, you have to gradually nozzle out to gather speed. Unless our speed is more than 550 kt [1018 km/h], we always fly with the flaps at mid position as it generates more lift."

Vertical landings

In the Royal Navy, the landings are divided into three different cases:
- Case 1: visual recovery
- Case 2: radar to visual recovery
- Case 3: carrier-controlled approach.

"If we come back as a four-ship formation, we join the 'wait' as two pairs, the wait being a 2-minute, 30-second race track at 1,000 ft [305 m] overhead the ship," explains Ian Sloan. "Exactly 2 minutes and 30 seconds before landing, we will slot in: flying at 600 ft [183 m] above deck on the starboard side of the ship, we will break across the bow into the circuit, with 20 seconds between the Leader and his No. 2, and one minute between each pair. We are usually downwind 1 mile [1.6 km] abeam the carrier and we turn finals when we are abeam the stern of the vessel. All landing checks are carried out between the break and the final turn: flaps down, gear down and water injection on." In the Sea Harrier, water injection gives an extra 5 per cent in power, the 50 Imp gal (227 litres) of demineralised water lasting 90 seconds. "When we roll out level after the final turn, we start looking at the 'meat ball', underneath the bridge, on the island. We will then literally correct ourselves to the 'roger' – that is the correct slope, at 5° off the stern (called Red 175 because it is 175° relative to the ship's direction). Once established on finals, we will select 60° of nozzle at 150 kt [278 km/h]. For the whole process, we have to be nice, tidy and expeditious.

"At 0.6 nautical miles [1.1 km] from the ship, we slam the nozzle lever to the hover stop. Generally, we will aim to use half brak-

To support Sea Harrier FA.Mk 2 training, 899 NAS acquired seven Harrier T.Mk 8s, produced by conversion of T.Mk 4As and 4Ns. The aircraft lacked radar, but had undernose Doppler added and a cockpit with FA.Mk 2 instrumentation.

ing stop to avoid coming to a complete halt as the ship is still moving forward. We have a very useful device called speed trim at our disposal: on the throttle, using the airbrake switch, we can fine-tune our position alongside ship – with 10° forward or aft increments of nozzle – without taking our hands off the controls. At this stage, a 'yellow coat' will be pointing at our allocated landing spot on the deck. Using the stick to control reaction controls at the wingtips, the tail and the nose, we then cross the deck.

"We aim to land on the right-hand tramline in order to have a safety margin against putting our left outrigger over the deck's edge. As soon as we are stabilised above the landing spot, we take a little bit of power off to initiate a descent before putting some power back on again to avoid accelerating downwards at an increasing rate of descent. When we hit the deck, we reduce power to idle straight away and select water off, flaps up, landing light off and nozzles aft." On an 'Invincible'-class carrier, as many as eight Sea Harriers can land in quick succession, with four taxied into the 'graveyard', at the bow of the ship, where the Sea Dart missile system was formerly located.

Night flying is more procedural, with the pilots paying total attention to the 'meat ball' and his instruments. At night, the LSOs play a much more active role. They give more information and are more proactive. "For any recovery, we will hold the ship on radar to get a good fix," says Ian Sloan. "Not having a radar at night is usually a no-go."

The withdrawal of the last Sea Harrier in April 2006 will mark the end of an era. Although the mighty aircraft has proved its worth on many occasions, from the Falklands to Bosnia, it will be the victim of budget constraints that will leave the Royal Navy with no air defence fighter until the delivery of the first multi-role Joint Strike Fighter.

Henri-Pierre Grolleau

By April 2006 the sight of Sea Harriers hovering over the deck as they return from a sortie will be consigned to history. Once the Sea Harriers have landed they can be quickly moved away to clear the deck for following aircraft.

Today INAS 300 has 15 single-seat Harriers available. A typical carrier deployment involves six to 10 aircraft, although the full complement can be carried in emergency situations. Weapons available include the MBDA Magic II air-to-air missile and Sea Eagle anti-ship missile, plus a range of free-fall bombs. 30-mm ADEN cannon are also routinely used.

The Indian Navy has reportedly rejected taking over ex-Royal Navy FA.Mk 2s as the lack of commonality between them and the existing aircraft would, in effect, add a new type to the maintenance/logistics effort. Furthermore, with the MiG-29K theoretically around two years away from service there is little need for the extra aircraft. Instead, a long-planned major upgrade for the FRS.Mk 51s is still believed to be part of the IN's planning.

According to press reports this is to be undertaken by HAL (previous overhauls and major repairs had been performed by BAE Systems) and will involve new avionics – including the installation of Elta EL/M-2032 multi-mode radar – and the integration of the Rafael Derby active-radar BVR missile. Upgrading the Sea Harriers would maintain their viability beyond the currently planned out-of-service date of 2010, although this schedule is largely dependent on the ability to get the new 37,500-tonne Air Defence Ship and INS *Vikramaditya* (ex-*Admiral Gorshkov*) 'on line' in time. The latter and its complement of MiG-29Ks is expected to be ready for service in 2008, while the keel for the ADS has been laid at the Cochin shipyard. Delays to either may extend the Sea Harrier's operational life beyond 2010 but, on the other hand, if the programmes stay tightly to schedule and Sea Harrier upgrade work does not begin soon, it may be deemed an unaffordable luxury given the short amount of time the modified aircraft would be expected to serve.

David Donald and Simon Watson

Right: *Shore base for the Sea Harrier fleet is INS* Hansa *at Goa-Dabolim.* Hansa *was originally commissioned as a land base at Sulur in 1961 but – following the ousting of the Portuguese from their Goa enclave in 1961 – moved in to Dabolim airfield in 1964. Today it is the Indian Navy's main base, housing helicopters, Il-38s, Do 228s and trainers, as well as the Sea Harriers. In the near future it is likely to be the main base for the Mikoyan MiG-29K, BAE Systems Hawk and ADA Naval LCA that the Indian Navy is expected to operate.*

INAS 551B is the Sea Harrier training unit, operating both single- and two-seat aircraft. Four of the latter are in use, consisting of two original Harrier T.Mk 60s (below) with 'needle' nose, and two ex-RAF Harrier T.Mk 4(I)s (below right) which retain their 'thimble' noses.

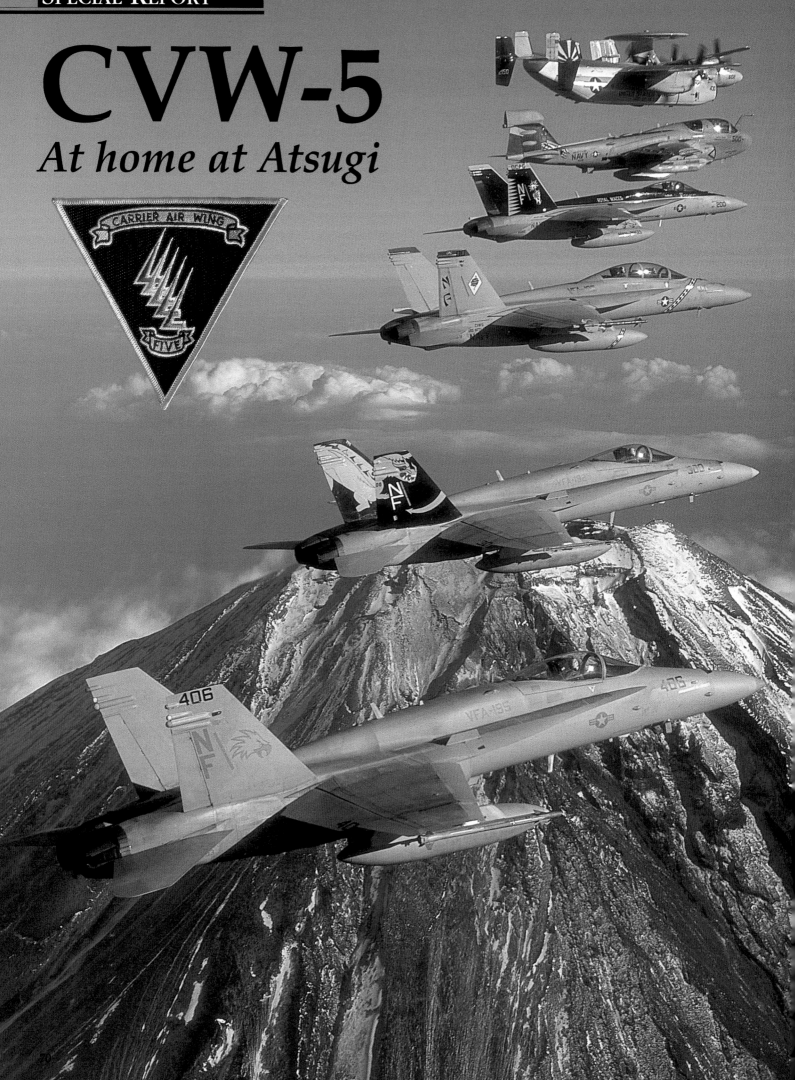

SPECIAL REPORT

CVW-5
At home at Atsugi

Carrier Air Wing Five

During World War II the United States was engaged in some of the fiercest fighting in its history, against the might of the Japanese military. In the aftermath, the United States was instrumental in the rebuilding of Japan. Sixty years have now passed and, over time, the bitterness of war has been replaced by friendship with America gaining a new ally as a result. Americans have been based in Japan since the end of the war, including a US Navy presence. Today the largest concentration of Navy aircraft in Japan is Carrier Air Wing Five (CVW-5), which is based at NAF Atsugi near Tokyo.

CVW-5 shares NAF Atsugi with the Japanese Maritime Self-Defence Force and its aircraft, including numerous P-3C Orions. When not at Atsugi, CVW-5 can be found prowling the oceans aboard USS *Kitty Hawk* (CV 63). Air Wing Five is unique among carrier air wings for several reasons. It is the only forward-deployed carrier air wing and, as it is located closer to many likely theatres of operation, it is able to get to most current hot-spots faster than other combat formations.

Atsugi is a large base situated in the densely populated greater Tokyo/Yokohama metropolitan district. The base is home to two squadrons of JMSDF Orions as well as CVW-5. Here a ShinMaywa US-1A can be seen on the ramp amid the various P-3s.

Unlike other carrier air wings, CVW-5 tenant units are all located at one facility, NAF Atsugi. This creates a cohesive environment, and when they deploy aboard *Kitty Hawk*, squadron members already know each other from life around the base. Typically, Navy air wing personnel and aircraft are scattered at different bases around the country, so the logistics of getting all the units together on a carrier in a crisis situation can take somewhat longer. One downside of CVW-5 being based at one location is that aircraft parts are not stocked in bulk

Air Wing Five is now an 'all-Hornet' organisation, having bade farewell to its Tomcats and Vikings in 2004. The maintenance requirements of the new Super Hornet are much less than for the Tomcat.

as they would be at the Stateside 'super bases', where large numbers of aircraft are concentrated in single-type communities.

All pilots selected for duty in Air Wing Five have to meet 'Pri-A' standard, meaning 'priority A' – top-notch naval aviators that are hand-selected. The Navy wants its best in Japan at the controls, ready to handle any emergencies

Special Report

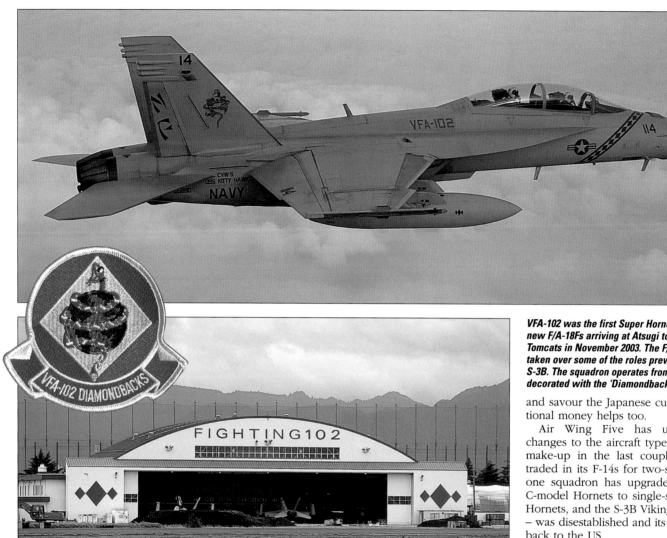

VFA-102 was the first Super Hornet unit in CVW-5, its new F/A-18Fs arriving at Atsugi to replace VF-154's Tomcats in November 2003. The F/A-18Fs have also taken over some of the roles previously assigned to the S-3B. The squadron operates from this hangar, suitably decorated with the 'Diamondbacks' markings.

Below: VFA-27 'Royal Maces' was originally a part of CVW-5 with the F/A-18C. The squadron deployed to Lemoore in California to undergo transition to the F/A-18E, and on its return completed the transformation of Air Wing Five to its new-look complement.

and savour the Japanese culture, and the additional money helps too.

Air Wing Five has undergone several changes to the aircraft types within the wing's make-up in the last couple of years. It has traded in its F-14s for two-seat Super Hornets, one squadron has upgraded from the legacy C-model Hornets to single-seat E-model Super Hornets, and the S-3B Viking squadron – VS-21 – was disestablished and its aircraft were flown back to the US.

The CVW-5 aircraft inventory has now stabilised and there are no planned major changes slated in the near future. The commander of Carrier Air Wing Five is Rear Admiral James D. Kelly, the Carrier Air Group Commander (CAG) is Captain Joseph Aucoin, and the Deputy CAG (DCAG) is Captain Garry Mace.

or problems that could arise. All Naval personnel based in Japan receive additional money to help offset the higher cost of living there. Almost all based at Atsugi prefer being located there due to having an opportunity to absorb

The CAG's view

Captain Joseph Aucoin has been in the Navy 24 years and was an F-14 RIO, serving in VF-33, VF-101 as an instructor, VF-84 and VF-41 as both a department head and commanding officer. He has about 4,500 hours of flight time and has been with CVW-5 as the CAG for the past two and a half years. When asked about Air Wing Five, he said, "This is the place where we are going to fight the next battle, being near to Asia and other hot areas. We have to be creative for training and we get realistic air wing training being close to the DMZ ranges, Okinawa and the Philippines.

"Since we are forward-deployed, it really buys the Navy a lot. We are in the middle of the action and we are subject matter experts on the Pacific AOR (Area Of Responsibility). While our air wing is always ready to go to battle, what limits us right now is the carrier, since the *Kitty Hawk* needs to go to the yard for maintenance periodically. We could deploy as an expeditionary air wing if need be, should the carrier not be available. Some logistics would have to be solved, but it is very feasible.

"Coming from east coast communities and then coming here, a noteworthy difference is – with all our squadrons based here at Atsugi – everyone is working together 365 days a year. This streamlines the wing tremendously since coordination and cooperation is an ongoing process. There are difficulties, too, as we have to provide maintenance to a variety of airframes at one base. We have some other challenges being based here that includes quality of life issues, limited ranges, limited deployment locations such as Guam, and some noise complaints.

VFA-27's CAG-bird carries colourful markings, including the 'Royal Mace' fin badge. The aircraft carries markings for the CAG on the port side, and for the Deputy CAG (whose name happens to be Mace) to starboard. Although the mace badge (right) has been a constant, the unit was renamed the 'Chargers' when it converted from the A-7E Corsair II to the F/A-18A Hornet in January 1991. It later received F/A-18Cs and became a part of Air Wing Five in Japan in the mid-1990s. The name changed again to avoid conflict with HS-14.

"However, this is the best job in the world and seeing how these motivated people do their job, day-to-day, and the long hours they put in on the ship is phenomenal. They are hard-working and a great resource – we make sure we take care of them!"

Deputy CAG

Captain Garry Mace has been DCAG of CVW-5 for a year now. He was previously the CO of both VFA-125 and VFA-27, was an F/A-18 instructor pilot in the FRS (Fleet Replacement Squadron), and was with VFA-195 prior to that, making this his third tour in Japan. He also flew with the CH-53E and A-4 communities in the past. He added, "With all CVW-5 assets all

Right: Air Wing Five uses the latest AIM-9X version of the trusty Sidewinder. This offers high off-boresight capability thanks to a thrust-vectoring motor and advanced seeker.

Below: VFA-192's CAG-bird is decorated with the golden dragon from the unit's nickname. The dragon insignia dates to 1950, when the squadron renumbered as VF-151 (from VF-15A). The current design was introduced in 1956 when the unit became VA-192. Along with the aircraft from VFA-195, the Hornets of VFA-192 have been a fixture at Atsugi as part of CVW-5 since November 1986. Both units subsequently upgraded to the F/A-18C. The assigned carrier was initially USS Midway (CV 41), which was followed by Independence (CVA 62) in 1991 and then the current assignment aboard Kitty Hawk in 1998.

Previously the 'Tigers', VFA-195 adopted the 'Dambusters' nickname in commemoration of the squadron's achievements in attacking the Hwachon Dam in Korea on 1 May 1951. The current badge, from which the eagle fin marking is taken, was adopted in 1985. The CAG-bird (below) – one of the most colourful in the Navy – carries the legend 'Chippy Ho!' on the spine.

being based at Atsugi, the synergy is evident. The various squadrons work together all the time, so we do not require typical basic air wing training that other wings receive during Fallon detachments.

"We can pick up and move out fast, and we deploy more than any other air wing. Thus we maintain our carrier proficiency by simply going to the boat in the normal course of business. This means that we do not have to spend as much time to re-qual on the carrier before leaving. All the junior officers hang out together, regardless of their aircraft community, so you don't see the various 'cliques' here. The

Above: CVW-5's resident Prowler unit is VAQ-136, whose CAG-bird is seen flying past Mount Fuji, close to the base at Atsugi. The 'Gauntlets' have been at Atsugi with Air Wing Five since 1980, initially assigned to USS Midway, which itself was moved to Yokosuka in 1973 to initiate the formal US carrier presence in Japan.

Below: Mount Fuji again provides the backdrop, this time for a VAW-115 E-2C Hawkeye. Like most of the Pacific Fleet early warning squadrons, the 'Liberty Bells' were formed on 20 April 1967 when the previous single Pacific unit, VAW-11, was sub-divided into new squadrons formed around its carrier detachments. In the 1990s the name 'Sentinels' was used.

CVW-5 units

Modex	Aircraft	Squadron	Nickname
100	F/A-18F	VFA-102	Diamondbacks
200	F/A-18E	VFA-27	Royal Maces
300	F/A-18C	VFA-192	Golden Dragons
400	F/A-18C	VFA-195	Dambusters
430	C-2A	VRC-30 Det. 5	Providers
500	EA-6B	VAQ-136	Gauntlets
600	E-2C	VAW-115	Liberty Bells
610	SH-60F, HH-60H	HS-14	Chargers

700 was assigned to VS-21 'Fighting Redtails' operating S-3Bs, but the unit was decommissioned on 1 February 2005 due to Super Hornets joining the wing

Above: VAW-115 flies standard Group II Hawkeyes with the original four-bladed propellers, although the unit is slated to upgrade to the Hawkeye 2000 at some point, and should receive new eight-bladed propellers in due course.

Right: Suitably bedecked in the Rising Sun, a C-2A Greyhound from VRC-30 Det 5 lands at Atsugi. Note the deployed tail bumper that protects the rear fuselage in the case of over-rotation. The motto from the unit's badge (below) sums up its primary task succinctly.

information sharing between units is superb and it makes for a really tight air wing.

"Here in Japan, we are within a quicker striking distance than the other carrier wings due to our proximity to an array of 'hot spots'. In my three tours here, we have made five unplanned cruises to the Arabian Gulf. One of the challenges of being based here is the lack of training ranges in the vicinity and not being able to fly live ordnance out of Atsugi due to Japanese regulations. We go on major detachments in order to accomplish our strike training. On the other hand, with all units based here, we do get peripheral additional training, so there are pros and cons, but overall the system is excellent.

"The local population here love us and, in fact, we even have fan clubs. Working in Japan is fun and the Japanese government shares the cost of forward-basing since we are a key component of their self-defence scheme. The quality of people here – the way they rise to challenges, the operational tempo, how they maintain and fly their aircraft – makes me proud. We have a non-standard training turn-around period, we have to be much more flexible, and we are prone to a very short-fuzed redirection in our deployment schedule. As always, the CVW-5 personnel step up to the plate and rise to the challenge."

Ted Carlson

A C-2A cavorts near Mount Fuji. Atsugi was the first base to which Greyhounds were deployed, VRC-50 beginning operations in December 1966. In the 1990s the US Navy adopted a revised concept in which Greyhounds were assigned directly to air wings in two-aircraft detachments rather than being theatre-based. VRC-50 was therefore disbanded, and support for the Japan-based carrier was transferred to VRC-30 at North Island. Despite the change of concept, the Greyhounds of Det 5 remain forward-deployed at Atsugi.

Rotary-wing support and protection for Air Wing Five is provided by HS-14 'Chargers', which was formed on 10 July 1984 and originally flew the Sikorsky SH-3H. The squadron was allocated to CVW-5 in 1994, replacing the disbanded HS-12 'Wyverns'. Today the unit flies the SH-60F 'Ocean Hawk' on inner-zone USW (undersea warfare) and SAR duties, and the HH-60H 'Rescue Hawk' on combat rescue, SAR and transport tasks. When fitted with AAS-44 FLIR/laser turret and M299 four-round launcher, the HH-60H is capable of firing the AGM-114K Hellfire missile. The photos on this page depict the squadron's SH-60F CAG-bird, which has a black tail and patriotic markings on the tailboom. In the photo above the AQS-13F dipping sonar is clearly visible.

AIR POWER ANALYSIS

US Department of Homeland Security

The Department of Homeland Security (DHS) was created on 19 November 2002 in response to the terrorist attacks that were carried out against the United States on 11 September 2001. Charged with preventing similar attacks, the new cabinet-level department was formally established on 1 January 2003. Its component agencies are tasked with analysing threats and intelligence, guarding borders and airports, protecting critical infrastructure, and coordinating the nation's response to future emergencies.

Tom Kaminski

The United States shares common borders with Canada and Mexico that, respectively, span 5,525 miles (8891 km) and 1,989 miles (3201 km). Additionally, its maritime border comprises 95,000 miles (152888 km) of shoreline, and an offshore exclusive economic zone (EEZ) covers an area of 3.4 million sq miles (8.8 million km²). During 2003 the responsibility for protecting these borders and enforcing immigration laws was assigned to the newly created Department of Homeland Security. The new department combined assets that had previously been assigned to various cabinet-level organisations including the Departments of Justice, Transportation and Treasury. Several law enforcement organisations were included in these transfers, comprising the US Coast Guard (USCG), the Immigration and Naturalization Service (INS) and the US Customs Service (USCS). Today, the Coast Guard reports directly to the Secretary of Homeland Security, whereas the aviation assets from the latter two organisations are assigned to the Directorate of Border & Transportation Security and report to US Customs & Border Protection. DHS is responsible for a fleet of more than 400 aircraft.

United States Customs and Border Protection (CBP)

Established on 1 March 2003, US Customs and Border Protection (CBP) is the Department of Homeland Security's unified border agency. Upon its creation, CBP assumed control of numerous agencies, including much of the former US Customs Service (USCS) and the entire US Border Patrol (USBP). The responsibility for the Office of Air & Marine Operations (AMO), which was previously a component of the USCS, was initially assigned to US Immigration and Customs Enforcement (ICE). However, on 31 October 2004, AMO was transferred intact to the CBP. A transition team is currently developing plans that will consolidate all of CBP's air and marine personnel, missions and assets under a single organisation. This will occur by the end of 2005. More than 40,000 employees manage, control and protect the nation's borders and official ports of entry. During 2004 CBP personnel were responsible for apprehending 1.15 million undocumented immigrants and made 56,231 drug seizures that accounted for 2.2 million lb (997920 kg) of narcotics.

CBP – Office of Border Patrol (OBP)

The United States Border Patrol transferred to the DHS on 1 March 2003, along with several other organisations, to form the US Customs and Border Protection (CBP). Established on 28 May 1924, the Office of Border Patrol (OBP), which is headquartered in Washington D.C., is the uniformed law enforcement arm of the DHS. The organisation had previously reported to the Department of Justice as a component of the Immigration and Naturalization Service. Since its transfer to DHS, the Border Patrol's primary mission has become preventing, pre-empting and deterring threats against the United States through ports of entry (POE), and controlling the boundary between POEs by interdicting illegal crossings.

Although a greater emphasis has been placed on preventing terrorists and weapons of mass destruction (WMD) from entering the country, the OBP still carries out its traditional duties of interdicting illegal (or so-called undocumented) immigrants, narcotics and other contraband. It carries out integrated land, air and – where required – marine patrols along the borders shared with Canada and Mexico, as well as 2,000 miles (3219 km) of coastal waters surrounding Florida and the territories of Puerto Rico and the Virgin Islands. The organisation's area of responsibility (AOR) is divided into 20 sectors and each is under the operational control of a Chief Patrol Agent or sector chief, who controls the ground, marine and air resources within the sector.

While much of the work conducted by the organisation is carried out by ground units, approximately 100 pilots and 115 aircraft support their operations. Aircraft assigned to the Border Patrol Air and Marine Operations department include both fixed- and rotary-wing types. Whereas most of OBP's fixed-wing aircraft were purchased new, more than half of its helicopters were originally operated by the US Army. Border Patrol aircraft currently operate from 23 locations throughout the country and, although operations along the Canadian border have expanded recently, the vast majority of the aircraft remain assigned to the nine sectors along the southwest border with Mexico. The Air and Marine Operations Center at El Paso International Airport, Texas, serves as the headquarters for the organisation and all training for Border Patrol pilots is conducted there.

Operations

Border Patrol aircraft and crews are tasked with a variety of missions in support of agents operating on the ground, as well as other local and federal law enforcement agencies. Their duties include surveillance and intelligence-gathering, and land and coastal patrols. Pilots often use disturbances in natural terrain conditions to locate, observe and track undocumented immigrants and interdict illegal drugs being smuggled into the country. These operations are commonly referred to as 'sign-cutting', whereas the patrols along the border are known as 'linewatch'. Aerial inspections of freight trains that have entered the US from Canada or Mexico are another task assigned to the aviation units. Tactical responses include transporting special response teams (SRT) and Border Search, Trauma & Rescue Teams (BORSTAR), and Border Patrol Tactical Unit (BORTAC) personnel. Illegal immigrants often try to cross the desert between POEs on foot. During the summer months the desert crossings become increasingly treacherous. As a result, the aircraft spend a significant amount of time supporting search and rescue (SAR) missions and the BORSTAR teams.

OBP requires that all of its pilots serve as Border Patrol Agents for a period of three years before applying for aviation duty. To be considered for flight duty each candidate must have a minimum of 200 flight hours and a commercial pilot's rating. Typically, once the pilot has been posted to an air operations sector he or she will fly fixed-wing aircraft until logging 1,500 flight hours. Even after the pilot transitions to rotary-wing types they will typically accumulate another 250 hours with an instructor before being cleared to operate with night vision goggles.

The number of personnel assigned to the OBP's helicopter crews varies between sectors. Within the San Diego sector, for example, two pilots generally fly the observation helicopters whereas the Tucson sector assigns a single pilot

Department of Homeland Security

A Border Patrol MD600N provides air cover for a ground agent who has taken a group of undocumented immigrants into custody in the desert to the east of San Diego, a typical operation for the OBP's air assets. Although immigration control is the main task in this region, OBP is also in the front line of the campaigns against terrorism and drug-trafficking.

Office of Border Patrol – Washington D.C.

Border Patrol Air and Marine Operations Center	
El Paso International Airport, Texas	C-28A, Cessna 210, OH-6A, MD600N, UH-1H
Blaine Sector Air Operations	
Bellingham IAP, Washington	MD600N, Cessna 182
Buffalo Sector Air Operations	
Niagara Falls IAP, New York	OH-6A, AS 350B3, Cessna 206
Del Rio Sector Air Operations	
Del Rio International Airport, Texas	Cessna 206, PA-18, AS 350B/B3, OH-6A, UH-1H
Detroit Sector Air Operations	
St Clair County IAP, Kimball, Michigan	OH-6A
El Centro Sector Air Operations	
Imperial County Airport, El Centro, California	OH-6A, MD500E, C182, PA-18, AS 350B3
El Paso Sector Air Operations	
El Paso International Airport, Texas	OH-6A, MD600N, Cessna 182, AS 350B, UH-1H
Deming Municipal Airport, New Mexico	MD600N
Grand Forks Sector Air Operations	
Grand Forks IAP, North Dakota	AS 350B3, Cessna 206
Havre Sector Air Operations	
Havre City-County Airport, Montana	Cessna 206, MD500E
Houlton Sector Air Operations	
Houlton International Airport, Maine	OH-6A, Cessna 206, MD500C
Laredo Sector Air Operations	
Laredo International Airport, Texas	Cessna 182, PA-18, UH-1H, OH-6A, MD600N
Marfa Sector Air Operations	
Marfa Municipal Airport, Texas	Cessna 206, PA-18, OH-6A
McAllen Sector Air Operations	
McAllen Miller International Airport, Texas	MD600N, Cessna 182/206, PA-18, OH-6A
Miami Sector Air Operations	
North Perry Airport, Pembroke Pines, Florida	MD600N, Cessna 206
New Orleans Sector Air Operations	
Lakefront Airport, New Orleans, Louisiana	MD600N
Ramey Sector Air Operations	
Rafael Hernandez Airport, Aguadilla, Puerto Rico	OH-6A
San Diego Sector Air Operations	
Brown Field Airport, San Diego, California	MD500E, MD600N, UH-1H
Spokane Sector Air Operations	
Felts Field, Spokane, Washington	AS 350B3, Cessna 206
Swanton Sector Air Operations	
Franklin County State AP, Swanton, Vermont	OH-6A, AS 350B3, Cessna 206
Tucson Sector Air Operations	
Tucson International Airport, Arizona	Cessna 182, PA-18, OH-6A, AS 350B3
Libby AAF, Fort Huachuca, Sierra Vista, Arizona	UH-1H, AS 350B3, OH-6A
Yuma Sector Air Operations	
MCAS Yuma/IAP, Arizona	AS 350B, OH-6A, UH-1H, Cessna 182

Above: This map shows Border Patrol sectors. Each has an air station to provide air assets to assist ground and marine forces.

Below: Only four MD500Es serve with the Border Patrol, this example being operated by the San Diego Sector Air Operations at Brown Field. Others fly with the units based at Havre, Montana, and El Centro, California.

Although the most numerous type in OBP service, the OH-6A's days are numbered as they are soon to be replaced by Eurocopter EC120Bs. This line-up is at Brown Field, San Diego. Note the incandescent spotlights.

Above: A pair of MD600Ns patrols in southern California. The MD600 is a stretched derivative of the MD500.

BORTAC agents demonstrate a fast-rope rappel from a UH-1H. The dozen upgraded Super Hueys operated by OBP are mainly based in the southwest desert, and are valued for their ability to insert large teams, including BORTAC special operations and BORSTAR rescue personnel. Some of the Hueys have been fitted with a weather radar in the nose.

with an observer/emergency medical technician. Newer pilots that have not yet been cleared to fly OBP helicopters primarily fly the fixed-wing types. As with the helicopters, the crew complements for fixed-wing aircraft vary between sectors.

Border Patrol rotary-wing aircraft

Border Patrol's oldest and most numerous helicopter is the OH-6A Cayuse, and approximately 30 remain in service. Referred to commercially as the Hughes 500 (Model 369A) and nicknamed the 'Loach', the OH-6A was originally built by Hughes Helicopter and all previously served with the US Army. In addition to the OH-6A, Air Operations flies a single Hughes 500C (Model 369HS), which is similar to the OH-6A, and a small number of McDonnell Douglas MD500E (Model 369E) helicopters. (Hughes Helicopters was sold to the McDonnell Douglas Corporation in January 1984 and was renamed McDonnell Douglas Helicopters Incorporated (MDHI). The company continued to operate under that name following the corporation's purchase by the Boeing Company in 1997. In February 1999 Boeing sold MDHI to the RDM Group in the Netherlands. RDM subsequently renamed the company as MD Helicopters Inc.).

The OH-6As have, however, been updated with the same Rolls-Royce 250-C20 turboshaft engine as the 369HS, making them equivalent to that model. The 278-shp (207-kW) engine offers an 11 per cent increase in power over the OH-6A's original T63A-5A (250-C10B). The 500E differs from the older models in being equipped with a five-bladed main rotor, 'T'-type tail, extended skids and a revised nose profile. It is also equipped with a more powerful 250-C20B that provides 375 shp (279 kW). Both types are mainly engaged in low-altitude, low-speed surveillance duties and security patrols. Although capable of carrying two crew members and up to three passengers, the light observation helicopters are normally flown with just the flight crew. Generally equipped for daylight observation duties, the OH-6As are only provided with a steerable bank of incandescent floodlights, whereas the MD500Es feature a 500-watt Spectrolab SX-5 xenon searchlight. During night-time operations, however, the crews regularly operate with night vision goggles. The 500E is capable of a maximum speed of around 150 kt (278 km/h) and has a maximum range of 260 nm (481 km) or an endurance of 2.7 hours. With a useful load of nearly 1,500 lb (680 kg), the aircraft can carry up to five crew and passengers. A turreted electro-optical/infrared (EO/IR) imaging system can be installed on the 500E.

USBP also operates 11 MD600N helicopters, which trace their roots to the OH-6A. Originally intended to replace the OH-6A, Border Patrol had planned to purchase up to 45 examples. However, the aircraft's inability to perform certain missions resulted in the purchases being cut short of that goal. Capable of carrying up to eight passengers and crew, the multi-mission MD600 offers a maximum speed of 152 kt (282 km/h) and has an endurance of 4.4 hours or a range of 423 nm (783 km). Equipped with the NOTAR anti-torque system and a six-bladed main rotor, the MD600N has a useful load of 2,000 lb (907 kg). A 600-shp (447-kW) Rolls-Royce 250-C47M engine powers the helicopter. In addition to observation duties, the MD600Ns are used for SAR missions, and in support of the BORSTAR and the special operations personnel assigned to BORTAC units. Night operations are enhanced by the installation of an EO/IR sensor turret and a 500-watt Spectrolab SX-5 xenon searchlight.

The Patrol also operates 13 examples of the single-engined Eurocopter AS 350B3 AStar (Ecureuil) that were delivered from July 2002. The acquisition followed the evaluation of four AS350Bs, which also remain in service. The AS350B3 is powered by the Turboméca Arriel 2B, which provides 847 shp (632 kW) of power, whereas the AS 350B is equipped with a 641-shp (478-kW) Arriel 1B engine. The former has a maximum range of 395 nm (732 km) and the range of the latter is slightly less at 357 nm (661 km). Equipped with both FLIR and powerful Spectrolab SX-5 xenon searchlights, the AStars are better suited for nocturnal operations than the earlier types, and their versatility allows them to perform a variety of missions. Capable of carrying up to six passengers and crew, the helicopter can loiter for more than three hours, and is capable of carrying BORTAC special operations personnel on external benches like the US Army's MH-6 'Little Birds'.

Largest helicopter in the Border Patrol inventory is the Bell UH-1H, of which 12 are currently in service. Primarily used to support airmobile units by inserting agents into areas that are heavily used by smugglers, the UH-1Hs are also used to transport the BORSTAR teams and BORTAC special operations personnel. Although Border Patrol evaluated five UH-1Hs that had been equipped with a variant of the LHTEC T800 engine, developed for the Army's RAH-66 Comanche, it determined that the benefits afforded by the new engine did not justify the expense the project would have required. The aircraft were returned to the Army in September 1996 at the conclusion of the 14-month evaluation. The UH-1Hs that are currently in service have all been updated to Super Huey configuration and are equipped with upgraded T53-L-703 engines, transmissions, gearboxes and rotors, along with improved avionics. OBP has replaced the Lycoming T53 in UH-1H N100PW (serial 68-15337) with a 1,692-shp (1262-kW) Pratt & Whitney Canada PT6C-67D engine that forms the basis of the UH-1H Plus programme. Developed jointly by DynCorp, Pratt & Whitney and Global Helicopters, the US$496,000 modification is known commercially as the 'Global Eagle'.

In October 2004 CBP announced it would purchase two new helicopters for Border Patrol. The American Eurocopter EC120B Colibri was selected as its new single-engined light observation or light 'sign-cutter' helicopter. The EC120Bs will replace the OH-6As and plans call for the purchase of 55 helicopters at a cost of US$75 million over a five-year period. The aircraft will be assembled at a new facility at the Golden Triangle Airport in Columbus, Mississippi. Powered by a 504-shp (376-kW) Turboméca Arrius 2F (TM319) turboshaft, the EC120 has a useful load of 1,653 lb (750 kg) and is capable of carrying two crew members and up three passengers. It has a maximum speed of 127 kt (236 km/h), a range of 367-383 nm (680-710 km) or an endurance of over four hours.

Border Patrol will also receive approximately 30 twin-engined Bell Model 430 helicopters under a US$163.9 million contract. First flown in 1994, the 430 – which will serve as a twin-turbine medium utility helicopter – is powered by a pair of 783-shp (584-kW) Rolls-Royce 250-C40B engines. The helicopter is capable of carrying a crew of two and up to eight passengers, depending upon its cabin configuration, and has a useful load of 3,975 lb (1803 kg) It has a maximum range of 353-367 nm (654-680 km) or an endurance of 3.8 hours when operating at its maximum gross weight.

Border Patrol fixed-wing aircraft

All but two of the nearly 40 aircraft in Border Patrol's fixed-wing fleet are single-engined types, comprising 14 two-seat Piper PA-18 Super Cubs and more than 20 Cessnas. The latter include the 182 Skylane, 206 Stationair and the 210 Centurion. Powered by a 150-hp (111-kW) four-cylinder Lycoming O-235 reciprocating engine, the Super Cub is primarily used as a proficiency trainer. With a maximum range of 400 nm (741 km) the Super Cubs are also useful for surveillance work. Capable of

Department of Homeland Security

Above: 104 was one of two Elbit Hermes 450 UAVs leased by the OBP between June and November 2004 as part of the Arizona Border Control Initiative. They operated from Libby AAF, Fort Huachuca.

Right: The slow-flying, high-wing Piper PA-18-150 Super Cub is mainly used for training, but also undertakes surveillance work. Various Cessna models provide the backbone of the fixed-wing patrol fleet.

carrying up to four passengers and crew, the Skylane is powered by a six-cylinder Continental O-470 reciprocating engine that provides 230 hp (172 kW) and gives the aircraft a range of 820 nm (1519 km). The larger Cessnas can carry four to six passengers and crew, and conduct surveillance missions that range from six to eight hours in length. The 206s and 210s are powered by either the Lycoming TIO-540 or Continental IO-520 six-cylinder reciprocating engines, rated respectively at 310 hp (321 kW) and 300 hp (224 kW). The only larger aircraft in the inventory are a pair of Cessna 404 Titans that had previously seen service with the US Navy under the designation C-28A. Prior to their delivery to the Border Patrol in June 2001, the 404s had been stored at the Aerospace Maintenance and Regeneration Center (AMARC) at Davis Monthan AFB, Arizona. The Titans are powered by a pair of 375-hp (280-kW), six-cylinder Continental GTSIO-520 reciprocating engines and can carry as many as 11 passengers and crew.

OBP's fixed-wing aircraft are limited to operations above 500 ft (152 m) and are used to direct agents working on the ground, much like a military forward air control platform. They are also used as observation aircraft and carry out some of the same missions assigned to the OH-6/MD500 helicopters. When required, the Cessnas are also used to transport agents between locations.

Unmanned aerial vehicles

Near-term Border Patrol plans call for the procurement of unmanned aerial vehicles (UAV), and the office's budget for 2005 includes US$10 million that has been earmarked toward their purchase and deployment. Between June and September 2004, OBP utilised a pair of Hermes 450 UAVs that were leased from Elbit Systems at a cost of US$4 million. The UAVs supported the Tucson Sector and the Arizona Border Control (ABC) Initiative, operating under the control of AMO's Air and Marine Operations Center (AMOC) in Riverside, California. They were flown from Libby Army Air Field at Fort Huachuca. By the time the experiment came to a close the UAVs had detected more than a dozen attempts to smuggle drugs across the border and were instrumental in some 800 apprehensions and numerous rescues. As a result of this success, OBP made plans to continue the UAV experiment utilising a pair of RQ-5A Hunter UAVs from the US Army's 305th Military Intelligence Battalion at Fort Huachuca. The air vehicles began operations along the border in November 2004 and were scheduled to operate through January 2005. During 2005 OBP will continue to refine its UAV plans and will conduct respective demonstrations over North Dakota and Puerto Rico during the winter and spring.

Above: An AS 350B3 AStar flies over sand dunes in southern Arizona, one of the busiest regions for the Border Patrol. It is fitted with the standard Spectrolab SX-5 Starburst xenon searchlight and FLIR Systems Ultra 7500 EO/IR turret.

Below: AStar N852BP was the first Border Patrol aircraft to fly with a new scheme, which includes the seal of the US Customs and Border Protection, the DHS organisation that parents both the OBP and Customs.

81

Air Power Analysis

CBP – Office of Air and Marine Operations (AMO)

Headquartered in Washington D.C., the Office of Air and Marine Operations (AMO) had been part of the US Department of the Treasury until transferred to the Department of Homeland Security on 1 March 2003. AMO was subsequently assigned as one of five divisions within US Immigration and Customs Enforcement (ICE) on 9 June 2003. ICE serves as the lead agency responsible for enforcing immigration and customs laws, and is the investigative arm of the Border and Transportation Security Directorate. In a move intended to consolidate DHS's air and marine assets, AMO was reassigned to US Customs and Border Protection on 31 October 2004.

The division's 1,000 personnel support the operation of a fleet of more than 130 aircraft that includes both civil and former military fixed- and rotary-wing types. Although the entire fleet is registered to the AMO and the Oklahoma City National Aviation Center (OCNAC) at Will Rogers World Airport in Oklahoma, the aircraft are assigned to individual aviation branches and units throughout the CONUS and its territories. OCNAC provides operational, administrative and logistical control over all of the aviation resources, and ensures that all field offices meet the same standards. It also provides aviation law enforcement tactics training and manages national fixed- and rotary-wing training contracts.

Each of the air units is assigned to an air, or air and marine, branch, and reports to the air branch chief. AMO's primary mission had been preventing smugglers from bringing illicit drugs and other contraband into the country. However, it is now tasked with the additional duty of preventing terrorist activity. During 2003 the AMO was directly responsible for the seizure of 38 tons of cocaine and nearly 170 tons of marijuana.

The Customs Aviation Program was established in 1969 and was merged with the Marine Interdiction Program to form the Customs Air and Marine Interdiction Department (AMID) in October 1999. AMO's three core competencies include airspace security, air and marine interdiction, and air and marine law enforcement. As well as its responsibilities of countering terrorism and smuggling, it is also tasked with supporting other law enforcement agencies and conducting counter-drug operations in foreign countries.

Over the years the organisation has operated a variety of aircraft including both civilian types and those bailed from the military. Examples of the latter included the Grumman OV-1 Mohawk and S-2 Tracker, and Bell UH-1 and AH-1. Fixed-wing aircraft currently operated by the AMO include the Cessna 206 Stationair, 210 Centurion and 550 Citation II, Lockheed Martin P-3A and P-3B Orion, Piper PA-42 Cheyenne III, and Raytheon C-12C King Air. Its rotary-wing fleet comprises examples of the Eurocopter AS 350B AStar, MD Helicopters MD500E and Sikorsky UH-60A Blackhawk.

AMO's assets support the operations of the Joint Interagency Task Force (JIATF) throughout the so-called source, transit and arrival zones. Its aircraft are primarily located within the arrival zones around the coastal waters near San Diego, the southwest border, the Gulf Coast, south Florida and the Caribbean. Additionally, air assets operate from three forward operating locations at Queen Beatrix International Airport, Aruba; Manta Air Base, Ecuador; and Comalapa Air Base, El Salvador. They also utilise forward operating sites in Peru, Costa Rica, Belize and the US Naval Station at Guantanamo Bay, Cuba.

Aircraft and crews stationed at these sites support joint inter-agency operations throughout the source and transit zones. The transit zones encompass the major smuggling corridors of Mexico, the Caribbean and the eastern Pacific Ocean. Air and marine operations in the Caribbean transit zone are under the tactical command of JIATF South in Key West, Florida. The exact location of operations in the eastern Pacific determines whether they are under the direction of the JIATF South or JIATF West, which is located on Coast Guard Island, Alameda, California.

AMO also works directly with Mexico, as part of Operation Halcon, monitoring and tracking suspected drug-smugglers heading north to the US through the Mexico transit zone. Normally two Citations are stationed in Mexico and operate from bases in Hermosillo and Monterrey. Within the Caribbean transit zone its assets support missions conducted under Operation Bahamas, Turks and Caicos, or OPBAT, which is a joint task force coordinated by the US Drug Enforcement Administration (DEA).

Air and marine fleets are also tasked in support of other federal agencies and, during special events such as the Presidential Inauguration and the Olympics, the aircraft provide perimeter security and are used to deploy tactical teams should the need arise.

Airspace security

As a result of the terrorist attacks that occurred on 11 September 2001, AMO has operated in partnership with the North American Aerospace Defense Command (NORAD) to provide airspace security. Its units were the first to augment NORAD operations in the wake of the attacks. In fact, P-3 aircraft operated by the Surveillance Support Branches (SSB) provided more than 25 percent of all domestic Airborne Early Warning (AEW) missions/coverage in support of Operation Noble Eagle in the initial month following the attacks. These operations subsequently continued over the northern border in support of Operation Liberty Shield.

AMO Citations and Blackhawks are tasked with providing airspace security over the National Capital Region (NCR) above Washington D.C. and patrol the restricted airspace surrounding the city. Its units have also been called upon to provide airspace security over special events, including the 2002 Winter Olympics, the 2003 and 2004 Superbowls, baseball's 2004 All-Star Game, and the Democratic and Republican Party national conventions in Boston and New York. Airspace security operations have also been conducted over New York, Las Vegas and Los Angeles during periods when terror alerts have been elevated. More recently, DHS designated the Presidential Inauguration as a National Special Security Event (NSSE). As a result, the 'no-fly' area around Washington D.C., which normally includes a 16-mile (26-km) radius around the city, was tripled. In response to the elevated security AMO increased the number of Citations and helicopters assigned to the National Capital Region Air & Marine Branch.

AMO has begun positioning additional assets along the northern border with Canada in support of CBP and other Federal agencies, and plans to expand operations along that border. Plans call for the activation of five new air and marine branches and the initial pair are located in Bellingham, Washington, and Plattsburgh, New York. The Bellingham Air and Marine Branch was activated on 1 September 2004 and the Plattsburgh Air and Marine Branch stood up on 8 October 2004. Although the ongoing integration efforts within CBP could alter these plans, the remaining three branches will be located in Detroit, Michigan, Grand Forks, North Dakota, and Great Falls, Montana. Eventually three aircraft and 28 personnel will be assigned to each of the new branches.

AMOC

Located at March Air Reserve Base, the Air and Marine Operations Center (AMOC) is in Riverside, California. The centre serves as a multi-agency coordination centre and compiles information collected from an array of civilian and military radar sites, tethered aerostats and other detection assets. The data is used to provide seamless 24-hour radar surveillance along the entire southern tier of the US, Puerto Rico and the Caribbean. AMOC provides command and control support to air, marine and ground units on patrol or engaged in special operations, and coordinates intercepts of suspected aircraft. The centre also integrates information systems with other domestic and international counter-drug centres and law enforcement agencies, and serves as a focal point for tactical coordination between government agencies.

NCRCC

In January 2003 AMO, along with the Department of Defense and other agencies, was tasked with creating a coordinated airspace security operation over the 23,000 square miles (59565 km^2) that comprise the National Capital Region around Washington D.C. In response to this requirement, the multi-agency National Capital Region Coordination Center (NCRCC) was activated in Herndon, Virginia. Located in a building owned by the Federal Aviation Administration, the NCRCC coordinates the identification and interception of aircraft that approach or enter the restricted area.

Operations

Each of the 470 pilots and 50 air enforcement officers (AEO) assigned to AMO has attended the 16-week programme at the Federal Law Enforcement Training Center (FLETC) in Glynco, Georgia. AMO requires that new pilots have a minimum of 1,500 hours of flight time, hold single- or multi-engined qualifications, and be instrument-rated or hold helicopter instrument ratings. P-3 pilots must also hold a multi-engined rating. Unlike OBP however, the

Department of Homeland Security

The AMO's latest purchase is the Pilatus PC-12/45, of which an initial two have been acquired, with more to follow for the MEA role. The PC-12s are outfitted with weather radar in a wingtip pod, and a retractable EO/IR sensor turret. In October 2004 the AMO was reassigned from ICE control to that of CBP.

The King Air 200 in the foreground is one of the Miami Air and Marine Branch's C-12Ms, which have maritime search radar and FLIR under the belly. They are used to track drug-smuggling boats and to control AMO interceptor vessels.

An AMO Beech C-12C taxis at North Island, home base of the San Diego Air & Marine Branch. This aircraft, like several in the AMO inventory, wears no markings to give away its identity, allowing it to operate closer to targets of interest without drawing attention.

Office of Air and Marine Operations - Washington D.C.

Unit	Location	Aircraft
Air and Marine Operations Center (AMOC)	March ARB, California	
Albuquerque Air Branch	Kirtland AFB, New Mexico	Cessna 210/550, Beech 200, AS 350B2, UH-60A
El Paso Air Unit	El Paso IAP, Texas	
Bellingham Air and Marine Branch	Bellingham IAP, Washington	Cessna 210, UH-60A (pending delivery of PC-12)
Caribbean Air and Marine Branch	Rafael Hernandez Airport, Aguadilla, Puerto Rico	C-12C/M, UH-60A, AS 350B2
Houston Air Branch	David Wayne Hook Memorial Airport, Tomball, Texas	Cessna 210/550, AS 350B2
Kansas City Air Unit	Wheeler Downtown Airport, Kansas City, Missouri	
Jacksonville Air and Marine Branch	NAS Jacksonville, Florida	PA-42, Cessna 210, AS 350B2, MD500E
Tampa Air Unit	St Petersburg-Clearwater IAP, FL	
Miami Air and Marine Branch	Homestead ARB, Florida	Cessna 550, C-12C/M, UH-60A, AS 350B2
New Orleans Air and Marine Branch	Hammond Municipal Airport, Louisiana	Cessna 210/550, C-12C, AS 350B2
Pensacola Air Unit	NAS Whiting Field, Milton, Florida	
Cincinnati Air Unit	Cincinnati Municipal Airport-Lunkin Field, Ohio	
Oklahoma City National Aviation Center	Will Rogers World AP, Oklahoma City, Oklahoma	Cessna 210/550, C-12C, AS 350B2
Plattsburgh Air and Marine Branch	Plattsburgh IAP, New York	Cessna 206, AS 350B2, TDY UH-60A (pending delivery of PC-12)
New York Air Unit	Long Island MacArthur Airport, Ronkonkoma, New York	
San Angelo Air and Marine Branch	San Angelo Regional Airport/Mathis Field, Texas	Beech 200, Cessna 206/550, AS 350B2
San Antonio Air Unit	San Antonio IAP, Texas	
San Diego Air and Marine Branch	NAS North Island, California	Cessna 206/210/550, C-12C, AS 350B2, UH-60A
Riverside Air Unit	March ARB, California	
Sacramento Air Unit	Mather Airfield, Sacramento, California	
Tucson Air Branch	Davis Monthan AFB, Arizona	Cessna 206/210/550, UH-60A, AS 350B2
Phoenix Air Unit	Williams Gateway Airport, Phoenix, Arizona	
Surveillance Support Branch West (SSB-W)	NAS Corpus Christi, Texas	P-3A/B, P-3B(AEW)
Surveillance Support Branch-East (SSB-E)	Cecil Field Commerce Center Airport, Jacksonville, Florida	P-3B, P-3B(AEW)
National Capital Region Air & Marine Branch	Ronald Reagan National Airport, Arlington, Virginia	TDY UH-60A, Cessna 550
National Capital Region Coordination Center	Herndon, Virginia	

Above: The single-engined Cessna fleet includes 206s and 210s. Most are fitted with an enlarged rear port window and 'barber chair' observation seat, like this 210M at North Island.

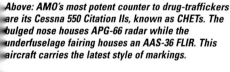

Above: AMO's most potent counter to drug-traffickers are its Cessna 550 Citation IIs, known as CHETs. The bulged nose houses APG-66 radar while the underfuselage fairing houses an AAS-36 FLIR. This aircraft carries the latest style of markings.

Right: Airspace security, especially during national events, is an important duty for the Citation fleet. This aircraft is seen over Houston's Reliant Stadium in December 2003 in the run-up to the January Superbowl XXXVIII in January.

Below: Among the CHET fleet is this Mexican-registered Citation II, assigned to the Tucson Air Branch at Davis-Monthan AFB.

personnel are not required to serve as agents on the ground before joining AMO, and detection systems specialists (DSS) assigned to the P-3s and the AMOC are not required to be Federal Law Enforcement Officers. Generally, the DSS have prior experience as military or civilian air traffic controllers.

AMO rotary-wing aircraft

Known as the Light Enforcement Helicopter (LEH), the short/medium-range Eurocopter AS 350B2 AStar is mainly used for aerial surveillance of stationary or moving targets. Powered by a single 732-shp (546-kW) Turboméca Arriel 1D1 and flown by a crew of two, it is regularly used in metropolitan areas. The AStar has a maximum range of 360 miles (580 km) or an endurance of three hours. The aircraft are equipped with a variety of systems, including the FLIR Systems Ultramedia digital camera system, Spectrolab SX-5 Starburst searchlight, gyro-stabilised binoculars and video downlink capability. AMO's fleet currently includes 23 AStars.

The McDonnell Douglas Helicopter (now MD Helicopter) MD500E (Model 369E) is used by the AMO on a limited basis as an LEH. Only three examples are in service, and all are assigned to the Jacksonville Air and Marine Branch in Florida. Assigned the same missions as the AStars, the aircraft has a range of 345 miles (555 km) and an endurance of three hours. Mission equipment includes an infrared detection set (IRDS), and a low-light level law enforcement camera that is combined with a laser.

AMO's 16 Sikorsky UH-60A Blackhawks, which are on loan from the US Army, serve as long-range, all-weather, tactical apprehension aircraft. The largest, most powerful helicopter used in airborne law enforcement anywhere in the world, the Blackhawks support a variety of missions including surveillance. They are, however, primarily tasked with supporting air, land and marine interdiction efforts as an apprehension platform. Additionally, the helicopters are used for airspace security, transporting evidence and inserting tactical teams. Equipped with an infrared detection set (IRDS) that comprises the FLIR Systems AN/AAQ-22 Star SAFIRE I thermal imaging turret and a Spectrolab SX-16 Nitesun searchlight, the UH-60As are tailored for night operations. Internal auxiliary fuel tanks installed in the aft cabin provide the helicopter with an endurance of 4½ hours or a range of 600 miles (966 km). It is normally flown by a crew of three comprising pilot, co-pilot and air enforcement officer.

AMO fixed-wing aircraft

Tasked with surveillance, tracking, photographic and intelligence-gathering missions in support of investigative and enforcement efforts, the Cessna 206 Stationair and the 210 Centurion are both used by AMO. They also have a secondary mission of providing logistics support, including relocating personnel, equipment or evidence that is time-critical to an investigation or enforcement action. Flown mainly outside the large metropolitan areas, the aircraft offer better range and endurance than helicopters, and blend more effectively with local traffic, thereby masking the presence of continuous air surveillance. Although capable of carrying up to six passengers and crew, the aircraft are normally flown by a two-man crew consisting of a pilot and observer. These short/medium-range aircraft are powered by different models of the six-cylinder Continental IO- and TSIO-520 reciprocating engines. The aircraft offer a maximum range of 650 miles (1046 km) or an endurance of 6½ hours. A large window is installed in the rear of the cabin, along with a so-called 'barber chair', to support the long-range/endurance observation missions. AMO operates six Stationairs and 18 Centurions, assigned to units throughout the United States and the Caribbean.

AMO's fleet of 15 Raytheon (Beech) Aircraft C-12Cs serve as medium-range, all-weather transports and are generally used to airlift six to eight enforcement personnel and equipment engaged in interdiction and investigative support missions. These duties often include relocating or evacuating tactical personnel, assets, prisoners and evidence. In conjunction with Cessna 550 interceptors, the aircraft have also been used to coordinate training between Colombian and Customs interdiction assets. The King Airs also operate as long-range apprehension aircraft and daytime maritime aerial search platforms. Although capable of carrying up to 13 passengers and crew, the C-12C is normally operated by a crew of two comprising a pilot and co-pilot. The King Air is capable of a range of 1,157 miles (1862 km) and has an endurance of 4½ hours.

A second variant of the King Air is equipped as a maritime patrol aircraft (MPA). Known as the C-12M, six of these radar- and FLIR-equipped, medium-range aircraft are assigned to branches in Florida and Puerto Rico, and are used to detect and monitor maritime smugglers in the so-called 'arrival zone'. The aircraft work closely with AMO's interceptor boats and, like the C-12C, are normally flown by two-man crews. The C-12C/Ms and King Air 200s are powered by a pair of 850-shp (634-kW) Pratt & Whitney Canada PT6A-41 turboprops.

Considered a long-range tracker and referred to as a Customs High Endurance Tracker (CHET), the Piper PA-42 Cheyenne III is used to intercept and track airborne drug smugglers and is equipped with an AN/APG-66 radar system and a FLIR. AMO's five Cheyennes are operated by three-man crews each comprising a pilot, co-pilot and sensor operator. The aircraft, which are powered by a pair of 850-shp (634-kW) Pratt & Whitney Canada PT6A-61 turboprops, have a range of 1,700 miles (2736 km) or an endurance of 6 hours. They are also capable of operating from remote landing strips.

A fleet of 26 modified Cessna 550 Citation IIs serves as AMO's primary CHET assets. The all-weather tactical jets are equipped with the same AN/APG-66 fire control radar installed in the F-16A, as well as an AN/AAS-36 infrared detection set (IRDS). These systems enable the crew to track and intercept airborne drug-smugglers and to enforce airspace security. Introduced into the US Customs Service interdiction efforts as a high-speed interceptor and tracker in 1976, the Citation is normally flown by a three-man crew comprising a pilot, co-pilot and sensor operator. It has a top speed of 358 kt (663 km/h), an endurance of 4 hours and a range of 1,594 miles (2565 km), and is capable of operating from remote locations. It has been used from forward operating bases in Panama, Honduras, Colombia, Peru, Venezuela, Mexico and Aruba.

The AMO operates two distinctive variants of the Lockheed Martin P-3 Orion. These include the P-3 Airborne Early Warning Detection and Monitoring (AEW D&M), and the P-3 Slick aircraft. Used to detect and monitor airborne drug-smugglers, the former is the only law enforcement AEW aircraft in the world. As a result of the commitment of DoD assets to other regions of the world the aircraft have recently assumed a greater role in the war on terror. Initially equipped with the AN/APS-125 radar, used by the Grumman E-2C Hawkeye, the fleet is now equipped with the more capable AN/APS-145 radar. AMO's fleet of eight aircraft is each operated by a crew of eight, including a pilot, co-pilot, flight engineer and five radar/sensor operators.

The AEW aircraft typically patrol the Caribbean and Gulf of Mexico in search of smugglers from Central and South America. Operating from bases at Cecil Field Airport in Jacksonville, Florida, and NAS Corpus Christi, Texas, the aircraft are assigned to AMO's two Surveillance Support Branches. In addition to the radar the Orions feature an array of communication equipment that allows them to receive and send information to the ground-based command, control, communications and intelligence centres, as well as other land, sea and air units. The P-3 AEW has an endurance of 12 hours and a range of 4,000 miles (6437 km). ICE's budget request for 2005 includes US$28 million that will fund an increase in P-3 flight hours from 200 to 600 per month.

The P-3 Slick aircraft is a long-endurance, all-weather, tactical turboprop aircraft that is also used to detect and monitor airborne drug-smugglers, and as a long-range tracker aircraft. Like the Citation it is equipped with the AN/APG-66 fire control radar, as well as electro-optical (EO) and infrared sensors. Capable of operating from most 8,000-ft (2438-m) runways, the Slick is supported by a ground maintenance team of only three people. AMO's eight Slicks often operate in tandem with the P-3 AEW aircraft, with the AEW used for detection and tracking while the long-range Slick intercepts, identifies and tracks suspected targets. The eight Slicks, which include four P-3As and four P-3Bs, are being updated with cockpit upgrades that will equip them with a new flight management system that includes an electronic flight instrumentation system (EFIS). The AEW model entered service in 1988.

New equipment

On 14 May 2004 AMO unveiled the Pilatus PC-12/45, its newest aircraft, at Ronald Reagan National Airport. Referred to as a medium or multi-role enforcement aircraft (MEA), two aircraft have been purchased at a cost of US$6.6 million each. Although the initial pair will be assigned to the new air and marine branches in Bellingham, Washington, and Plattsburgh, New York, additional examples will eventually replace AMO's fleet of C-12s and King Air 200s. The PC-12 combines the Cessna 210's slow-flight capabilities with the high-speed and payload features of the C-12. The aircraft will be tasked with aerial surveillance and logistical support duties, and will also be able to conduct air interdiction missions. Equipped with retractable EO/IR sensors and removable

The best-known of the Customs Service aircraft now assigned to AMO are the fleet of Orions. The scene above from 2004 shows a pair of P-3 AEW ('Domes') and a single P-3 Slick on the ramp at Cecil Field, Florida, one of two permanent bases used by the Orions. The Slick (right) has APG-66 radar in a radome above the flight deck and an EO/IR turret forward of the nosewheel bay for tracking targets. The P-3 AEW (below) has APS-145 radar for plotting air traffic, identifying and tracking suspect aircraft, and for vectoring CHET aircraft or P-3 Slicks into intercept or tracking positions. Suspects are followed until they land, at which point ground and heliborne forces move in.

Right: This Piper PA-42-720R Cheyenne III is operated by the Jacksonville Air and Marine Branch in the CHET role against suspect aircraft. Like the Citation CHETs, it carries APG-66 radar and AAS-36 IRDS.

consoles, the PC-12 will be flown by a crew of three comprising a pilot, co-pilot and sensor operator. The PC-12 is capable of short take-offs and landings requiring just 2,300 ft (701 m) and 1,800 ft (549 m), respectively. It has a range of 2,260 miles (3637 km), and can attain a speed of 270 kt (500 km/h) at an altitude of 30,000 ft (9144 m).

During March 2004 AMO purchased a pair of multi-role patrol aircraft (MPA) at a cost of US$50 million. Based on the Bombardier Q200-series turboprop, and referred to as Dash 8Q-202 MPA, the aircraft will be delivered in 2005. Mission equipment installed in the MPA will include Raytheon's SeaVue maritime surveillance radar, L-3 Wescam's MX-15 EO/IR sensor, and an integrated sensors and display

Many former US Customs Service aircraft have adopted 'Police' titles to re-inforce their law enforcement mission, while '121.5' is worn by the UH-60s and CHET aircraft to signify the VHF frequency upon which intercepted aircraft can talk to their DHS trackers. Sixteen UH-60As – colloquially known as 'Pot Hawks' – serve with AMO. The standard night operations equipment is a 1600-watt Spectrolab SX-16 Nitesun searchlight and AAQ-22 Star SAFIRE I EO/IR turret.

Arizona Border Control Initiative

OBP and AMO operations in Arizona have expanded greatly as part of the so-called ABC Initiative. Begun on 16 March 2004, the initiative was intended to detect and deter the number of illegal incursions by human- and drug-smugglers from Mexico. It permanently added 200 Border Patrol Agents to the Tucson Sector, and 60 agents were assigned on a temporary basis. The latter group was specially trained in search, rescue and remote tactical operations.

AMO has flown additional hours patrolling the area with helicopters and fixed-wing aircraft. The AMO and the Border Patrol have used the additional aircraft to rapidly transport search, rescue and enforcement assets to remote areas of the Arizona desert. The Border Patrol assigned four additional AS350 helicopters to the sector. As part of this effort AMO and OBP air operations units have scheduled their flights to maximise the use of air assets in specific areas. In addition, OBP BORSTAR personnel and Tactical Teams have been assigned to AMO flights in support of interdiction and rescue operations.

system (ISADS) developed by ATK Mission Research Group. The latter company serves as the prime contractor for the MPA project. Field Aviation in Toronto, Canada – which previously delivered five similar aircraft to Australia's Customs Service – will incorporate the structural changes required to install the sensors and mission equipment. Once that work is complete, the aircraft will be delivered to ATK Mission Research for final sensor and systems integration. ICE holds an option for a third MPA and could purchase as many as 12 examples.

Future plans call for the purchase of additional aircraft, including medium-lift and light enforcement helicopters, and medium enforcement and interceptor aircraft. ICE conducted an 18-day test with a USAF Predator unmanned aerial vehicle (UAV) during November 2003. However, at this time AMO has no firm plans to add UAVs to its inventory.

During May 2004 AMO issued a request for information associated with the purchase of 20 new sensor-equipped Light Enforcement Aircraft (LEA). The requirements for the new single-engined LEA include a high-wing, provisions for two flight crew and a minimum of one sensor operator. Its useful load must be a minimum of 1,800 lb (816 kg) and be capable of loitering for four hours at an altitude of 7,500 ft (2286 m), and a speed of 80 kt (148 km/h) while operating 60 nm (111 km) from its base. Mission equipment will include an EO/IR sensor equipped with a laser rangefinder and a digital recording system.

United States Coast Guard

Headquartered in Washington D.C., the US Coast Guard had been a component of the Department of Transportation until its transfer to the Department of Homeland Security on 1 March 2003. The full-time military organisation, which is the world's largest coast guard, includes nearly 49,000 active-duty, reserve and civilian personnel, and more than 200 aircraft. Its fleet of surface vessels includes three ice-breakers (WAGB), 12 high-endurance cutters (WHEC) and 28 medium-endurance cutters (WMEC) that are capable of supporting helicopter operations. Coast Guard units regularly operate in support of the Department of Defense (DoD) and specific Unified Combatant Commanders. Coast Guard assets are regularly tasked in support of US Pacific and Southern Commands (USPACOM and USSOUTHCOM) via Joint Interagency Task Force East and West. In times of national emergency the service would be attached directly to the Department of the Navy.

Led by the Commandant of the Coast Guard, a four-star Admiral, its organisation includes Headquarters, Atlantic and Pacific Area commands, and other support commands. Nine Coast Guard Districts, assigned to the Area commands, are responsible for Groups, Sectors, Air Stations, Air Facilities and other support units. Among the units assigned directly to headquarters are the Aviation Training Center (ATC), the Aircraft Repair and Supply Center (ARSC), and Helicopter Interdiction Tactical Squadron Jacksonville. The USCG's primary mission areas include maritime safety, protection of natural resources, mobility, maritime security, and national defence.

As the only military service with the authority to conduct law-enforcement operations, the Coast Guard is charged with enforcing federal maritime laws, regulations and international treaties. It is tasked with interdicting illegal narcotics and migrants at sea, protecting fisheries and marine resources, and enforcing environmental protection regulations. It is also responsible for maintaining aids to navigation, search and rescue (SAR), supporting the international ice patrol, environmental protection and response, and military operations.

The Coast Guard is the lead federal agency for Maritime Homeland Security (MHS), and is responsible for marine inspection, port safety and security. The MHS duties have taken on a greater importance since 2001 and, accordingly, become more visible. In fact, the service's four major national defence missions now include maritime intercept operations, deployed port operations/security and defense (DPOSD), peacetime engagement, and environmental defence operations. Several of these missions were carried out in the Persian Gulf, in support of US Central Command (USCENTCOM) and Operation Iraqi Freedom, when the Coast Guard deployed 11 vessels, including a single high-endurance cutter with an HH-65A helicopter. During 2003 the Coast Guard conducted 36,000 port security patrols and 3,600 air patrols.

A key player in combating the flow of illegal drugs to the United States, the Coast Guard is the lead federal agency for maritime drug interdiction. Responsibility for the air interdiction mission is shared with the US Customs and Border Protection's Office of Air & Marine Operations. Interdiction operations are also conducted in conjunction with other federal agencies and regional foreign governments, targeting the air and sea routes used by smugglers within the so-called transit zone. This 6 million-square mile (15.54-million km^2) area spans the Caribbean Sea, Gulf of Mexico and eastern Pacific Ocean. The seizures made annually by the Coast Guard account for nearly 25 per cent of the drugs intercepted by the US government each year.

Since the early 1960s the USCG has been the lead federal agency tasked with interdicting illegal immigrants that attempt to enter the United States by sea. These operations are similarly coordinated with foreign countries and other federal agencies, including those assigned to the DHS. Coast Guard units were responsible for the interdiction of 71 vessels, the seizure of 121 tons of cocaine, 13 tons of marijuana and made 326 arrests during 2004. It also interdicted 10,899 undocumented immigrants.

Both cutters and aircraft are used to enforce international fishing laws and treaties, and laws protecting marine mammals and endangered species throughout the entire US Exclusive Economic Zone (EEZ). This area extends 200 miles (322 km) from the US coastline and encompasses more than 90,000 miles (144841 km) of coastline, totalling 2.25 million square miles (3.62 million km^2).

Above: This impression shows how the Bombardier Dash 8Q-202 MPA is expected to look in AMO service. It is equipped with 360° SeaVue maritime search radar and a Wescam EO/IR turret.

Right: A UH-60A crew practises using the helicopter's downwash to persuade a speedboat to stop. This technique is one of many used against the drug-smuggling 'go-fast' boats.

AMO operates two helicopter types in the LEH role: three MD500Es (left) fly from Jacksonville, while the majority of AMO sectors have Eurocopter AS 350B2 AStars (below). The AStar is seen during operations over Houston in January 2004 to provide security cover for the Superbowl, a typical AMO operation.

Left: The Dassault HU-25 Guardian once flew from 10 air stations, but is now operated only at Cape Cod, Corpus Christi, Miami and Mobile. The latter is the main training base for the Coast Guard, where a few unmodified HU-25As remain in service, mainly for training duties.

Above and below: Equipped with APG-66 radar in a lengthened nose and WF-360 FLIR under the fuselage, the HU-25C was used for interceptor duties, tracking suspect aircraft. The aircraft are now in HU-25C+ configuration, with an upgraded APG-66(V)2 radar and AAQ-35 FLIR.

Air Power Analysis

Search and Rescue (SAR) is one of the Coast Guard's oldest and most publicised missions, and the agency is tasked as the maritime coordinator in support of the national SAR plan. To fulfil this mission, aircraft, cutters and patrol boats are required to be under way within 30 minutes of receiving a distress signal, and on scene in the search area 90 minutes after that. Stations are maintained along the Atlantic, Pacific and Gulf coasts of the 'lower 48', and in Alaska, Hawaii, Guam and Puerto Rico, as well as along the Great Lakes and principal inland US waterways.

In addition to SAR activities, the Coast Guard supports the International Ice Patrol (IIP) by searching for and tracking icebergs in the North Atlantic shipping lanes in order to provide vessels with their location. The missions are carried out annually between February and July from St. John's in Newfoundland by radar-equipped HC-130H aircraft, operated by a reconnaissance detachment (RECDET) deployed from Coast Guard Air Station Elizabeth City, North Carolina. The RECDET flies the ice patrol missions in the area south of Newfoundland known as the Grand Banks.

Polar ice-breaking missions are conducted in both the Arctic and Antarctic Oceans in support of scientific, environmental and military operations. The missions are assigned to three large ice-breakers (WAGB) that are supported by deployed aviation detachments (AVDET) equipped with HH-65Bs. These are provided by the Mobile-based POPDIV.

Maintenance of offshore navigational aids is generally carried out by sea-going buoy tenders and other vessels. However, rotary-wing aircraft are also used in support of large offshore navigation aids and installations.

Direct-reporting units

Aircraft Repair and Supply Center, Air Station Elizabeth City, North Carolina
Though located at Air Station Elizabeth City, the Aircraft Repair and Supply Center (ARSC) is assigned directly to Coast Guard headquarters. Its five divisions, which include aircraft engineering and aviation repair, provide the aviation force with complete logistic support including maintenance, repair, modification and periodic depot maintenance (PDM).

Coast Guard Aviation Training Center, Mobile, Alabama
Located at Bates Field in Mobile, the Aviation Training Center (ATC) is the service's largest air unit. Reporting directly to Coast Guard Headquarters, its eight divisions include operations, training, polar operations and aviation engineering. The Operations Division's (OPDIV) HU-25A aircraft stand alert duty in support of the Eighth Coast Guard District. The Training Division (TRADIV) provides Coast Guard aviators with initial and recurrent training on the HU-25, HH-65 and HH-60. This includes annual week-long refresher training in ATC's flight simulators. The Polar Operations Division's (POPDIV) HH-65s are assigned to Aviation Detachments (AVDET) that deploy aboard three ice-breakers in support of scientific, logistic and search and rescue (SAR) missions.

Helicopter Interdiction Tactical Squadron (HITRON) Jacksonville, Cecil Field, Florida
The only designated squadron within the Coast Guard operates eight MH-68A Sting Ray helicopters from the former NAS Cecil Field near Jacksonville, Florida. HITRON was created with the sole purpose of interdicting illegal narcotics, however, since 2001 the squadron has taken on additional duties in the war on terrorism. Recently, the squadron deployed aircraft and crews to New York and Alaska, where they carried out port security missions while threat levels were elevated. The helicopters are normally embarked aboard medium- and high-endurance cutters deployed to the Caribbean Sea and western Pacific Ocean.

C-130J Aircraft Project Office
Assigned directly to headquarters and located at Air Station Elizabeth City, the C-130J APO is tasked with supporting the acquisition and missionisation of the service's newest aircraft. The six aircraft are all currently assigned to the APO.

Commander Atlantic Area (LANTAREA)/Maritime Defense Zone Atlantic (MDZA)

Headquartered in Portsmouth, Virginia, the LANTAREA includes five Coast Guard Districts. Its area of responsibility (AOR) spans 40 states, borders 29 foreign countries, and contains more than 5 million square miles (12.9 million km²) of ocean, inland waterways and tributaries. Resources include five Coast Guard Districts, 13 air stations and three air facilities. The command is also responsible for the International Ice Patrol and the Maritime Defense Zone Atlantic.

1st Coast Guard District
Headquartered in Boston, Massachusetts, the 1st Coast Guard District is responsible for the northeast United States from the Canadian border to northern New Jersey. A single air station is assigned.

5th Coast Guard District
Headquartered in Portsmouth, Virginia, the Fifth Coast Guard District controls three air stations. Its AOR includes the states of Delaware, Maryland, Virginia, North Carolina, and parts of New Jersey and Pennsylvania.

7th Coast Guard District
Headquartered in Miami, Florida, the district's AOR includes South Carolina, the Georgia coast, part of Florida and the entire Caribbean. Four air stations and a single air facility are assigned. The district also detaches aircraft to a pair of remote operating locations in the Bahamas in support of anti-narcotics operations as part of Operation Bahamas, Turks and Caicos. The joint operation, which is known as OPBAT, incorporates assets from the US Army, US Drug Enforcement Agency (DEA) and Bahamian law enforcement agents.

8th Coast Guard District
Headquartered in New Orleans, Louisiana, the 8th Coast Guard District is responsible for an area that includes 26 states along the Gulf Coast and across most of the Midwest. The district covers some 1,200 miles (1931 km) of coastline and 10,300 miles (16576 km) of navigable inland waterways that include the Illinois, Mississippi, Missouri and Ohio Rivers. Three air stations are located along the Gulf coast. The Coast Guard Aviation Training Center's Operations Division also supports the 8th district.

9th Coast Guard District
Headquartered in Cleveland, Ohio, its boundaries encompass the shores of the Great Lakes states of Illinois, Indiana, Michigan, Minnesota, New York, Ohio, Pennsylvania and Wisconsin. This area includes 6,700 miles (10783 km) of coastline and includes two air stations and a single air facility, all located in Michigan.

Commander Pacific Area (PACAREA)/Maritime Defense Zone Pacific (MDZP)

Headquartered on Coast Guard Island, Alameda, California, the PACAREA commander is responsible for four Coast Guard Districts. PACAREA's AOR spans 10 states, but contains more than 73 million square miles (190 million km²) of territory across the Pacific Rim. Resources assigned to PacArea include four Coast Guard Districts, 11 air stations, and a single air facility. The command is also responsible for the Maritime Defense Zone Pacific.

11th Coast Guard District
Headquartered in Alameda, California, the 11th district controls three air stations and a single full-time air facility. All five of the district's air stations are located in California, although its AOR also includes the states of Arizona, Nevada and Utah.

13th Coast Guard District
The district, which is headquartered in Seattle, Washington, is responsible for the entire northwest district of the United States, including Idaho, Montana, Oregon and Washington. Three air stations and single air facility are assigned to the district.

14th Coast Guard District
Although the 14th district, headquartered in Honolulu, Hawaii, is assigned only a single air station, its AOR comprises most of the central and western Pacific, including Guam and American Samoa.

17th Coast Guard District
Headquartered in Juneau, Alaska, the 17th's two air stations, both located in the southern portion of Alaska, cover America's largest state, the northern Pacific Ocean and the Bering Strait.

Based at Jacksonville, the HITRON operates all of the MH-68A Sting Ray AUF (armed use of force) helicopters. They are armed with a high-power rifle and a machine-gun to stop 'go-fast' smuggling boats with force if needs be, the gunner aiming for the boat's engine(s). Intercept techniques are being practised in the scene above, while the view at right highlights the undernose EO/IR turret and searchlight used to aid night-time operations.

Headquarters US Coast Guard – Washington D.C.

HITRON Jacksonville	Cecil Field AP, Jacksonville, Florida	MH-68A
CG AR&SC	Pasquotank County RAP, Elizabeth City, North Carolina	(aircraft depot)
C-130J APO	Pasquotank County RAP, Elizabeth City, North Carolina	HC-130J

Coast Guard Aviation Training Center (CGATC) – Bates Field/Mobile Regional Airport, Alabama

Operations Division (OPDIV)	Bates Field/Mobile RAP, Alabama	HU-25A
Polar Operations Division (POPDIV)	Bates Field/Mobile RAP, Alabama	HH-65B
Training Division (TADIV)	Bates Field/Mobile RAP, Alabama	HH-65B/C, HH-60J, HU-25A

Commander Atlantic Area (COMLANTAREA)/Commander US Maritime Defense Zone Atlantic (COMUSMARDEZLANT) – CGSC Portsmouth, Virginia

1st Coast Guard District – CGSC Boston, Massachusetts

AS Cape Cod	Otis ANGB, Falmouth, Massachusetts	HU-25C+, HH-60J

5th Coast Guard District – CGSC Portsmouth, Virginia

AS Atlantic City	Atlantic City IAP, New Jersey	HH-65B/C*
		*transition to HH-65C will be complete by Jun 05
AS Washington.	Ronald Reagan National Airport, Virginia	C-37A, VC-4A
AS Elizabeth City	Pasquotank County RAP, Elizabeth City, N.C.	HC-130H, MH-60J

7th Coast Guard District – CGSC Miami, Florida

AS Miami	Opa Locka Airport, Florida	HU-25D, HH-65B
AS Clearwater	St Petersburg-Clearwater IAP, Florida	HC-130H, HH-60J
AS Savannah	Hunter AAF, Georgia	HH-65B
AF Charleston	Charleston Executive Airport, South Carolina	HH-65B
AS Borinquen	Rafael Hernandez Airport, Puerto Rico	HH-65B

8th Coast Guard District – CGSC New Orleans, Louisiana

AS New Orleans	NAS JRB New Orleans, Louisiana	HH-65B
AS Corpus Christi	NAS Corpus Christi, Texas	HH-65B, HU-25C+
AS Houston	Ellington Field, Houston, Texas	HH-65B

9th Coast Guard District – CGSC Cleveland, Ohio

AS Traverse City	Cherry Capital Airport, Michigan	HH-65B
AF Waukegan	Waukegan Regional Airport, Illinois	HH-65B
AS Detroit	Selfridge ANGB, Michigan	HH-65B
AF Muskegon	Muskegon County Airport, Michigan	HH-65B

Commander Pacific Area (COMPACAREA)/Commander US Maritime Defense Zone Pacific (COMMARDEZPAC) – CGSC Coast Guard Island, Alameda, California

11th Coast Guard District – CGSC Alameda, California

AS Los Angeles	Los Angeles IAP, California	HH-65B
AS San Diego	San Diego IAP/Lindbergh Field, California	HH-60J
AS San Francisco	San Francisco IAP, California	HH-65B
AS Sacramento	McClellan Airfield, Sacramento, California	HC-130H
AS Humbolt Bay	Arcata Airport, California	HH-65A*
		*to transition to HH-65B in May 05

13th Coast Guard District – CGSC Seattle, Washington

AS Astoria	Astoria Regional Airport, Oregon	HH-60J
AS North Bend	North Bend Municipal Airport, Oregon	HH-65B
AF Newport	Newport Municipal Airport, Oregon	HH-65B
AS Port Angeles	CGAS Port Angeles, Washington	HH-65A*
		*to transition to HH-65B in Apr 05

14th Coast Guard District – CGSC Honolulu, Hawaii

AS Barbers Point	Kalaeloa Airport, Oahu, Hawaii	HC-130H, HH-65B

17th Coast Guard District – CGSC Juneau, Alaska

AS Sitka	Sitka/Rocky Gutierrez Airport, Alaska	HH-60J
AS Kodiak	Kodiak Airport, Alaska	HC-130H, HH-60J, HH-65B

Above: 2001 is the first of six HC-130Js for the USCG, initially assigned to the C-130J Aircraft Project Office at Elizabeth City. In 2005 the new Js will begin to fly full operational patrol missions.

Backbone of the long-range patrol effort is the HC-130H Hercules. The aircraft above is assigned to AS Clearwater, while the aircraft below is from AS Kodiak, seen on the icy ramp at Valdez. Both are fitted with the CASPER equipment that includes a Wescam ASX-4 20-in (51-cm) diameter EO/IR turret and a palletised cabin equipment. Note the enlarged forward cabin window for long-range visual searches.

US Coast Guard aircraft

The service's fixed-wing fleet includes four different aircraft types. However, only the Dassault HU-25 and Lockheed Martin HC-130 support operational missions. The remaining types include single examples of the Grumman VC-4A (G-159) and the Gulfstream Aerospace C-37A, which are used in support roles.

Tasked with a wide range of missions, the HC-130H is the Coast Guard's primary long-range surveillance (LRS) aircraft. Its duties include search and rescue (SAR), law enforcement, fisheries and environmental protection, narcotics interdiction, airlift and support of the International Ice Patrol. A crew of seven comprising two pilots, flight engineer, navigator, radio operator and two dropmasters normally operates the Hercules. During airlift missions one of the dropmasters is replaced by a loadmaster.

The HC-130H has a range of 4,500 nm (8334 km) and it is capable of remaining airborne for 14 hours. It is equipped with advanced navigation systems including the AN/APN-215 weather radar and the AN/APS-137 inverse synthetic aperture radar (ISAR), which is capable of providing radar images from distances greater than 50 miles (80 km).

Beginning in 1999, the HC-130H fleet was modified to carry the C-130 Airborne Sensor with Palletized Electronic Reconnaissance (CASPER) system, which comprises an AN/ASX-4 (Wescam Model 20TS) forward-looking infrared (FLIR) turret assembly, and the airborne tactical workstation (ATW). When installed, a sensor system operator (SSO) and a tactical system operator (TSO) are added to the crew. A small number of aircraft are also equipped with an AN/APS-135 side-looking airborne radar (SLAR) pod. The system mainly supports the International Ice Patrol mission over the North Atlantic but can also be used to detect and map oil spills. For airlift duties the HC-130s can carry up to 70 passengers or cargo.

The operational fleet currently includes 22 HC-130Hs, and the first of six C-130Js was delivered on 31 October 2003. The latter aircraft, which are assigned to the C-130J Aircraft Program Office (APO), have not yet received the required long-range maritime patrol modifications. They are currently stationed at the Aircraft Repair and Supply Center in Elizabeth City, North Carolina, and limited to flying airlift missions. The first aircraft is scheduled to receive these modifications during 2005. Two examples supported operational evaluations at Air Station Kodiak, Alaska, during July 2004.

Developed from the Falcon/Mystère 20G, the HU-25 Guardian serves as a medium-range surveillance (MRS) aircraft. The service originally purchased 41 HU-25As from Dassault-Breguet, outfitted by the contractor's subsidiary Falcon Jet Corporation in Little Rock, Arkansas. The modifications included the installation of search windows in the fuselage, a drop hatch in the floor and AN/APS-127 radar. The Guardians have been tasked with SAR, narcotics interdiction, law and treaty enforcement, and marine environmental protection. Since delivery of the final aircraft numerous modifications have resulted in the HU-25B, HU-25C, HU-25C+ and HU-25D models. No longer in service, the HU-25B was equipped with the AN/APS-131 AIREYE sensor pod and used for tracking oil spills. The pods were retired in October 2002. Tasked with identifying and tracking sea and airborne targets, the HU-25C interceptor was equipped with a modified AN/APG-66 radar system in place of the APS-127, and a Texas Instruments WF-360 FLIR system was installed under the fuselage. In addition, the cockpit lighting was optimised for using Night Vision Goggles (NVG), and a SATCOM communications suite was installed.

Between 2001 and May 2003 Northrop Grumman delivered six HU-25Ds and nine HU-25C+s to the Coast Guard. The modified HU-25As and HU-25Cs are equipped with the AN/APG-66V(2) air intercept radar, an AN/AAQ-35 (Wescam MX-15) FLIR system and a new tactical work station (TWS) that allows the SSO to control the sensors and process the data. Additionally, the HU-25Ds are equipped with the Telephonics AN/APS-143V(3) inverse synthetic aperture airborne radar. A small number of HU-25As also remain in service but are primarily used in support of training.

Normally operated by a crew of five, comprising two pilots, an avionicsman or SSO, dropmaster and observer, the HU-25A can carry up to six passengers with a reduced crew. The HU-25A has a maximum range of approximately 1,700 nm (3148 km) or a maximum endurance of 5¾ hours. These numbers are slightly reduced for the HU-25C+ and HU-25D. The remaining HU-25s will undergo an avionics modernisation beginning in late 2004 that will update the navigation equipment and flight instrumentation, and allow the aircraft to remain in service until at least 2010.

Based on the Gulfstream V business jet, a single C-37A provides transportation for Coast Guard and DHS officials. Operated by a crew of four, including two pilots, the aircraft has a maximum range of 6,500 nm (12038 km) or a maximum endurance of 14 hours. The C-37A is configured to carry up to 12 passengers.

A single VC-4A, which had until recently been assigned to the Seventh Coast Guard District and used for operational support duties, was recently transferred to Air Station Washington. The Gulfstream I, which now operates alongside the C-37A, is capable of carrying up to 11 passengers. It is operated by a crew of three and has a maximum range of 1,000 nm (1852 km). The VC-4A had previously been operated by NASA, and was transferred to the Coast Guard to replace an older VC-4A.

Coast Guard helicopters

The Coast Guard's rotary-wing fleet includes more than 130 helicopters comprising the HH-60 Jayhawk, HH-65 Dolphin and MH-68 Sting Ray. Among its fleet are 43 aviation-capable ships consisting of high- and medium-endurance cutters, and ice-breakers. Although both the HH-60 and HH-65 are cleared to operate from these vessels, due its smaller 'footprint' the HH-65 is preferred for flight-deck operations. The Jayhawks are normally stationed ashore.

Based on the Aérospatiale (now Eurocopter) SA 366G, the first of 95 HH-65A Dolphins reached initial operational capability (IOC) in 1984. The short-range recovery (SRR) aircraft are mainly tasked with SAR, narcotics interdiction, environmental protection, law and treaty enforcement, airlift and support of polar ice-breaking operations. Equipped with a Rockwell-Collins CDU-900G automated flight management system, and a Bendix RDR-1300C radar, the Dolphin is capable of flying search patterns automatically. This system can automatically place the HH-65 in a hover at an altitude of 50 ft (15 m). The aircraft is equipped with an electronic flight instrumentation system (EFIS), and a Northrop Sea Hawk forward-looking infrared (FLIR) system.

In 2000 the service began equipping the fleet with a new cockpit under a US$34 million programme that upgrades the control display units and incorporates new flat-panel multi-function displays (MFD), capable of displaying FLIR video. The programme also updates the flight management software and reduces the aircraft's empty weight by 100 lb (45 kg). Once updated, the aircraft are redesignated HH-65B. In addition to the radar system, the aircraft are equipped with a searchlight and a 600-lb (272-kg) capacity rescue hoist. The crew of four consists of two pilots, an aircrewman and a rescue swimmer for SAR missions. The Dolphin is capable of a maximum range of 300 nm (556 km) or a maximum endurance of 3½ hours. The last of 84 operational HH-65B modifications will be delivered in May 2005.

Although early problems with the Lycoming LTS-101 engines were corrected, the service conducted evaluations of the Turboméca Arriel 2C1 and 2C2 engines during 2002 and 2003 and has embarked on a project that will install the latter engine in its Dolphins. The first Dolphin equipped with Arriel 2C2-CG engines, which have been tailored specifically for the service, was modified at the Aircraft Repair and Supply Activity (ARSC) in Elizabeth City, North Carolina. The re-engined aircraft (6510) made its first flight on 27 August 2004. The redesignated HH-65C was subsequently transferred to the Aviation Training Center in Mobile at the conclusion of its flight test programme. It is currently supporting the development of a new training syllabus at Mobile. The maximum take-off weight for the HH-65C is increased by 280 lb (109 kg) compared with that of the HH-65B, but limitations on its landing gear still restrict its weight to 8,900 lb (3130 kg) for shipboard operations.

As part of the upgrade programme, the HH-65C is also equipped with a new heat shield and a more powerful transmission, like that fitted to the AS 365N4 Dauphin. Integrated Coast Guard Systems (ICGS), a joint venture between Lockheed Martin and Northrop Grumman, recently signed a US$124 million contract on behalf of the USCG with American Eurocopter that covers the production of re-engining kits for 95 Dolphins. The contractor will deliver the kits to the ARSC, which will install them in the helicopters. The service plans to re-engine 84 additional aircraft over the next two years and all 95 examples will be updated to HH-65C configuration by May 2007. The initial examples were assigned to Air Station Atlantic City, New Jersey, in January 2005. Under current plans the Dolphins will remain in service through 2014.

Above: Having been flown by the USCG since 1981, the HH-60J Jayhawk fleet is due for an upgrade. The HH-60T programme will modernise the aircraft's cockpit, systems and sensors from 2005. The Jayhawk is used for medium-range missions, including armed interdiction, and occasionally operates from cutters, although the Dolphin is preferred for shipborne duties.

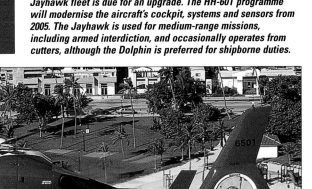

Above: A ski-equipped HH-65A prepares to land on a 'Polar'-class ice-breaker during operations in the Antarctic. The Dolphins used for such missions are parented by the Polar Operations Division (POPDIV), part of the Mobile-based Aviation Training Center.

Left: An HH-65A flies along Miami's beachfront during a routine patrol. The most numerous of the Coast Guard types, the Dolphin is dispersed widely around the Coast Guard stations, and is used on a wide variety of roles. Among these is SAR, for which a hoist is fitted above the starboard side cabin door.

By May 2005 the operational Dolphin fleet will have been fully upgraded to HH-65B standard, with new cockpit instrumentation. These examples are from AS Miami (above, with Wescam 12DS/TS200 EO/IR 'ball') and AS Savannah (right, landing on CG Cutter *Harriet Lane*).

Members of the Coast Guard's Pacific Area Tactical Law Enforcement team practise fast-roping from an HH-60J at the Aviation Training Center at Mobile during a vertical insertion exercise. The ATC provides instruction for HH-60, HH-65 and HU-25 crews while Hercules training is undertaken by the USMC.

An HH-60J from AS San Diego picks up a 'survivor' during a SAREX. Both Dolphin and Jayhawk fly SAR missions with a crew of four, including an aircrewman who works the hoist and calls instructions to the two pilots, and a rescue swimmer.

Sikorsky's HH-60J Jayhawk, which entered service in 1981, is tasked with the medium-range recovery (MRR) role. The HH-60J's missions include SAR, narcotics interdiction, law enforcement, supporting aids to navigation and environmental protection. Its configuration is similar to the US Navy HH-60H combat rescue helicopter. It is equipped with AN/APN-217 Doppler search and weather radar, AN/AAQ-15 forward-looking infrared (FLIR), secure communications equipment, and its cockpit is NVG-compatible. Like the HH-65, it carries a Nitesun spotlight and is equipped with a rescue hoist. The HH-60J is able to operate up to 300 nm (556 km) from base, and remain airborne for as long as 6 hours. During SAR missions it is normally operated by a crew of four and is capable of carrying up to six passengers or survivors.

Beginning in 2005, the service's 41 surviving HH-60Js will receive updated avionics, based upon equipment developed for the US Army's special operations helicopters. Known as the Common Avionics Architecture System (CAAS), the update will initially provide the Jayhawk with five LCD multifunction displays, a new control display unit and dual GPS receivers. In the longer term the aircraft will also receive new radar and EO/IR sensors. The first HH-60J will receive the CAAS modifications in 2005. The modified Jayhawk, which will be redesignated HH-60T, is due to fly in 2006. The remainder of the fleet will receive the CAAS modifications, along with other improvements intended to extend service life, during depot maintenance between 2007 and 2011.

As the result of its successful airborne use of force (AUF) experiment, the Coast Guard leased eight Agusta A109E helicopters in 2000. Intended to combat the flow of illegal narcotics aboard so-called 'go-fast' boats, the service initially leased two Boeing MD-900 Explorer helicopters before ordering the Sting Rays. Assigned the designation MH-68A, the helicopter entered service in 2001. The MH-68A is equipped with a 0.50-in (12.7-mm) Robar RC-50 precision rifle, and a 7.62-mm M240 machine-gun. When required, these weapons are used to physically stop the smugglers and allow surface personnel to board the boats. They are normally deployed aboard 378-ft (115-m) 'Hamilton'-class and 270-ft (82-m) 'Famous'/'Bear'-class high- and medium-endurance cutters. These ships, unlike the older 210-ft (64-m) 'Reliance'-class medium-endurance cutters, are equipped with hangar facilities. Initially leased for a period of two years, the service subsequently extended that lease and will retain the MH-68A through at least 2008.

Traditionally tasked with humanitarian and law enforcement duties, the USCG only initially employed the Agusta MH-68A Sting Ray for AUF missions. In response to its newly expanded duties under the DHS, the service has now developed weapon installations for its HH-60 and HH-65 helicopters. Five HH-60Js have already been equipped with 7.62-mm M240 machine-guns and Wescam 12DS EO/IR sensor turret, civil radios and revised exterior lighting. Those Jayhawks modified with this equipment packages have received the multi-mission designation MH-60J. Similarly, HH-65Cs that are equipped with the AUF installations will be redesignated as MH-65B/Cs. During 2005 the Coast Guard will invest $1.8 million in modifying its helicopters to carry out airborne use of force (AUF) missions.

Force modernisation

On 25 June 2002, the Coast Guard announced the award of the largest acquisition programme in its history to Integrated Coast Guard Systems (ICGS). The joint-venture partnership, which comprises Lockheed Martin and Northrop Grumman, is tasked with managing the USCG's Integrated Deepwater System (IDS) programme. IDS will provide the Coast Guard with an integrated, multi-mission and highly flexible system of so-called Deepwater assets. Once the IDS has been fully implemented in 2022, the service will have acquired three new classes of cutters and associated small boats, new fixed-wing aircraft, a combination of new and upgraded helicopters, and both cutter- and land-based unmanned air vehicles (UAVs). Each of these systems will be linked with Command, Control, Communications and Computers, Intelligence, Surveillance and Reconnaissance (C^4ISR) systems. Under the IDS the Coast Guard will also retire older equipment that has become more costly to operate and maintain.

Initially, however, the contractor team is upgrading the so-called 'legacy' aircraft, including the HC-130 and HH-60, and finalising the designs for the first of the new aircraft that will be built. Eventually, Deepwater will replace several of the legacy aircraft with a combination of UAVs and more capable manned aircraft.

The Bell Helicopter HV-911 Eagle Eye has already been selected as the Vertical Take Off and Landing UAV (VUAV), and in February 2003 the contractor was authorised to begin concept and preliminary design work. Testing of the initial examples is scheduled to begin in 2005. Ultimately the service hopes to buy 69.

ICGS has also selected the EADS/CASA CN235-300M aircraft for the MRS mission. Referred to as the CN-235 Medium Range Surveillance Maritime Patrol Aircraft (MRSMPA), the initial pair will be delivered in early 2007. The MRSMPA will also assume some of the long-range surveillance and transport duties now assigned to the HC-130. Although ICGS initially identified the need for 35 MRSMPAs, the actual number that will be ordered has not yet been formalised. The project is currently in the system development and demonstration (SDD) phase.

The Agusta-Bell AB-139 VTOL recovery and surveillance (VRS) helicopter was initially planned as Deepwater's medium-range recovery aircraft, to replace the HH-60T from 2014. However, in April 2005 it was announced that the AB-139 had been dropped from the Deepwater plans as it could not undertake all of the Jayhawk's duties. A number of upgrades that are planned for the HH/MH-60J/T will allow it to remain in service through 2022.

An upgraded version of the MH/HH-65C short-range recovery (SRR) helicopter, known as the multi-mission cutter helicopter (MCH), will be capable of operating from flight deck-equipped cutters (FDEC), including the NSC and OPC. As part of the programme 93 aircraft will be recapitalised and undergo a service life extension (SLEP). They will be equipped with new landing gear, fenestron (shrouded tail rotor) and a longer nose. Additionally, the avionics installed in the so-called HH-65X will be common with those of the VRS. The Dolphin's re-engining programme had earlier been part of the HH-65X project, but the service made the decision to update the fleet prior to conducting the SLEP.

In the longer term, a high-altitude endurance UAV (HAUAV) will provide long-range surveillance over large areas for extended periods of time, assuming part of the long-range surveillance capability now conducted by manned HC-130s. Seven HAUAVs will be deployed in 2016. The UAVs will be capable of transmitting EO/IR imagery to cutters, MPAs, and shore-based Command and Control (C^2) centres. The Northrop Grumman RQ-4 Global Hawk and General Atomics Mariner are candidates to fill this role.

Department of Homeland Security air bases

Key to map symbols

- ✪ Coast Guard Air Stations and Facilities
- ■ AMO aircraft-operating locations
- ▼ OBP aircraft-operating locations

The majority of DHS bases are in the south of the country, reaching from Florida to California to protect the southern approaches to the US. They guard the main arrival zones for both the narcotics smugglers and undocumented immigrants. The eastern and western seaboards, and the Great Lakes region, feature a string of USCG bases to maintain SAR and maritime cover. Permanent bases that are outside the 'lower 48', and not in the area of the map, are: AS Borinquen/Rafael Hernandez AP, Puerto Rico (USCG/AMO/OBP), AS Barbers Point, Oahu, Hawaii (USCG), AS Sitka, Alaska (USCG) and AS Kodiak, Alaska (USCG).

Above: The Bell HV-911 Eagle Eye has been selected for the USCG's VUAV mission.

Left: The Coast Guard has operated a single VC-4A (Gulfstream I) on transport duties since the 1960s. The original aircraft has recently been replaced by an ex-NASA machine, and now serves alongside the C-37A at AS Washington (Ronald Reagan NAP).

Right: 'Coastguard 01' has been a Gulfstream product for many years. Today this C-37A Gulfstream V (G500) is used.

Below: This impression shows how the EADS-CASA CN235-300M will look when it enters USCG service in 2007.

SPECIAL FEATURE

Tucson Testers

ANG/AFRC Test Center

While the latest aircraft types undergo tests at Edwards, Nellis and Eglin, important work goes on in southern Arizona to maintain the combat viability of some of the older aircraft in the USAF inventory. Equipped with early F-16Cs, the AATC is tasked with developing systems to keep the reservist fleet fully able to take its place alongside active-duty elements on global operations, as well as the testing of some systems for all elements of the Air Force.

These two F-16Cs from the AATC both carry the Lockheed Martin Sniper XR pod, slated to supersede the LANTIRN and Litening II pod as the primary USAF targeting system. The aircraft also carry AIM-120 AMRAAMs, which have supplanted the AIM-9 Sidewinder as the F-16's wingtip-mounted weapon. When AIM-9s are carried in addition to AMRAAMs, they are mounted on the outboard underwing pylon, as here.

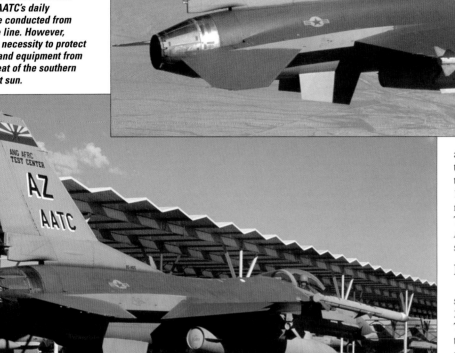

Right: Four of the AATC's six single-seat F-16s formate for the camera, led by the commander's aircraft. The use of the Block 25 F-16C reflects the widespread distribution of this version (and similar Blocks 30/32) throughout the ANG and AFRC.

Below: Thanks to Tucson's fine weather, the AATC's daily operations are conducted from outside on the line. However, shelters are a necessity to protect ground crew and equipment from the searing heat of the southern Arizona desert sun.

Located in the southern part of Arizona, Tucson has long been associated with the place where US military aircraft are retired and put out to pasture. The large Aerospace Maintenance and Regeneration Center (AMARC) facility, located at Davis Monthan AFB, is already home to several hundred of the oldest F-16s that have been retired from service. There is a smaller and lesser known facility located just across town responsible for keeping those older F-16s still in service as capable and reliable as possible through the end of their airframe lives. That facility is the Air National Guard/Air Force Reserve Command Test Center (AATC). It is co-located with the Air National Guard's 162nd Fighter Wing (FW) at Tucson International Airport.

The 162nd has a long association with the F-16 as a large wing with multiple squadrons dedicated to training F-16 pilots. The wing operates many variants of the F-16, including the F-16A/B/C/D/E/F. The AATC, a tenant unit of the 162nd FW, operates only the Block 25 F-16C+, with six C-models and a single D-model on strength with the unit.

The AATC is assigned to the National Guard Bureau as detachment 11 with the unit's commander, Colonel John Mooney, reporting directly to a General Officer at Guard headquarters. The AATC also maintains ties with several other commands within the United States Air Force. It is an operating location for Air Force Reserve Command (AFRC), as well as an Air Combat Command Test Center (like the Air Warfare Center at Nellis AFB, Nevada). Though not part of the 162nd FW, AATC maintains a very close relationship with the host wing. One of the most important parts of this relationship is in the maintenance organisation. The AATC has approximately 50 personnel slots allocated to its maintenance operation. Rather than operating an independent logistics group, the AATC has provided those positions to the 162nd FW in return for the wing providing all maintenance support for AATC's seven 'Vipers'. The rest of AATC's personnel are made up of ANG, AFRC, active-duty, government civil service and civilian contract personnel.

Functions of the AATC

AATC's largest and most important role is the support of the Active, Reserve and Guard Block 25/30/32 F-16 fleet, but it is not its only tasking. The test centre has numerous other functions that relate to the support of guard and reserve aircraft. The two major commands in the Air Force reserve component use the AATC to keep platforms like the A-10 relevant to the overall force structure. To support the ANG and AFRC A-10 communities, the AATC uses a single, test-coded A-10 assigned to the 917th Wing at Barksdale AFB, Louisiana. This aircraft is usually deployed to Davis Monthan AFB to fly test missions in support of various projects. Crew chiefs and maintainers from the 917th Wing support this A-10 while undergoing test sorties at Davis Monthan. When it comes to other aircraft types in the Air National Guard and Air Force Reserve Command, the AATC looks for volunteer units within the reservist organisation to provide assets and personnel to support projects on an as-needed basis.

The area around Tucson provides the AATC with good ranges and excellent weather for most of the year – a perfect environment for both training and test work. The desert mountains provide some challenging terrain for low-level operations.

The Block 25 was the first of the F-16C versions, and was initially delivered without AMRAAM capability, which arrived with the Block 30/32. The latter have a common engine bay for either GE F110 (Block 30) or P&W F100 (Block 32) engines but with generally similar avionics to the Block 25.

Some projects assigned to the AATC are not specific to a single platform, rather they are applicable to multiple aircraft types. One of the more recent successful programmes run by the AATC was one which implemented the Situation Awareness Data Link (SADL). It was initially installed and integrated in the F-16 Block 25/30/32, but has more recently been integrated into other aircraft such as the A-10 and HH-60. As SADL is installed on more aircraft, it creates a larger force that is networked together in the battlespace. Other projects have included the Color Advanced Video Tape Recorder integration into the A-10, F-15 and F-16, as well as affordable night vision goggle lighting for multiple platforms.

Occasionally, the AATC receives an outside tasking to support other programmes and projects. Recently, the test centre has done much work on the operational evaluation of the advanced Lockheed Martin Sniper XR targeting pod. This tasking came from the Air Force Operational Test and Evaluation Center and was done in support of the upcoming deployment of the new pod on F-16s to replace the legacy LANTIRN targeting pods. The first ANG unit slated to become operational with Sniper XR will be the 138th FW at Syracuse, New York, planned for 2005. Sniper XR – with 640 x 512-pixel high-resolution third-generation FLIR coupled to a laser designator and CCD-TV – provides the 'Viper' with a greater lethality when carried. The AATC previously supported the Litening II integration effort prior to F-16s using the pod in Operations Enduring Freedom and Iraqi Freedom. The Litening II enjoyed much success as it provided a leap in technology and capability to the F-16, and because of that success has been integrated onto other platforms like the A-10, B-52 and F-15E, as well as AV-8Bs and F/A-18Ds of the USMC.

AATC commander

In speaking with Colonel John Mooney, the current commander of the AATC, it is quite obvious one of the main goals of AATC is doing more with less. By keeping programmes low-cost and low-risk, his organisation continues a tradition of success. Colonel Mooney came to the AATC after serving as the Operations Group Commander of the 120th FW at Buckley AFB, Colorado. While there, Colonel Mooney was selected to be the group commander of the large Air Expeditionary Wing that was formed and deployed just prior to Operation Iraqi Freedom. The wing deployed to an undisclosed location in the Middle East and Colonel Mooney led 3,000 airmen from several ANG and AFRC F-16 units, and personally flew nine combat missions. He said that he has always been intrigued by the AATC operation as he has seen what the test centre has provided to the Guard and Reserve F-16 community in recent years. Now as commander, he said he expects AATC to remain busy because more legacy equipment in the future will require some growth. He also pointed out that one of the biggest advantages of the AATC is its small size, so growth can be limited to what is necessary.

Most of the day-to-day operations and taskings of the AATC revolve around the F-16. The bulk of the assigned personnel are in support of these aircraft and all are full-time Reservists and Guardsmen. The F-16 pilots on staff are all Fighter Weapons School graduates and some were instructors at the school, which is a tribute to their skill and proficiency. Experience is high, with average F-16 flight time being close to 3,000 hours. The current Director of Operations for the AATC is serving his second tour with the Center. Lieutenant Colonel Eric Overturf was originally on active duty before coming to the AATC in 1995, and is now back after spending several years as a full-time Reservist with the 301st FW at NAS Fort Worth, Texas. 'Turf', as he is known to his peers, made the point that the current ANG/AFRC F-16 fleet is ageing. The goal of AATC is to provide improvements to those aircraft through their full operational lifespans. As new aircraft, like the F-35, are procured by the USAF in the next decade, the Guard and Reserve are going to receive newer versions of the F-16, and the goal of improving those assets as they are handed down will continue. The AATC will likely get newer F-16s in coming years as those assets are shifted from active-duty units. Currently, the responsibility for improving newer 'Vipers' lies with the Air Warfare Center at Nellis, which has the 422nd TES and 85th TES to accomplish this. Additionally, even though the A-10 will start retiring from active service in the next decade, the Guard and Reserve will continue to operate the Warthog. This continued operation could see an increased role of the AATC in A-10 capability and improvement upgrades.

Like those of the parent 162nd Fighter Wing, the AATC aircraft wear the colourful state flag of Arizona on the fin. The 162nd FW has three F-16 training squadrons assigned: the 148th, 152nd and 195th Fighter Squadrons.

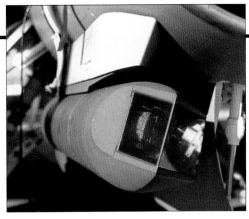

As carried by three of the quartet at right, the Sniper XR targeting pod was AATC's main development priority in 2004. The pod has a rotating head, and employs the latest FLIR and laser technology. The pod is available in downgraded form for export as the Pantera, and it forms the basis of the internal targeting suite for the F-35.

Currently, the AATC supports all of the active-duty, Guard and Reserve Block 25/30/32 F-16 software capability upgrade (SCU) programmes. The latest software release fielded is SCU-5 and the final version was released in the fall of 2004. SCU-5 has been fielded in various forms since early 2003 when an emergency edition (SCU-5E1) was fielded just prior to the start of OIF. This release provided a much-needed JDAM capability to the F-16s deployed to the Middle East for action over Iraq. Next on the agenda for the AATC is SCU-5.1. This software will support new colour displays being installed, as well as the capability to employ the latest member of the Sidewinder family, the AIM-9X. It is targeted for release in mid-2005 in conjunction with start of the planned installation of new colour displays in the Block 25/30/32 fleet. Generally, major software releases are made in two-year increments, with some smaller updates made in between. The next major update planned is SCU-6, which is targeted for release in April of 2007. Several major improvement planned in conjunction with SCU-6 include a new stores management system, a new up-front control and Advanced Identification Friend or Foe, as well as full-colour integration for the 'Viper' cockpit. SCU-7 will follow in April 2009 and, while elements of this update are still being defined, this could be the last major software upgrade for this portion of the 'Viper' community. By the time SCU-7 is implemented, most of the fleet will be close to 25 years old and airframe hours will be approaching their limits.

The many recent successes of the AATC led to its selection for the 2004 Air Combat Command International Test and Evaluation Special Achievement award. Colonel Mooney said the nomination stemmed from the innovative solutions provided by the test centre prior to the start of OIF. AATC was tasked to develop, test and implement a plan to establish a capability to counter any theatre ballistic missile threat in western Iraq. Personnel at the centre formed a team as the need for a command, control, communications and intelligence (C3I) datalink was identified, as was the need for datalink connectivity between the F-16C+ (Block 30), which was the primary fighter asset tasked, and Link 16-equipped platforms. The team took the identified requirements and proceeded to develop, test, refine and validate a time-sensitive targeting C3I employment concept. They tested the concept at Nellis with F-16C+, F-15E, A-10 and F-16CJ units that eventually deployed for missions in OIF. The key piece of hardware added to the F-16C+ was the Battlefield Universal Gateway Equipment (BUG-E). This equipment, married to the previously installed and integrated SADL system, provided a connection to other datalinked assets (like the Link 16-equipped F-15E) in theatre, as well as a connection to the Combat Air Operations Center and the Joint Force Air Component Commander. Not only were theatre commanders able to see the Link 16 air picture of western Iraq from distant locations, but they were able to send command and control messaging/targeting information to air assets flying combat missions as well. All of this was done concurrently with implementation of the quick-reaction, emergency release version of SCU-5E1 that was required for the F-16C+ to have a necessary JDAM capability in theatre. The integration of the Litening II targeting pod on the A-10 and B-52 was also done in the same timeframe, along with supporting efforts to integrate the pod on the F-15E. All of the team's efforts produced positive results in combat operations.

Andy Wolfe

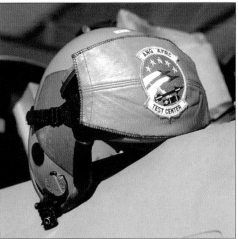

Though small, the AATC is a vital part of the test and evaluation community within the Air Force. The highly experienced personnel have a record of always being prepared for the next emergency tasking, whenever the call comes. They continue to be 'there, willing and able' to provide the innovative solutions that have given them the outstanding reputation they enjoy today.

VARIANT FILE

Boeing 707
Military variants

The story of the military Boeing 707 consists of two quite separate and contradictory strands. On the one hand, the 707 is widely viewed as an obsolescent anachronism in its original airliner form, and operators who can afford to do so have discarded them in favour of more modern wide-bodied jetliners. But the decline of the 707 airliner contrasts starkly with the continuing, pivotal importance of some of its derivatives, which are among the most modern and most capable aircraft in their class. The USA's 'air power infrastructure' is based upon the E-3 Sentry and E-8 JSTARS, which have dominated recent real-world operations, and which have been largely responsible for providing the 'air dominance' upon which US forces have been able to rely.

58-6970 was the first military 707, a Series 153 delivered to the USAF's 1254th Air Transport Wing on 4 May 1959 for VIP transport, including that of the President (although it was not the official 'Air Force One'). It received TF33 turbofans in 1963 and was withdrawn from use in 1996.

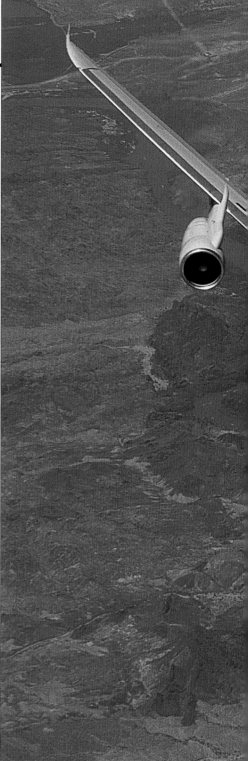

For many years the Boeing 707 was the biggest selling commercial jetliner in the world. Though best known as a civilian jet airliner, the Boeing 707 has also seen extensive military service, even if one discounts the dedicated military C-135 family. Because they are so similar, and because they were both derived from the same original Model 367-80 design, the relationship between the 'civil' 707 and the 'military' C-135 is often misunderstood, and many have assumed that the USAF C-135 was a military version of the 707 airliner. This is far from being the case, and if anything, the aircraft was originally driven by military requirements.

In the immediate post-war world, leading air forces quickly re-equipped with jet fighters and jet bombers, which left their transport and tanker fleets looking increasingly archaic and anachronistic, and making it increasingly difficult to support rapid overseas deployments. The USAF, in particular, clearly needed a jet transport to move groundcrew, spares and support equipment wherever the jet fighters and jet bombers deployed, while the KC-97 tankers were also demonstrably too slow to support Strategic Air Command deployments.

The Model 367-80, initially known as the Jet Stratoliner, was therefore aimed principally at the US Air Force, and was intended as a replacement for the Model 367 (C-97 Stratofreighter) in the tanker and transport roles, though the aircraft also had obvious potential as a civilian airliner, and Boeing hoped that it would be a competitor to the newly announced Douglas DC-8 jetliner.

The 367-80 model number was not used because the aircraft really had any commonality with the Stratocruiser, but because Boeing hoped that Douglas would be lulled into a false sense of security. Boeing hoped that its competitors would assume that the new project was simply a warmed-over Model 367, in just the way that Boeing's early Model 473 jetliner designs (drawn up during the late 1940s) had been. In fact, the new aircraft was a much more advanced, 'clean sheet of paper' design, and was known internally as the Model 707 when design studies were launched in 1951. The new jetliner owed very little to the ancient Stratocruiser, but inherited much from Boeing's B-47 and B-52 jet bombers. The 707 had a similar lightweight highly swept wing, similar integral fuel tanks and pod-mounted engines.

Classic layout

Though early Boeing jetliner studies looked more like the B-52, with a shoulder-mounted wing and a fuselage-mounted 'bicycle' undercarriage, the company soon arrived at a more conventional low-wing arrangement, with a 35° swept wing (at quarter-chord), podded engines underwing, and a tricycle undercarriage whose main units retracted into the centre section. The 707-6, which formed the basis of the 367-80 (707-7) was 126 ft 4 in (38.51 m) long, with a 130-ft (39.62-m) wingspan, and differed from the 367-80 prototype only in having its J57 engines paired in B-52 type pods. Before cutting metal, Boeing looked several times at

Above: Going to war without AWACS cover would be unthinkable for the USAF. This E-3B, in the latest Block 30 configuration, carries the 'OK' tailcode of the 552nd ACW, which parents the USAF's entire E-3 fleet.

alternative powerplant options. The 707-7-27, for example, was an American 'Bear', using four separate T34 turboprop engines overwing, while the 707-7-39 had six J57s – the four inboard engines paired in B-52 type pods and the two outboard engines in single pods. Both of these designs had streamlined F-104 type wingtip tanks, but both were rejected in favour of a more conventional layout with four separately podded, underslung engines spaced out along the leading edge.

Whereas the competing DC-8 used hydraulically boosted flying controls, Boeing chose to use manually operated, aerodynamically balanced controls, with spring tabs. The new Boeing airliner also used multi-surface flight controls, with inboard ailerons augmented at low speed by outboard ailerons. The latter were locked in the neutral position when the flaps were retracted.

Boeing built the 367-80 prototype as a private venture, spending $16 million (225 per cent of the company's 1951 profits) and hoped

Below: Resplendent in the special 'Air Force One' scheme introduced by sister-ship 62-6000, 72-7000 was the second of two VC-137Cs (707-353Bs) procured especially for Presidential transport.

Above: The E-3 Sentry was the first 707 variant to be developed for a specific military purpose: that of replacing the EC-121 in the AWACS role. The aircraft above is the first of two radar evaluation aircraft, which were known initially as EC-137Ds. The aircraft were hasty conversions of 707-320 airframes, hence the evidence of cabin windows.

Right: The flight test apron at Seattle is headed by the two pre-production E-3As (73-1674/1675) and the first production machine (75-0556). Also in the line is a 707 tanker for the Imperial Iranian Air Force, next to an IIAF 747.

that separate but related designs would be ordered for military and civil use. When Mrs Boeing, wife of the Chairman, named the aircraft when it rolled out on 1 May 1954, it was christened 'Jet Stratotanker' and 'Jet Stratoliner'. The aircraft made its first flight on 15 July.

When the USAF evaluated the Model 367-80 prototype, they decided that its 132-in (3.35-m) cabin was too narrow, and insisted that the production military version (then known as the Model 367-138B and later as the Model 717 and 739) should have a cabin that was 12 in (30.5 cm) wider. (Some 117 KC-135As had already been ordered as interim tankers, pending the resolution of the KC-X tanker contest). Boeing initially hoped to use the same fuselage cross-section for both civil and military versions, but when early airline prospects ordered the DC-8 in preference to the 707, the company was forced to widen the cabin of the civil Model 707 by a further four inches (to 148 in/3.76 m), and to lengthen it by 10 ft (3.05 m). The original cabin width had been adequate for five-abreast, tourist-class seating, but was not sufficient to allow six-abreast seating, and was certainly not sufficient to compete with the 147 in (3.73 m) wide DC-8. Interestingly, the original 707-1 design study had used a 122-in (3.10-m) cabin – three inches 'slimmer' than the DC-7, and the original 132 in cabin was selected to match the width of the Stratocruiser's upper deck. The decision was

The final aircraft from the 707 line were powered by fuel-efficient CFM56 (F108) turbofans. The US Navy bought the type as the E-6A to perform the TACAMO (Take Charge And Move Out) submarine communications relay role using trailing wire antennas.

then taken that the 'short-lived' military tankers would be built using lightweight 7178 aluminium alloy, while the airliners would use 2024 alloy. This further reduced commonality between the two aircraft types (which were originally to have been built using common tooling) to a mere 20 per cent.

The Model 367-138B (Model 717) spawned a vast number of sub-variants, and the C-135 family fulfilled a wide variety of military roles, but these aircraft were described in *International Air Power Review* Volume 10 and lie outside the scope of this article, which covers the military use of variants based on the slightly later Model 707 airliner airframe.

707 variations

The 707 was built in a number of basic variants during the course of its production life, and many of these saw military use, or formed the basis of dedicated military versions. The initial 707 airliner was known as the 707-120, though the last two digits of the designation suffix varied according to the customer. This baseline aircraft had a 138-ft 10-in (42.32-m) fuselage, 13,486 US gal (51050 litres) of fuel, and was powered by four JT3C engines. Sixty were built.

The 707-138 was a longer-range version, with a 10 ft (3.05 m) section removed aft of the wing, with 17,286 US gal (65434 litres) of fuel, and was powered by 11,200-lb (49.84-kN) thrust JT3C-6 engines (these produced 13,000 lb/57.85 kN with water injection). Only seven were built, for Qantas, designated as 707-138s. A taller tailfin, a ventral tailfin/tail bumper, and a hydraulically-boosted rudder were all retrofitted to these early 707s, and most received new Krüger flaps inboard of the engines.

Only five 707-227s were built, and none of these long-range JT4A-3 powered aircraft saw military use. Some 70 examples of the 707-320 (briefly known as the 707 Intercontinental) were built, with more powerful 16,200-lb (72.09-kN) thrust JT4A engines, increased wing area and fuel tankage, a strengthened landing gear, increased span tailplanes and new flaps. The fuselage was stretched by 6 ft 8 in (2.03 m), and passenger capacity was increased from 96 to 104 first-class or 165 to 180 economy-class passengers. The 707-420 was identical, except in being powered by Rolls-Royce Conway R.Co.5 engines.

Whereas the 707-320 and 707-420 were scaled-up, long-range derivatives of the baseline 707-120, the 720 was a reduced-capacity, short-range aircraft, with a 130-ft 10-in (39.88-m) fuselage, an extended leading-edge glove, and lower weight. The JT3C-7/-12 powered aircraft carried 88 first-class or 141 economy-class passengers.

The need to compete with the Convair 990 led to the development of a number of 707 models with Pratt & Whitney JT3D turbofan engines, resulting in the 707-120B, 707-320B and 720B. The first of these flew on 22 June 1960, and the turbofan-powered aircraft soon entirely supplanted the earlier turbojet-engined aircraft on the production line. From 1963, Boeing offered a turbofan-engined convertible passenger/cargo aircraft with a main deck side cargo door, and a reinforced floor with tie-downs, using the designation 707-320C.

Aiming to provide the same level of command and control for the ground war as the E-3 provides for the air war, the E-8C JSTARS carries a huge synthetic aperture radar under the forward fuselage. The 17 operational aircraft were produced by conversion.

Later in its career, as the first wide-bodied airliners entered service with their high-bypass ratio turbofan engines, the 707 and other first-generation jetliners appeared noisy and anachronistic. Boeing developed and test flew new, quieter nacelles for the JT3D in 1973, but these attracted little customer interest, and did not enter production. A few years later, when ICAO rules threatened to ground unmodified 707s, a variety of companies provided Stage 2 and Stage 3 'hushkits' for the type.

Most of these basic 707 versions saw some military use, jet- and turbofan-powered, passenger and Convertible, 707-120, 707-320 and even 720. Their military versions are detailed later.

Military use

During the piston-engined airliner era, Boeing's commercial transports had faced stiff competition from Lockheed and especially Douglas. However, in the post-war period, building on the success of the B-47 Stratojet, Boeing gained an almost-unassailable lead in jet airliner design. With the virtual collapse of Britain's rival de Havilland Comet (actually the first jet airliner in productive commercial service), the Boeing 707 enjoyed staggering success, and its performance soon attracted the attention of a number of military operators. Though dwarfed by the C-135's 820 aircraft production total, military use of the 707 totalled about 250 aircraft, making it one of the most successful post-war military jet transports.

For many military customers, the 707 made much greater sense than the C-135, even in the tanker role. With its wider cabin the 707 was more versatile, and because it was in widespread commercial service across every continent, the 707 was easier and cheaper to support, and often offered some commonality with aircraft operated by the national (and sometimes nationally-owned) airline. In addition, the aircraft was available second-hand (whereas the KC-135 was not), and about one half of military 707s were 'pre-owned'.

Thus, while the USAF procured some 48 dedicated transport versions of the KC-135 tanker (though most of these were relatively quickly replaced by C-141 StarLifters) it also procured significant numbers of military Boeing 707s. On the export market, the 707 has been much more successful than the dedicated military C-135. Whereas the -135 family retained the -135 designation regardless of its role, purpose-built models of the 707 built (or purchased) for the USAF have used a bewildering array of designations, and export aircraft have often been referred to by their civil model numbers.

Some 37 new 'stock' 707s and one new Model 720 were simply sold to military customers for transport use, together with about 116 'pre-owned' former airline 707s and Model 720s. It is difficult to determine exactly how many military 707s have been delivered, since some government-owned and operated aircraft wear civil registrations, and sometimes even civilian airline colour schemes.

Some of the military 707s have subsequently been converted to fulfil other roles and requirements, some of them modified by Boeing and some by other companies. Some aircraft have been converted to a number of tanker configurations, while others have seen service in the Sigint and Elint roles. The 707 has also formed the basis of a number of dedicated military special duties derivatives (including the E-3 and E-6), most of them newly-built, and indeed it was mainly these 93 aircraft (and the aircraft rebuilt as E-8s) that kept the 707 in production in its later years.

With their CFM56 turbofan engines, the E-6s and export E-3s were among the most modern 707s flown, echoing the features of the 707-700, a 1979 one-off prototype, further production of which was cancelled in case it took orders away from the new 757.

Further 707s were paid for by military funds when civil aircraft were chartered for trooping (especially during the long involvement in Vietnam) and when large numbers of redundant airliners were purchased by the USAF for their JT3D (TF33 engines) and wide-span tailplanes, which were retrofitted to 156 KC-135A tankers to turn them into KC-135Es, as well as to four KC-135Ds and 21 special mission aircraft.

707 at war

While the military KC-135 and RC-135 saw extensive combat service in Vietnam, the military 707's combat debut came rather later. In US hands, the E-3 played a part during the invasion of Grenada in 1983, and went on (with the E-8 JSTARS prototypes) to enjoy a pivotal role in Operation Desert Storm. Since then, E-3s and E-8s have formed a key element in the air forces used in successive operations over the Balkans and in the Middle East.

A small number of 707s have been used by the United States on clandestine duties, usually without national insignia and with either spurious serials or, as here, an out-of-sequence serial based on the aircraft's construction number. Designated EC-137D, this aircraft flew with the 2nd Special Operations Flight.

Taking the lead from its smaller KC-135 stablemate, the 707 has been widely used as a tanker. Canada modified two of its CC-137s with wingtip Beech 1800 pods, here streaming drogues for CF-18 Hornet receivers.

The basic 707 has also seen plenty of action, perhaps most often in Israeli hands. The IDF/AF leased two or three 707-131s from IAI during the 1973 Yom Kippur war, using the aircraft in the transport role. The 707 proved a massive success, and the IDF/AF quickly acquired more examples of the type, which was soon in the headlines again.

When Israel launched Operation Jonathan (also known as Operation Thunderbolt), its mission to rescue Israeli hostages held at Entebbe airport, Uganda, the force included four C-130 Hercules and two Boeing 707s. One of the 707s was fitted out as an airborne command post. The second was configured as a hospital aircraft, and flew to Nairobi, Kenya, where it sat ready to take on any casualties.

The mission was a signal success, resulting in the death of all eight terrorists and the freeing of all but two of the hostages (who were killed in the crossfire). Only one of the rescue force (the commander) was killed.

The IDF/AF used another 707 command post to support the strike against Iraq's Osirak nuclear facility on 7 June 1981. Boeing 707s played a more active part during the Israeli invasion of the Lebanon in 1982. At least one EW-configured aircraft was used during the 9 June 1982 operation against Syrian air defences, operating in conjunction with Standard- and Shrike-armed Phantoms to blind Syria's AD radars and disrupt its air defence communications, laying the groundwork for the IDF/AF's methodical destruction of the Syrian fighter force in the air.

Following a number of terrorist outrages, the Israeli cabinet authorised a retaliatory strike against the PLO's beachfront HQ in Tunis. This necessitated a 1,280-mile (2060-km) trip to the target, flown entirely over water. Eight F-15s (plus two air spares) flew the strike on 1 October 1985, supported by a 707 tanker.

Refugee airlift

When Israel mounted an airlift to recover 14,000 Ethiopian Jews from Addis Ababa to Israel, just days before the capital fell to rebel troops, an El Al Boeing 747 grabbed the headlines by carrying an astonishing 1,200 refugees. But during Operation Solomon this aircraft was augmented by eight more El Al airliners (two more 747s, four 767s and two 757s), 18 IDF/AF Hercules and eight Boeing 707s. The latter had their seats removed and rubber mattresses were placed on the cabin floors, allowing each aircraft to carry 500 passengers. The airlift ran from 24-25 May 1991.

Just as USAF RC-135 regularly flew 'ferret' missions around the peripheries of the USSR, so the IDF/AF used its Sigint-configured Boeing 707s to try to detect, classify, locate and identify Syrian air defence radars, measure response times and monitor communications traffic. From the late 1980s, the IDF/AF flew missions over the Mediterranean, just outside Syrian airspace. Syrian fighters were usually scrambled, but usually contented themselves with shadowing the IDF/AF Boeings, which were normally escorted by pairs of F-15s.

All this changed on 14 September 2001, when two MiG-29s turned towards the Israeli aircraft and increased their speed, a clearly aggressive move. The F-15 leader ordered the Boeing to withdraw and activate its ECM systems, and called for reinforcements. Six more F-15s, six F-16s and a Boeing 707 tanker were scrambled, but before they could arrive, the engagement was concluded. After warning the Syrian MiG-29 pilots to change course (using the international distress frequency) with no response and no compliance, the Eagles attacked, shooting down one MiG with a Python IV and the other with an AIM-9M.

More recently, IDF/AF Boeing 707s played a vital role in the seizure of the *Karine A*, a 4,000-tonne ship carrying 50 tons of weapons and ammunition intended for the Palestinian Authority. The commandos were dropped from Black Hawk and CH-53 Yasur helicopters that had been refuelled by KC-130 tankers, with F-15s providing top cover. The operation was supported by 707 tankers and Sigint aircraft, and controlled by the IDF's Chief-of-Staff Shaul Mofaz, IAF commander Dan Halutz, Navy commander Yedidia Ya'ary and IDF Intelligence Branch commander, Aharon Zeevi, from a command post 707 that left Ovda at midnight on 2 January 2002.

On 28 November 2002, IDF/AF Boeing 707s hit the headlines again, when two Boeing 707 Re'ems, one carrying a medical crew and one a command post, flew to Kenya to assist with the treatment of the wounded following the Mombasa terrorist attack. Israeli-modified 707s also saw extensive service in southern Africa, as SAAF Elint and jammer 707s supported offensive operations in Angola.

707 tanker wingtip HDUs

This is the Sargent Fletcher 34-000 hose-drogue unit of a South African tanker. Israeli tanker conversions have the pod mounted further inboard than Boeing-designed installations.

Brazilian KC-137s use the Beech 1800 pod. This differs from the other options by having the hose trailing from an arm that deploys down away from the wingtip. Power comes from the nose-mounted turbine.

Fitted to an Australian tanker by IAI conversion, this is the Flight Refuelling Mk 32 pod. Fuel flow director lights for receiver pilots are incorporated in the lower side of the pod.

Boeing 707 military variants

Iran's Peace Station tankers had a Boeing Flying Boom installed to refuel receivers equipped with USAF-style receptacles. These ranged in size from fighters to the Boeing 747 (above). Following the Islamic revolution the 707 continued to play a big part in the IRIAF, and an example is seen at right with an F-5B, RF-4E and F-14A. Some of the aircraft have had their tanker equipment removed and have been used by civil operator Saha Air.

Perhaps the best known combat use of the 707 came in 1982, when Argentina's three 707s were pressed into use. Immediately before the war the aircraft mounted an intensive series of flights to Britain, France, Spain, Italy and Israel to pick up spares and weapons. When Argentina invaded the Falklands the 707s could not participate directly, as they were too large to use Port Stanley's runway. They were, however, used to transport men and material within Argentina.

Once Britain despatched its Task Force to recover the islands, the Argentine navy requested that the FAA should locate the British ships. Though they lacked any role equipment, the 707s did have the necessary range for the task, and their weather radar gave them a rudimentary search capability. A 707 captained by Vicecomodoro Jorge Ricardini found and photographed the Task Force on the first mission, flown on 21 April. The aircraft was intercepted by a Sea Harrier almost immediately after passing a radio report back to HQ.

Subsequent 707 missions were intercepted further out from the Task Force ships, and on 22 April one aircraft was intercepted and 'boxed in' by three Sea Harriers, in an effort to underline the 707's vulnerability. But Argentina realised that Britain was loathe to shoot down an unarmed airliner over international waters, and the 707s continued their missions until 23 April, when Britain advised (through diplomatic channels) that the aircraft would be fired upon in future, if they breached the British Maritime Exclusion Zone south of 35° S.

Dodging SAMs

Long-range reconnaissance missions had resumed by 19 May, when two RAF Harriers from *Hermes* narrowly failed to intercept a prowling 707. On 22 May a 707 operating north of the MEZ was fired upon by HMS *Bristol* and HMS *Cardiff*, each of which launched two Sea Dart SAMs. The 707 took violent evasive action, and escaped by diving almost to sea level. The final 707 sortie took place on 7 June, when an aircraft located the damaged HMS *Argonaut* as it limped home.

Boeing 707 airframes were arguably most critical to the success of a military operation during Operations Desert Shield and Desert Storm, and during subsequent operations in the Balkans and the Middle East. In Desert Shield USAF (and Royal Saudi Air Force) E-3 Sentries immediately established a 24/7 radar screen which allowed deployed fighters to defend vital coalition bases in Saudi Arabia and the Gulf against any possibility of Iraqi air attack. The E-3s flew more than 400 missions during Desert Storm, clocking up more than 5,000 hours on-station in sorties which lasted between 16 and 18 hours, usually with two crews aboard each aircraft. The Sentries provided senior military and political leaders with accurate and timely information about the enemy's actions and movements, and provided radar surveillance and control for more than 120,000 coalition sorties (at a rate of about 3,000 sorties per day). Their Fighter Controllers assisted in or directed 38 of the 40 air-to-air kills scored by coalition fighters.

A few 707s have been used as pilot training and deployment support aircraft for the operational E-3 and E-6 fleets. NATO (above) has operated a total of five aircraft in the TCA (Trainer Cargo Aircraft) role, although not simultaneously, the aircraft latterly being designated CT-49. The USAF Sentry fleet was supported by a pair of TC-18Es. The initial pair was retired in favour of two more aircraft (demodified EC-18s, right) before they, too, were withdrawn from service.

707s have been popular with governments to provide VIP transport. In the Gulf region such aircraft were operated by Qatar (below), Saudi Arabia and the United Arab Emirates (left).

During these operations, the 707-based E-3 Sentry airborne warning and control system (or AWACS), has been arguably the most important tool in achieving air dominance. In previous wars, the USAF has been able to establish 'air superiority' – a situation under which a commander has sufficient control of the air to prevent effective enemy air activity against his own ground forces. Sometimes, US commanders have even enjoyed 'air supremacy' – in which allied aircraft can operate largely unmolested, and in which the defeated enemy can achieve only the most limited effect against allied operations.

With 'air dominance' a commander enjoys total control of theatre airspace, and is able to entirely prevent the enemy from undertaking air operations at all.

Just as the E-3 has allowed air commanders to dominate theatre airspace, another 707-based platform, the E-8 JSTARS has started to provide commanders with a similarly impressive picture of the ground war, with a similar effect on allied dominance.

Secret 707s

Sometimes the military has made more clandestine use of the Boeing 707 airframe, and in the US, in particular, the line between civil and military has sometimes been blurred. During the 1970s and 1980s, Southern Air Transport (sometimes described as formerly a CIA proprietary airline) carried out military transport operations. In November 1985, for example, two Southern Air Transport 707s flew US-supplied parts for F-4 fighters and helicopters, and Israeli-supplied Hawk anti-aircraft missiles, from Tel Aviv to Tehran as part of the 'arms for hostages' programme. Four further shipments were made on 15 February, 22 May, August and October 1986. The May shipment, by two Boeing 707s flying to Tehran (via Tel Aviv) from Kelly AFB, consisted of 90 tons of weapons, including 500 TOW missiles and an unspecified number of Hawk surface-to-air missiles. Though US sources acknowledged five shipments (10 'plane-loads') during 1985 and 1986, Iranian sources suggest that the total was actually 20 'plane-loads'. This may be explained by the fact that some shipments went from Kelly AFB to Ramstein AFB in Germany on board USAF C-141 StarLifters, and were then flown from Ramstein to Ben Gurion IAP on board Boeing 707s. The money paid by Iran for these weapons allegedly found its way to the Contras in Nicaragua.

Southern Air Transport was by no means the only airline 'used' by the CIA and other US government agencies. Between 1976 and 31 December 1988, for example, Crittenden Air Transport received $571,350,000 and 15 aircraft from the CIA (including at least one 707 and no fewer than seven C-130s). Special codes were reportedly used during air traffic control proce-

A small number of 707s have been converted for signals intelligence-gathering. Argentina (left) contracted IAI to modify one of its aircraft for the role, while Israel itself has operated a number of Sigint machines under the RC-707 designation (below). In IDF/AF service the RC-707 is known as the Barboor (Swan).

Following the Nimrod AEW fiasco, the RAF turned to Boeing to replace its antediluvian Shackleton AEW fleet by ordering seven Sentry AEW.Mk 1s (E-3Ds). They were to be among the last 707 airframes built: subsequent prospective customers for special mission aircraft would be offered the 767 as the basic airframe.

dures that were used to alert US customs to ignore the aircraft and not make any customs inspections. On one occasion a Crittenden Air Transport aircraft flew a load of American M16 rifles from Manila to Moscow, where it loaded AK-47 assault rifles, which were flown to San Salvador. Denver Ports of Call and Evergreen International were also allegedly associated with these activities.

Out of production

The 707 airliner programme drew to a close during the mid-1970s. While rival Douglas continued to develop and stretch the DC-8, the development of advanced 707 versions ceased as Boeing's attention was taken by its ill-fated 2707 SST, and by the giant 747 Jumbo Jet. As the oil crisis provoked what some called 'the worst downturn in the US airline industry since the 1930s', Boeing was left without a direct competitor to the stretched DC-8. Having failed to update the 707 since the 707-320B, civil orders dried up, though military customers continued to demand new 707 airframes.

The last 707 built for an airline was a 707-3F9C (s/n 21428) delivered to Nigeria Airways on 30 January 1978, but the military 707 (in E-3 and E-6 form) remained in production until 28 May 1992, when the final RAF E-3D (ZH106 s/n 24999, the 1,011th 707 built) was delivered. By 1992, the 707 airframe was showing its age, even when newly built and fitted with modern high-bypass ratio turbofan engines, and Boeing began to offer the Model 767 as the basis of future large military aircraft. The Boeing 767 offered a larger cabin than the 707/E-3, was 6 ft (1.83 m) longer, with 50 per cent more floor area and double the cabin volume, and the newer aircraft also offered superior speed, altitude and endurance.

IAI Elta fitted its Phalcon AEW radar equipment into a 707 to produce arguably the most radical alteration to the airliner's basic shape. Chile acquired the only aircraft to be sold to date, naming it Condor in FACh service. The nose and cheek fairings house arrays of electronically steered transmit/receive radar modules.

Thereafter, new customers had to choose between rebuilt, refurbished and converted second-hand 707 airframes (as used for the E-8) or new-build 767 airframes. The Japanese Air Self Defence Force, for example, left it too late to order E-3s, and instead opted to buy 767s fitted with the same mission systems and equipment, while the USAF opted for the conversion of second-hand Boeing 707s to form the basis of its E-8 JSTARS aircraft.

The last new-build E-3 and E-6 airframes have proved extremely successful, whereas the use of refurbished ex-airline airframes has highlighted a number of problems, and has effectively killed off the development of further new 707 derivatives. The biggest long-term issue for the 707 will be corrosion. It has been suggested that the cost of repairing airframe corrosion pushed the price of the E-8 refurbishment close to that of building new airframes, even though the USAF was able to 'cherry-pick' the very best 707-320B airframes for the E-8 JSTARS programme. It has been suggested that a typical US $65 million refurbishing/rectification price per airframe is now more than the cost of a good used Boeing 767 or Airbus A310.

The JSTARS programme highlighted the 707 airframe's corrosion problems, which were concentrated in four 'hot spots', around the nosewheel well and forward fuselage, the lower fuselage, at three points aligned with the leading edge of the wing, and in front of and behind the main undercarriage wells. Significant corrosion was also found in the wings, especially near the engine pylons, and in the upper part of the wing roots. Other operators discovered corrosion in other areas, with the RAAF's 707s having problems with the upper wing skins, fuselage top and tail surfaces.

Surprisingly large numbers of military Boeing 707s remain in service, though many have been retired, often because they can no longer comply with modern noise regulations, a problem that drove many commercial 707s out of service during the late 1970s and early 1980s. Some 707s have been replaced because they no longer convey a sufficiently prestigious image to be used for VIP or 'Head of State' transport duties, while others have been replaced by aircraft which offer greater capacity, performance or range. Many military 707s have been replaced by wide-bodied airliners rendered superfluous by the recession in the airline industry. The advantages conferred by the 707's four engines have largely been negated by the improved reliability and safety characteristics of modern twin-engined airliners, which have become the backbone of the world's airlines, and which are increasingly replacing remaining 707s in military use. With modern aircraft like the Boeing 767 and A310 sitting unwanted in storage in the desert, ageing 707s are cheap to replace.

While transport 707s are something of an endangered species, the USAF's E-3s and E-8s, and the Navy's E-6s, can look forward to decades of productive front-line service, while a Boeing 707 testbed named Paul Revere is today playing a crucial role in shaping the very future of US air power infrastructure.

Jon Lake

Boeing 707 military variants: Part 1

VIP and personnel transports

VC-137A

The three new-build VC-137As (58-6970 to 6972) delivered to the USAF in 1959 were probably the best-known Boeing 707s in the world, becoming the first jets routinely used by the US President, and then going on to see prolonged service as Presidential and VIP transport aircraft, in what was known as the 'Special Air Missions' role. The VC-137A was based on the 707-120 airframe, and used the Boeing designation Boeing 707-153 (with the customer designation suffix 53 indicating the US Air Force). The first aircraft (58-6970, sometimes affectionately known as 'Queenie') was purpose-designed for Presidential duties, so differed from the other two.

The first aircraft, which flew on 7 April 1959, had sleeper accommodation and airborne conference facilities in the centre cabin, supported by a special communications centre. There was an eight-passenger cabin forward of the wing with 14 reclining passenger seats behind, and the aircraft had three galleys and two dressing rooms. Without the knowledge of the then-President Eisenhower, the CIA fitted 58-6970 with secret reconnaissance cameras for a planned trip to Moscow, though this visit was cancelled in the aftermath of the Gary Powers incident. The aircraft was the first jet to be used as 'Air Force One', though the primary Presidential Aircraft continued to be a VC-121E and a VC-118A.

The two remaining VC-137As flew in a more standard, if rather luxurious, passenger fit, seating 36-40 passengers, along with 7-18 crew members.

Seen in their original schemes, these are the first (top) and third (right) VC-137As. The second two aircraft were used for general staff transport.

VC-137B/C-137B

The three VC-137As were re-designated as VC-137Bs after being re-engined with TF33 (JT3D-3) turbofan engines in 1963, which effectively brought them up to a similar standard to the newly delivered VC-137C. The aircraft were repainted in the colour scheme designed for the VC-137C, and 58-6970 became the 'back-up' for the VC-137C, which was the primary aircraft used as 'Air Force One'. The smaller VC-137B had a very different interior layout to the C model, however, and a much shorter range, and so was never completely suitable as a reserve aircraft.

The three VC-137B aircraft were later re-painted in a more standard scheme during President Carter's administration, losing their V- for VIP designation prefix to become simple C-137Bs. Some sources suggest that the designation change marked relegation to 'cargo-only' flying, but this is believed to be incorrect.

58-6970 retired to Seattle's Museum of Flight on 18 June 1996, 58-6971 retired in 1997 and ended its days at the Pima County museum (adjacent to the AMARC boneyard) in 1999, while 58-6972 was scrapped at Boeing's Wichita, Kansas, plant in 1996, suffering from severe corrosion.

C-137B 58-6971 is seen late in its operational career, having been re-engined with TF33 turbofans. It wears the scheme adopted as standard by the 89th Airlift Wing.

VC-137C

The piston-engined VC-121 and VC-118 Presidential Aircraft were finally officially replaced by a jet when the first VC-137C (62-6000) was delivered on 10 October 1962. The new aircraft was based on the longer, bigger-winged Boeing 707-320B airframe, with JT3D (TF33) turbofan engines. The new aircraft used the Boeing Company designation Model 707-353B. It was fitted with a Presidential sleeper/HQ suite in the rear fuselage, with staff accommodation forward, so that passenger capacity increased to 49.

The VC-137C was delivered in a smart silver, white and blue colour scheme designed by the famous stylist Raymond Loewy, whose portfolio also included the Ritz cracker logo, the Studebaker Starlight automobile, and the smart livery used by the Pennsylvania Railroad. The new colour scheme was famously designed in collaboration with Jackie Kennedy, the wife of the then President. Tremendous attention was paid to detail, and a new shade of blue was used on the national star and bar insignia in order to tone better with the rest of the scheme!

A second VC-137C (72-7000) was delivered in December 1972, and both aircraft remained in use with the 89th Airlift Wing until replaced by C-32s in 1998-99, though they lost their Presidential tasking in September 1990, when the 747-based VC-25 entered service. After their replacement as 'Air Force One', 62-6000 and 72-7000 lost their distinctive dark blue forward fuselage top and cheat line, and were repainted in the similar standard 89th Wing VIP scheme.

While used for Presidential duties the VC-137A 58-6970 and VC-137Cs (62-6000 and 72-7000) were maintained and operated by a separate Presidential flight within the 89th Wing, with all personnel requiring an 'extended background investigation.' VC-137 62-6000 went to the USAF Museum at Wright-Patterson AFB in May 1998 while 72-7000 flew its final presidential flight in August 2001, and was subsequently retired to the Reagan Presidential Library at Simi Valley, California, where it is now displayed. The latter aircraft had carried every President from Nixon to George W. Bush.

Jackie Kennedy's 'mission' to bring glamour to the White House included overseeing the scheme applied to 62-6000, the first 707 purchased specifically for Presidential transport.

72-7000 is seen in July 2001 during its last days of Presidential service. By that time the 'Chief' himself was flying on the VC-25A (747) seen in the background.

C-137C

The VC-137s were augmented by a pair of pre-owned 707-320Bs during 1985, and these were designated as C-137Cs. The two aircraft (85-6973 and 6974) were a Boeing 707-396C and a Boeing 707-382B that had reportedly been confiscated amid allegations of drug-running. These ex-airline aircraft were used by the 89th Airlift Wing to transport the Vice-President and other government officials, alongside a handful of VC-135 'Tubes'. The more lavishly appointed C-137Cs were sometimes known as 'Super Tubes'. They retired in August 1998. One (85-6973) subsequently became an E-8C JSTARS platform.

C-137C 85-6974 was flown by the 89th AW until replaced by a C-32 (Boeing 757).

Boeing 707 military variants

West Germany acquired four 707s to equip the long-range transport arm of the FBS. Two of them continued in military service in Germany, flying as NATO TCA trainers, while the airframe of another was used as the basis for an E-8C.

C-137 (Germany)

When the Luftwaffe acquired four 707-307Cs through USAF channels in late 1968, the aircraft were designated as C-137s (despite using a freight door) and were allocated the USAF serials 68-11071 to 11074, though they actually wore Luftwaffe codes 10+01 to 10+04. The aircraft were replaced by A310s during the 1990s.

CC-137 Husky

The Canadian Armed Forces took five Boeing 707-347Cs from a cancelled Western Airlines order (hence the customer suffix 47), receiving all five aircraft between February and May 1970. This brought to an end a particularly long and tortuous procurement programme. This had started when the Canadians expressed an interest in acquiring KC-135s, but they were unable to place an order before the production line closed. Canadian interest then switched to the C-141 StarLifter, but the production line again shut down before an order could be finalised.

Canada's air force used the 707 mainly for long-range staff/cargo transport, notably on the scheduled flights to Europe to support Canada's contribution to the NATO garrison.

The five CC-137s (13701 to 13705) replaced a larger fleet of turboprop Canadair CC-106 Yukons, and served with No. 437 Squadron until replaced by the CC 150 Polaris (Airbus A310) from 1994. The last two aircraft were retired in April 1997, having been upgraded to tanker (see later). Four of the five aircraft were converted to E-8C JSTARS standards.

EC-137D and US government 707s

The EC-137D designation was used to describe two quite different configurations. The best known were a pair of EC-137D aircraft which served as prototypes for the E-3, and which are described separately. The other was a single ex-airline 707-355C (N707HL) purchased in 1992, also designated EC-137D (sometimes erroneously reported as EC-137E). The aircraft was modified by E-Systems as an 'electronic support aircraft' and is believed to have had inflight refuelling receiver capability and comprehensive command post capabilities, justifying its EC- designation prefix. The aircraft operated without titles or national insignia, and may sometimes have worn spurious civil registrations or serials. Externally, the aircraft had a small number of 'extra' communications antennas, and had a circular port on each side of the fuselage, just in front of the wing leading edge.

Uniquely this aircraft was assigned an out-of-sequence serial number (67-19417) based on its construction number (19417). The aircraft ostensibly flew with the 310th Airlift Squadron from Robins AFB, Georgia, and perhaps later from MacDill, and later with the 2nd Special Operations Flight, Air Force Special Operations Command, and has been described as being allocated to C-in-C AFSOC. Other sources suggest that the aircraft was possibly used by the CIA and by other government agencies, perhaps including the Foreign Emergency Support Team (FEST).

The shortcomings of the Boeing 707 as a platform for the FEST were allegedly underlined in April 1999, following Al Qaeda's bombing of the US Embassies in Nairobi and Dar-es-Salaam. The 36-year old aircraft went unserviceable at Rota, Spain, during a refuelling stop, and delayed the arrival of the inter-agency crisis management and counter-terrorist team by some 13 hours. The aircraft had already been criticised for being unreliable and too small, and an inter-agency working group (drawn from the Departments of State and Defense, the CIA, and the FBI) had already recommended to the National Security Council (NSC) that two specially modified Boeing 757s should be purchased for the FEST.

67-19417 was retired to AMARC as FP0076 in October 2002, after replacement by a similarly anonymous and shadowy 757. A second aircraft, an ex-airline Boeing 707-321B, was ostensibly operated by Air Force Material Command as N2138T. This aircraft wore an almost identical colour scheme to 67-19417, and was fitted with the same ports in front of the wingroot, but did not have the same antenna fit. It later became the 'Paul Revere' MC²A-X testbed.

Some sources suggest that a third CIA/FEST aircraft was a 707-327C (c/n 19529) notionally operated by Asnet Inc. and reportedly using the spurious military serial 80065 and/or 86005. Another serial noted on a USAF/Government-operated 707 was 31044.

The FEST aircraft are thought to have operated from McGuire AFB, New Jersey, possibly maintained on four-hour standby to transport an inter-agency counter-terrorist team and to support the CIA and NSA. Other sources have speculated that the parent or 'cover' unit was the 486th Flight Test Squadron at Wright Patterson AFB.

Right: This aircraft was allegedly tasked with transporting FEST agents. The serial '31044' is spurious.

Below: The single EC-137D (67-19417) may also have been used for FEST transport, although its stated role was transport for the C-in-C USSOCOM.

C-18A

Although the USAF's designation system had changed in 1962, with new transport aircraft designations 'starting again' from C-1, new 707 airliners acquired after that date continued to use the basic C-137 designation until January 1982, when nine former American Airlines 707-323C airliners were purchased by the USAF. These aircraft were allocated to Wright-Patterson AFB, where they were designated as C-18As (81-0891 to 0899, though 81-0899 was scrapped after the discovery of serious corrosion), perhaps because they featured the port side cargo door, unlike the earlier USAF VC- and C-137s.

It was probably always the intention that the aircraft would subsequently be converted for special duties as EC-18s, and there are some grounds for believing that they were originally purchased with the express intention of converting them to E-7 standards, though this designation was subsequently dropped.

The aircraft were stripped of their airline equipment and were initially used for training and transport duties, though in 1985 four (81-0891, 892, 894 and 896) were converted as EC-18B ARIA platforms.

81-0895 is seen in unmodified C-18A state, employed as a transport by the 4950th TW prior to modification to EC-18D status.

There are conflicting reports as to what happened to the remainder. 81-0893 lost its passenger windows and later gained a bulbous nose radome, becoming an EC-18D cruise missile control aircraft like 81-0895. 81-0897 was broken up for spares following the discovery of severe corrosion. 81-0898 (and 81-0893) became TC-18E pilot trainers for the E-3 fleet.

107

Variant File

Miscellaneous 707 transports

Some 32 new-build 707 transports were delivered to military/government customers, but these have been outnumbered by second-hand ex-airline aircraft. The majority of military 707 transport aircraft operate in standard airliner 'fit' (perhaps with a VIP interior or minor communications upgrades) and are effectively standard 707-120s or 707-320s. Aircraft converted for the tanker/transport and electronic warfare roles are described separately.

Angola
Angola added an ex-Chilean air force 707-321B (D2-MAY) to its existing -321B (D2-MAN) and -3J6B (D2-TPR). The new aircraft was sent to IAI for a Sigint conversion, but this may not have been completed.

Argentina
Before 1982, Argentina operated three 707 transports (TC-91 to TC-93). The first of these was ordered as a military VIP aircraft (707-3F3B, initially serialled T-01) though all three were actually completed as -387Bs for Aerolineas. Following the Falklands War, Argentina procured four more Boeing 707s (T-94, T-95, T-96 and LV-WXL), and converted one of the original aircraft (TC-93) to operate in the photo-reconnaissance and Elint roles. This aircraft is more fully described in the IAI electronic combat aircraft section. Two aircraft (TC-92 and T-96) have been withdrawn after accidents, and as many as three more (T-94, T-95 and LV-WXL) may be in storage. There is reportedly funding for procuring one further 707, and for converting one aircraft to tanker standards.

Australia
The Royal Australian Air Force received four ex-Qantas 707-338Cs (A20-623, -624, -627 and -629) and three ex-Saudia -368Cs (A20-103, -261 and -809). The -338Cs were converted as tanker transports, -103 crashed, and -809 was used for spares, but the remaining -368C is still used in the transport role. Australia's 707s are more fully described in the IAI tanker section.

Benin
Benin used a 707-321 (TY-BBW) and a 707-336B (TY-BBR) as government VIP aircraft. The -336B was lost in 1985, and the -321 was retired in 1995 and languished at Ostend for a number of years before finally being sold as a cabin crew trainer, parked near Ghent.

Chile
Though Chile's first 707 (901) was sold to IAI and then on to Angola, the FAC retains a -351C transport (902) and an IAI-converted -330B tanker transport (903), as well as a -385C based Phalcon AEW aircraft (904). All serve with Grupo 10, and the tanker and Phalcon are described separately.

Colombia
Colombia took one aircraft (707-373C FAC 1201) as a Presidential/VIP transport, and this has since been given a tanker capability by IAI. A second aircraft (707-358C ARC 707) served with the Army, but its present status is unknown.

Democratic Republic of the Congo
The single 707-138B used as a VIP transport (9Q-CLK) by the Democratic Republic of the Congo is now in storage.

Egypt
The Egyptian Government used a single civil-registered 707-366C VIP transport ('Egyptian 01'), but this is believed to be for sale.

India
India's two 707-337Cs (K2899 and K2900) were originally used in the transport and support roles by the Air Research Centre, but they have been converted for Sigint, probably by IAI.

Indonesia
Indonesia's sole 707-3M1C VIP aircraft (A-7002, originally built for Pertamina Oil) was withdrawn from service with Skwadron Udara 17 in late 2004 and sold to Omega for possible conversion to a tanker.

Iran
The Imperial Iranian Air Force took delivery of a 707-368C and a -370C configured as VIP transports (1001, the Shah's personal aircraft, and 1002) and three 707-125 transports (5-8103, 5-8104 and 5-8107). Some of these may remain in use with 13 Transport Squadron at Tehran-Mehrabad, alongside survivors from the batch of 14 'Peace Station' 707-3J9Cs. Though most of the latter were tankers, 5-8301, 5-8304, 5-8307, 5-8311 and 5-8312 were delivered as 'vanilla' transport aircraft, and the final aircraft (5-8314) was delivered as a VIP transport aircraft, using the civil registration EP-NHA. An unknown number of the aircraft originally delivered as transports were subsequently converted to tanker configuration, while some tankers have later been demodified for transport use.

Israel
The IDF/AF briefly operated three or four Boeing 707-131 transports leased in from IAI, but they were soon returned. They may never have worn full IDF/AF markings or serials. Israel subsequently took delivery of three 707-124s and five 707-131s, and though these were used mainly in the transport role, they were relatively soon withdrawn from service. The IDF/AF also received two turbojet-engined 707-321s, six 707-328s and three 707-329s. These remained in service for rather longer, one (140) became a tanker and about six were converted for use in the EW and Elint roles. Three of the latter (120, 128, and 137) remained in use until recently (sources differ as to their current status). Plans to re-engine the three surviving EC/RC-707s and KC-707 with JT3D engines were cancelled after 128 was converted.

The backbone of the IDF/AF Boeing 707 fleet today is formed by turbofan-engined 707s. Eleven such aircraft have been delivered, and one more has been converted to turbofan power. At least two of these aircraft have been withdrawn from use, six have been converted as tankers or tanker/transports, and at least one more operates in the electronic combat role. Only one (272) lacks a refuelling boom or an antenna array indicative of an EW/ABCP role, and there are some question marks over this aircraft's exact role (since it does have an unusual dorsal 'saddle' antenna).

It is entirely possible that no IDF/AF 707s remain in standard 'Re'em'

The Fuerza Aérea Argentina has operated seven 707s over the years, including one which retained a civil registration. This ETA V line-up includes the Sigint aircraft.

902 is the Fuerza Aérea de Chile's transport 707, seen here supporting a visit to Nellis AFB. This smart scheme has now given way to an all-over light grey.

The Colombian air force's single 707 was given tanker capability but is seen here 'clean'. It is operated by the Escuadrón Presidencial.

SU-AXJ has been Egypt's Presidential transport since 1974. It carries its callsign ('Egyptian 01') on the fin.

Indonesia's presidential 707 was initially built for a civil operator but was transferred to the government. It is seen here prior to delivery.

For several years Israel's 707s operated in a mock airline scheme and carried quasi-civil registrations (4X-JYQ). Note the IDF/AF serial (242) on the nose.

The PAF's two 707s were acquired from the national airline in 1986 and 1989. This aircraft has a VIP passenger fit while the other has a more standard interior.

Boeing 707 military variants

Pride of the Fuerza Aérea Paraguaya's fleet is FAP-01, the presidential 707.

Portugal operated the 707 only briefly in the early 1970s, the main task being to support forces in the African colonies as they transitioned amid violence to independence.

(Buffalo) transport configuration, though it is likely that most of the tankers, and perhaps some of the other aircraft, retain a transport capability.

Liberia
The Liberian Government used a 707-321B and a -351SCD as VIP transport aircraft.

Libya
Libya's VIP 707-3L5C 5A-DAK was operated in civil markings by Libyan Arab Airlines.

Malaysia
Two 707-3L6Bs were delivered to Malaysia in 1974 and 1975, and were used as government VIP transports.

Morocco
Morocco's 707-3W6C tanker (CN-ANR) was sold to the IDF/AF, but the air force still operates a single 707-138B (CN-ANS), perhaps as a tanker/transport. The aircraft is though to serve with the VIP squadron at No. 3 Air Base, Kenitra.

Pakistan
No. 35 Composite Air Transport Wing at Chaklala includes No. 12 Squadron, which operates a 707-351B (68-19635, 'Pakistan 1') and a 707-340C (19866).

Paraguay
The Fuerza Aérea Paraguay's sole 707-321B (FAP-01) is operated by the Escuadrilla Presidencial of the Grupo Aéreo de Transporte Especial at Asunción-Silvio Pettirossi IAP.

Peru
The Fuerza Aérea del Peru has now retired its two transport 707-351Cs (OB-1400 and 1401, FAP-320 and 321) and retains only a single 707-323C (FAP-319, formerly OB-1371). The aircraft has been converted to three-point tanker configuration with three FRL refuelling drogues, and serves with Escuadrón de Transporte 841, a mainly C-130 equipped squadron within Grupo Aéreo de Transporte 8 at Lima-Callão, though some sources suggest that the aircraft is now stored at Lima.

Portugal
The Força Aérea Portuguesa operated two 707-3F5Cs (8801 and 8802) bought from Boeing in 1971. In 1976 one went to the Portuguese government, which leased it out to several civil carriers, including the state airline, TAP, which bought the other aircraft. They were sold to Aeritalia for conversion into tankers for the Italian air force.

Qatar
Qatar's VIP Boeing 707s have been replaced by more modern aircraft types. The Boeing 707-3P1C (A7-AAA) was sometimes known as 'Amiri One', and wore a smart colour scheme, with a broad crimson cheat line and tailfin. A second, back-up aircraft (707-336C A7-AAC) has also been sold.

Romania
Romania operated a pair of Boeing 707-3K1Cs in the VIP transport role. YR-ABD reverted to state airline Tarom, but YR-ABB was still in government use last year (2004), confounding reports of its disposal and sale.

Saudi Arabia
The Saudi Government has retired one of its VIP-configured Boeing 707s (707-138 HZ-123) to storage at Southend, though two more (707-338Cs HZ-HM2 and -HM3) serve the Royal Flight in Saudia markings.

Sudan
Sudan briefly used a single 707-368C as a government VIP transport aircraft.

Togo
Togo has used a Model 720 and two 707s for VIP transport duties. The first 707 (-312B 5V-TAG) crashed, but the hush-kitted 707-3L6B probably remains in use.

United Arab Emirates
The United Arab Emirates employed a variety of 707s for VIP transport duties. These included 707-330B A6-UAE, 707-330C A6-DPA, 707-3L6B A6-HPZ and 707-3L6C A6-HRM. A6-HRM and a 720, A6-HHR, served with the Dubai Air Wing, while A6-UAE was used by Abu Dhabi.

Yugoslavia
Yugoslavia seized an alleged gun-running Uganda Airlines 707-324C on 31 August 1991 and placed it in service with the air force as 73-601. It was soon sold and is now stored at Ostend, owned by an Afghan company.

Zaïre
707-382B 9T-MSS served as President Mobutu's personal aircraft, with Republique du Zaïre titles. The aircraft has been stored at Lisbon since the late 1980s.

The West African state of Togo has had three 707s to transport its government officials since 1980, when a 720 was bought. This was later sold and a 707 acquired (5V-TAG, seen above). This crashed in 2000 and was replaced by another 707 a year later.

707-126/226/326

The customer designation suffix 26 was reserved for the USAF's Military Air Transport Service, and from November 1955 Boeing drew up a number of military freighter versions of the Boeing 707. Some of these featured a side cargo door on the port side, while others had a new rear fuselage incorporating clamshell doors and a downward-hinging pressure bulkhead, or a vertically hinged tail. Some featured wingtip fuel tanks and some had a weight-saving 8-ft 4-in (2.54-m) reduction in fuselage length (to 137 ft 2 in/41.81 m) to compensate for higher density military freight. However, the USAF preferred to wait for a more optimised transport, with high wings and truckbed-height loading. This eventually emerged as the Lockheed C-141 StarLifter.

Model 720

There was very little military use of the reduced-capacity, shorter-range Model 720. Apart from one aircraft used as an ASW test and trials aircraft by Boeing, and one used by IAI Bedek as a trials aircraft for the Lavi fighter's radar, only three were delivered to air force/ governments. In 1971 Taiwan procured an ex-Northwest Orient Airlines 720-051B as a presidential transport for Chiang Kai Shek, and used this VIP-configured aircraft until 1991. An ex-Western Airlines 720-047B had a shorter military career, serving the Togolese Government as a VIP transport between November 1980 and March 1983. The Dubai Air Wing used Boeing 720-023B A6-HHR as a VIP transport.

Pratt & Whitney Canada used two 720-023Bs (C-FETB and N720PW) as engine testbeds, and these aircraft are still in use, while Honeywell's 720-051B (N720H) may still fly from Phoenix. The Hughes Aircraft Corporation also used a Boeing 720-060B (N7381) as a testbed, but this aircraft was scrapped in October 2003.

IAI's 720 Lavi radar testbed is now at the Hatzerim museum on outdoor display.

Model 738

Though the Model 367-80 was designed with side cargo doors, they were inadequate to allow the fullest exploitation of the aircraft's cabin for cargo carriage. In order to provide better, less restricted access, Boeing designed a number of 'swing-tail' options, some hinging to the side (as on the CL-44) and some swinging upwards. These 'swing-tails' formed the basis of two dedicated 1958 cargo versions, the commercial Model 735 and the military Model 738 (which was offered as a 707 or 717/C-135 derivative).

109

Variant File

Tanker conversions

Boeing KC-137 proposals

After the last military C-135 was delivered on 14 June 1966, and the production line closed, Boeing began to offer tanker versions of the Model 707 airliner (which was still in production) to military customers, both as a new-build aircraft or as a conversion, using redundant 707-320C airliners. Such aircraft were refurbished and modified with military avionics, including TACAN, IFF and dual UHF (with direction-finding), and with provision for engine thrust reversers, strengthened outer wings and new wingtips. Extra fuel tanks were installed underfloor, allowing the aircraft to retain full transport capabilities.

Boeing offered a variety of one-, two- and three-point 707 configurations (at a time when the KC-135 was single-point only), while tanker 707s were also produced by IAI. Boeing's options included a Boeing flying boom and boom operator's position or a Sargent Fletcher HDU on the centreline, with or without Sargent Fletcher HDUs or Beech 1800 refuelling pods underwing, or with just the underwing stores in a two-point tanker configuration. All aircraft were also offered with the option of a centreline refuelling receptacle or probe above the cockpit.

In 1965 Boeing used a 707-385C (N6857) to demonstrate its proposed tanker conversion, with Beech 1800 pods slung directly below the wingtips. The definitive configuration moved these just inboard from the tips, below the outer wing panels. The aircraft proved the modifications made to the first generation of 707 tankers, and was then restored to airliner configuration and sold to LAN-Chile in December 1969.

Later, in October 1982, Boeing leased a 707-331C from TWA to serve as a demonstrator for a drogue-equipped three-point tanker, with a Beech 1800 or Sargent Fletcher 34-000 HDU under each wing, and with a FRL FR600B or Sargent Fletcher HDU under the rear fuselage. Boeing subsequently bought the aircraft.

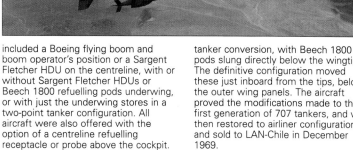

Boeing's second Tanker-Transport demonstrator refuels an F-105 Thunderchief from the starboard wingtip Beech 1800 pod. The port wingtip carried a 34-000 HDU, made by Sargent Fletcher who also provided the centreline HDU.

Boeing's first tanker demonstrator, seen here with the company's Sabre chase/photo aircraft, did not have a fuselage HDU but had Beech pods on the wings. This is the first installation with pods on the extreme tips.

CC-137 tanker conversion

When Canada ordered its five CC-137 transports, the contract provided for the subsequent conversion of two aircraft to tanker/transport configuration at a later date. The third and fourth aircraft (13703 and 13704) were therefore returned to Boeing in 1971 and 1972 to have Beech 1800 aerial refuelling kits installed. Installation on 13703 was completed by 19 May 1972, and on 13704 by October 1972. These two aircraft thus became the first 'production' 707 tankers. No new designation was allocated to the two two-point tanker-capable aircraft following their conversion.

The Canadian DND requested authorisation for two more tankers in 1977, and again in 1987, but funding was not forthcoming, and the two aircraft remained Canada's only tankers. They were the last CC-137s in service, retiring in April 1997. Both have since joined the E-8 JSTARS programme.

Canada's two Beech 1800-equipped tankers were modified by Boeing.

707-3J9C Peace Station

The next customer for the 707 tanker was Iran, which ordered 14 new Boeing 707-3J9Cs under Operation Peace Station, augmenting four ex-Eastern, ex-TWA 707-125s, one of which was sold back to TWA before conversion, and two other VIP transports.

Seen wearing its initial serial, this Peace Station three-point tanker was later reserialled as 5-8310.

The Iranian 707-3J9Cs were delivered in two batches, the first consisting of six aircraft. The first and fourth new-build Iranian aircraft (5-8301 and 5-8304) were delivered as 'vanilla' transport aircraft, but the remaining four new aircraft (5-8302 to 5-8306) were delivered as single-point tankers, with just a KC-135 type boom fitted.

Within the second batch the 7th, 11th and 12th aircraft (5-8307, 5-8311, and 5-8312) were transports, while the next three aircraft (5-8308 to 5-8310) and 5-8313 were tankers, but also fitted with underwing Beech refuelling pods. The final aircraft (5-8314) was delivered as a VIP transport aircraft, and used the civil registration EP-NHA. The first four tanker aircraft were quickly brought up to the same three-point tanker standard as the later tankers.

Some of the first trio of ex-TWA aircraft (5-8103, 5-8104 and 5-8107) and the other transports (5-8301, 5-8304, 5-8307, 5-8311 and 5-8312) have subsequently been converted to tanker configuration, while other aircraft have been stripped of their refuelling gear and used purely in the transport role. 5-8301, in particular, has been confirmed as a three-point tanker. One aircraft (5-8305) has been fitted out as an Elint/EW platform.

The aircraft serials detailed in the entry above were those used in service – they had been changed immediately before delivery.

This view across the Boeing Seattle flightline shows one of the Peace Station aircraft with its boom extended. 5-242 (later 5-8302) was one of the aircraft initially delivered without Beech wingtip pods, but was later modified so that it could carry them. The boom was used mainly for refuelling F-4 Phantoms.

KC-137 Brazil

The only aircraft to use the KC-137 designation officially are the Brazilian air force tanker aircraft. These four aircraft were obtained in 1986 from the Brazilian airline Varig and serve with 2° Esquadrão 'Corsario' of 2° Grupo de Transporte at Galeão (part of the Força Aérea Brasileira's V Air Force). The aircraft were converted as two-point tankers with Beech 1800 HDU pods underwing. They were fitted with an APU, extra fuel tankage and folding air stairs, and three (2402-2404) were painted in a low-viz tactical grey paint scheme.

The first of the KC-137s (2401) is maintained in a highly polished silver and white VIP colour scheme and has an airliner-type interior. It was sometimes used for VIP and Presidential transport duties, though it is being replaced in the VIP role by a new presidential A319CJ ordered for the Special Transport Group at Brasilia.

As well as its tanking duties, 2401 is used for VIP transport, for which it wears this smart scheme.

There is a requirement to replace the tanker 707s with up to four new tankers, with both Airbus and Boeing submitting proposals.

Boeing 'KC-137'

Apart from producing the tankers for Canada, Iran and Brazil, Boeing converted or assisted in the conversion of only seven further tanker aircraft, for Spain, Italy and Morocco, and manufactured eight new-build KE-3A aircraft for Saudi Arabia.

Spain
Boeing's three-drogue tanker demonstrator (707-331C N792TW) was eventually sold to Spain in November 1998, with its fuselage-mounted HDU removed, but retaining two Sargent Fletcher 34-000 HDUs underwing. Designated T.17 in Spanish service, the aircraft was joined by a second aircraft during 1991. The two T.17 tankers and one T.17 VIP aircraft served with 451 Escuadrón at Torréjon, though two aircraft are now believed to be in storage. Spain also operates an EW Boeing 707, converted by IAI and described separately.

Italy
Since 1992 the AMI has operated four ex-TAP 707-382Bs, converted to Boeing 707 T/T tanker configuration by Alenia and Sargent Fletcher, with SF 34-000 wing pods and a centreline Flight Refuelling Ltd 480C HDU. Two are

configured as Combis, with seating for 66 passengers and space for seven pallets adjacent to the cargo door. The other two are in a passenger transport fit, with a VIP cabin and seating for 110 'ordinary' passengers. The aircraft are operated by 14° Stormo 'Sergio Sartof', 8° Gruppo Volo, at Pratica di Mare.

High operating costs, noise and emissions prompted Italy's July 2001 decision to acquire four new Boeing 767 tanker/transports to replacing its ageing Boeing 707 tankers.

Morocco
When Morocco received the former 707-700 CFM56 engine testbed (CN-ANR, re-engined with JT3D engines and re-designated as a 707-3W6C) there were reports that it was delivered as a tanker. This was not the case. However, it was subsequently converted to tanker configuration in country by Boeing and AMIN Maroc.

There have been reports that Morocco's other 707 (a 707-138B, CN-ANS, c/n 18334) has also been converted to tanker status.

Mediterranean tankers – Boeing-assisted tanker conversions entered service with Spain (top), Italy (left) and Morocco (above). The Moroccan aircraft has passed to Israel, via Omega, where it is now in service as a boom-equipped tanker.

KE-3A

Boeing used the KE-3 designation for eight new-build tanker aircraft for Saudi Arabia, ordered in 1987 under the Peace Sentinel and Peace Shield programmes that also saw the Kingdom obtaining five E-3 Sentry AWACS aircraft, ground-based surveillance radar and I-Hawk SAMs. Serialled 1811 to 1818, the aircraft were also allocated USAF serials 82-0071 to -0076, 83-0510 and 83-0511.

The Saudi tankers were delivered to a similar standard to the Kingdom's E-3As, based on the 707-320C airframe with CFM56-2 engines. They were equipped with a centreline boom, but also had provision for underwing Beech 1800 refuelling pods. Initially flown in USAF markings in the USA, the aircraft flew in three-point tanker configuration before delivery, but in service have often been used as single-point tankers, without the underwing pods.

The penultimate aircraft, 1817, became the sole RE-3A Elint platform, and was re-serialled as 1901. It is uncertain as to whether the aircraft ever flew as a KE-3A, or whether it was delivered in its present service 'RC-135 Lite' configuration.

Saudi Arabia's KE-3As are the only tanker 707s with CFM56 power. Wing pods are rarely seen on operational aircraft.

Omega 707 tanker/transport

Omega Air, Inc., is an international company involved mainly in second-hand aircraft sales, leasing and in the supply and sale of parts. It has operated a large fleet of Boeing 707s. With an obvious shortfall in military tanker capacity, and increasing demand, the company saw contract air refuelling as a useful means of using some of its 707 capacity, and decided to convert one aircraft (Boeing 707-321B, N707AR c/n 20029) as a military tanker/transport. In 1997 Omega Air commissioned the Naval Air Warfare Center Aircraft Division (NAWCAD) to assist the company in producing what would be the first jet-engined commercial aerial refuelling tanker.

NAWCAD participated in and then reviewed Omega Air's test plan and oversaw contractor flight and ground tests, before conducting its own air and ground trials of the aircraft's compatibility with US Navy receiver aircraft. NAWCAD developed rendezvous and refuelling procedures, and provided initial baseline training for industry and experienced reserve

Omega's 707 tanker lands at Prestwick in Scotland, from where the company routinely supports US Navy exercises in the North Atlantic.

aircrew to operate the new tanker.

Omega's 707 is configured as a redundant single-point tanker, using the 'probe and drogue' method used by NATO and the US Navy and Marine Corps, rather than the 'boom system' used by the US Air Force. The aircraft has two Sargent Fletcher HDUs installed in the rear fuselage, side-by-side on each side of the aft centreline. Only one HDU can be used at one time, with the other available as an immediate spare should it be required due to malfunction or damage.

Omega's fuel system is basically the same as that used by the KC-135, albeit without forward and aft body tanks. Two fuel pumps were added to the aircraft centre wing fuel tank to pump fuel to the hose and drogue assemblies, with system controls and video monitoring devices fitted in the cockpit. Like the KC-135, Omega's 707 has two reserve tanks, two outboard main tanks, two inboard main tanks and a centre-section tank. The Omega's wing and fuselage tanks are larger than those of the KC-135, but without body tanks, the Omega aircraft's total fuel capacity is less. Because the 707's air refuelling pumps are in the centre-section tank, all offloaded fuel has to drain to the tank, just as the KC-135 has to drain offloaded fuel from the wings to the body tanks. The aircraft is fitted with a military communications suite, with two VHF/UHF radios, two HF radios and air-to-air TACAN.

In 2004 Omega said that its first tanker was flying more than 1,200 hours per year and was marketing a fleet of tankers to the United States, Canada, South Korea and NATO air forces. The company's existing tanker is usually leased by the flying-hour rate, and the fuel offloaded by the tanker is allocated and charged to the appropriate squadron in the same way that fuel received anywhere is allocated and charged, whether it is uploaded at other operating bases, government-approved civilian FBOs, or from USAF tankers. Omega does not pay for its fuel, but instead uses the same government credit card that military units use to pay for fuel, and buys fuel only from US government-approved sources.

A second Boeing 707 (the ex-Indonesian Air Force 707-3M1C c/n 21092) is being converted to tanker configuration as N707LG, but the company's interest has largely switched to the DC-10 as a tanker platform, and it recently purchased the JAL fleet of 20 DC-10-40s for tanker conversion.

The Omega tanker has two fuselage HDUs for redundancy (far left). A TV camera on the centreline provides imagery to allow monitoring of the refuelling process in the 707's cockpit, a Tomcat being shown here (left). The 707 has a complex fuel system control panel.

IAI KC-707 Saknayee (Pelican)

An unknown number of Israel's 23 707s (which included 17 707-320s) have been converted to tanker configuration, with various sources differing widely on the precise number of conversions. As far as can be ascertained, aircraft 140 (4X-JYT c/n 17625), a turbojet-engined 707-329, was the first KC-707 conversion, known as Salat Yarok (Green Salad) 1, entering service as a tanker in 1983. Some sources have also suggested that 707-124 4X-JYA/006 was converted to three-point (three-drogue) tanker configuration, but if so, this was almost certainly only for test and development purposes, in support of export programmes. Another Israeli 'mystery tanker' was 4X-AOV, noted in 1985 with underwing HDUs, but was probably an export aircraft.

Subsequent KC-707 conversions by IAI's Bedek Group were reportedly completed with the delivery of the seventh aircraft in January 1997, though only five tankers have been identified from this batch. Two further tanker conversions were 290 and 264, delivered in January and June 2004, respectively. Apart from 140 (and 006), all of the tanker conversions seem to have used JT3D turbofan engines.

Serial numbers 140, 248, 250, 260, 264, 275 and 290 have been positively identified as tankers, while 242 was reportedly undergoing conversion in 1999. Some of these aircraft have been fitted with more extensive antenna arrays, perhaps indicating a multi-role or secondary role capability. Some reports suggest that only four or five of the tankers are actually classed as KC-707s, and two or three may be officially regarded as boom-equipped, tanker-capable EC-707s.

Aircraft 248 and 250 have been photographed in two different three-point configurations (both with Sargent Fletcher 34-000 pods underwing, one with a centreline boom and one with a centreline HDU) and probably also flew with Flight Refuelling HDUs underwing too. It is believed that these three-point configurations were flown only when the aircraft were in use as demonstrators/trials aircraft supporting export 707 tanker programmes. When fitted, the underwing HDUs are mounted slightly further inboard than on Boeing 707 tanker versions. The IDF/AF's own 707 tankers are inevitably

This IAI-converted tanker has three HDUs, a configuration that has not been noted in IDF/AF service. It was probably a demonstration for export.

photographed in a single point configuration, with a Boeing-type boom on the centreline, and a remotely controlled EO camera turret for the boom operator on the flight deck. They have enhanced communications, IFF, and defensive aids systems, and are fitted with GPS and air-to-air TACAN.

The IDF/AF Boeing 707 tanker fleet is operated by No. 120 Squadron at Lod (Bacha 27), the name given to the military enclave of Ben Gurion IAP. Plans to replace the JT4A-11 turbojets of aircraft 140 with the P&W JT3D-3B turbofans used by the IDF/AF's other 707 tankers were cancelled, and the aircraft has been withdrawn from use and put up for sale, leaving six or seven tankers in service.

The majority of the IDF/AF's combat aircraft (F-15s and F-16s) are equipped for boom refuelling, so air force tankers are usually seen in single-point configuration. Some have additional electronic equipment.

IAI 707 tanker conversions

Israel supplied four tanker/Elint aircraft to South Africa (plus a single freighter) and one tanker, one freighter and one Phalcon AEW aircraft to Chile, and has undertaken – or helped to undertake – 707 tanker conversions for Australia, Colombia, Peru and Venezuela.

Australia
Within No. 84 Wing (the RAAF's Air Lift Group) at Richmond, NSW, No. 33 Squadron flies the surviving five 707-338/-368 aircraft from the six aircraft originally delivered. Four of these (the -338s) are configured as two-point tankers, having been converted to this configuration by IAI and Hawker de Havilland in Melbourne during the early 1990s. The other aircraft (A20-261) flies in the long-range transport role. The tanker aircraft are fitted with four submerged, cross-fed J.C. Carter fuel pumps in the centre-section fuel tank and have underwing FRL Mk 32 HDUs. The refuelling operator can monitor receiver aircraft using a steerable camera in a remotely-controlled lower fuselage turret. They do not have a centreline boom and therefore cannot refuel the RAAF's F-111s, although a third HDU (or boom) could easily be added. Nor do the aircraft have lower deck auxiliary fuel cells, since the aircraft would not be able to take off from most RAAF runways with such heavy loads.

The tanker conversion included a major avionics upgrade, with a dual-redundant Litton LN-92 ring laser gyro navigation system, and a Collins 150 air-to-air TACAN system with AN/ARN-118 and APN-139 subsystems. The aircraft also received Collins SIT-421 dual-redundant IFF transponders, and a forward-facing Hazeltine AN/APX-76B(V) IFF interrogator. This saw new IFF dipole antennas being mounted on the existing weather radar antenna. The aircraft were also fitted with a Magnavox AN/ARC-164 UHF transceiver and a Collins DF-301E/F UHF DF set.

Handicapped by escalating support and operating costs, corrosion and structural problems, the RAAF's 707 tankers are being replaced by Airbus A330 MRTTs under Project Air 5402.

Chile
The Fuerza Aérea de Chile still operates three Boeing 707s, though only one of these (a 707-330B, 903) is a tanker, known locally as a KB-707. All three serve with the II Brigada Aérea's Grupo 10 at Pudahuel, Santiago. The aircraft is fitted with Sargent Fletcher HDU pods underwing, and may also have a third HDU in the fuselage, though this cannot be confirmed.

Colombia
Colombia's ex-Korean Airlines 707-373C (FAC 1201, once the flagship of the Escuadrón Presidencial at Base Aérea 'Camilo Daza', El Dorado, Bogotá) was converted to tanker configuration in Israel during 1991. It has Flight Refuelling Ltd HDUs underwing only.

Peru
The Fuerza Aérea del Peru operates a single Boeing 707-323C (319/OB-1371) in the tanker/transport role. The aircraft is a two- or three-point tanker using Flight Refuelling Ltd HDUs and serves with Escuadrón de Transporte 841 at BA 'Jorge Chavez' IAP at Lima-Callao.

South Africa
When South Africa originally procured the first of its 707s in March 1982, it did so to fulfil a requirement for inflight refuelling tankers, though an EW capability was added before the aircraft entered service. The aircraft were converted to three-point tanker configuration by IAI, with underwing FRL HDUs and a further FRL HDU in the fuselage. No. 60 Squadron was re-activated to operated the new aircraft on 16 July 1986, reforming at Air Force Base Waterkloof, and flew its first operational EW sorties before the year's end. The squadron flew its first inflight refuelling sortie on 5 March 1987. Two of South Africa's five Boeing 707s (615 and 617, and perhaps 623) have a tanker capability, but since these are also configured to allow rapid conversion to the Elint role, they are more fully described in the Electronic Combat Aircraft section. Some sources suggest that the remaining two or three SAAF 707s can be flown as two-point tankers, with underwing HDUs only.

Venezuela
The Fuerza Aérea Venezolana operates two Boeing 707-346Cs (6944 and 8747) in the tanker/transport role. The aircraft, which were delivered in 1990, have a centreline boom and Sargent Fletcher underwing HDUs, and serve with the mainly C-130 equipped Escuadrón T1 of Grupo Aéreo de Transporte No. 6 at Base Aérea 'El Libertador' at Palo Negro. Though converted by IAI and not Boeing, the aircraft are known locally as KC-137Es.

Although most IAI tankers use SF 34-000 pods, Australia specified the FRL Mk 32. The RAAF's four-strong fleet is now painted in dull grey.

Chile has a single 707 tanker with Sargent Fletcher pods serving with Grupo 10.

Peru's single tanker is seen in transport configuration, without the wing FRL Mk 32 pods. However, the pylons for attaching the pods are still fitted.

Venezuela's two 707 tankers are each equipped with a boom for refuelling the FAV's F-16s, and wingtip HDUs (not installed here) for refuelling Mirage 50s.

Electronic warfare, reconnaissance and special mission aircraft

EC-18B ARIA

Four of the ex-airline 707-323C aircraft acquired as C-18As were further modified to EC-18B ARIA (Advanced Range Instrumentation Aircraft) configuration to augment eight older, similarly modified EC-135Ns serving with the 4950th Test Wing at Wright Patterson AFB as flexible airborne telemetry data recording and relay stations, filling gaps in the coverage offered by land- and ship-based stations. Their job was to track spacecraft during re-entry, and to support missile tests by

The EC-18Bs supplanted hard-worked EC-135s in the ARIA role. Note the black-painted port wing and engine nacelles, a measure taken to reduce glare for optical tracking instruments.

The bulbous nose of the ARIA houses a steerable tracking antenna for telemetry receipt, not a transmitting radar as is commonly thought.

acquiring, tracking, recording and retransmitting telemetry signals, primarily in the S-band (2200-2400 MHz) and the C-band (4150-4250 MHz). With additional modifications, ARIA could receive and record L-band and P-band signals.

The 707-based EC-18B is larger than the EC-135N, and was therefore able to carry a larger payload and also had better performance characteristics, including the ability to operate from shorter runways. This allowed the EC-18s to forward deploy more easily, and the type actually flew its first mission in January 1986 from Kenya.

The four aircraft were (81-0891, 81-0892, 81-0894 and 81-0896), each of which was fitted with a large, bulbous, slightly drooping 10-ft nose, a probe antenna on each wingtip, extensive tracking and communications systems, and an array of four fixed staring cameras designed to photograph re-entries. Each camera had its own optical quality, air-heated window, manufactured by Perkin Elmer from Schott BK-7 type glass. Each of the cameras could be operated at the rack or from a remote operator station, via a video monitor. On most aircraft the port upper wing surface was painted matt black to avoid reflections. The nose radome housed a 7-ft (2.13-m) steerable parabolic dish antenna for capturing, recording and rebroadcasting telemetry, though the ARIA dish antenna did not lend itself to simultaneous tracking and telemetry-gathering from multiple data sources without sacrificing data signal quality. The aircraft all had worldwide HF voice communications and real-time UHF satellite data retransmission capabilities, and all had two 14-track wide-band recorders to record, monitor and play back mission data. Two aircraft (892 and 894) also had horn antennas for scoring, and 894 was fitted as an inflight refuelling receiver. Several of the ARIA aircraft were upgraded with 4-MHz Racal Storehorse recorders and Microdyne S-Band, C-Band, P-Band superheterodyne receivers.

In 1994, the ARIA fleet was relocated to Edwards AFB, California, as part of the 452nd Flight Test Squadron, 412th Test Wing. 81-0896 'Pegasus' was later modified for use as a testbed for the SMILS/optics programme, leaving three ARIA EC-18Bs. The last of these was 81-0891, which performed the last official ARIA mission in January 1986.

From June 2000, the EC-18Bs were used as trainers for US Navy E-6 flight crews, who rotated through Edwards AFB for some six EC-18B sorties before moving to Tinker to transition to the operational E-6. 81-0891 flew the last Cruise Missile support sortie on 24 August 2001, before being transferred to the JSTARS modification programme in 2001.

EC-18B CMMCA Phase 0

The two dedicated EC-18D Advanced Cruise Missile Mission Control Aircraft were augmented by three EC-135Ns with CMMCA Phase 0 modifications, which gave them the ability to receive, record and retransmit cruise missile telemetry, though they were unable to track the missiles in the same way, since they did not have APG-63 radar. The CMMCA Phase 0 equipment was subsequently transferred to two EC-18Bs (81-0891 and 81-0894).

EC-18D ACMMCA

Two more of the C-18As (81-0893 and 81-0895) were modified to EC-18D configuration for use as Advanced Cruise Missile Mission Control Aircraft. Looking very similar to the EC-18Bs, tasked with tracking and telemetry work, and often referred to by the same ARIA (Advanced Range Instrumentation Aircraft) designation, the EC-18Ds were very different animals.

The extended nose of the ACMMCA aircraft accommodated a considerably smaller 3-ft (0.91-m) diameter tracking antenna, and also the antenna for an AN/APG-63 fighter radar used as a tracking/surveillance radar for stand-alone operations, and a small weather radar antenna. Inside the aircraft a dedicated ACMMCA console, with three operator's positions, allowed the aircraft to monitor the missile's systems, and to take control or destroy it if necessary.

Typical cruise missile test support missions were flown from Edwards AFB, with the B-52 launch aircraft departing its home base several hours before the ACMMCA took off. The ACCMCA joined the B-52 about 90 minutes before launch and acquired telemetry from the missile as the formation flew to the launch area. Mission control on the ground used the telemetry data provided from the ACMMCA to evaluate the missile's status. The ACMMCA continued to track the missile after launch, receiving and relaying L-band telemetry data from the missile, and relaying UHF voice transmissions from the chase and camera aircraft to mission control. Missions could entail continuous automatic tracking for more than five hours.

The two EC-18Ds looked outwardly similar to the EC-18Bs also operated by the 412th TW. The radome housed a fighter radar and small tracking antenna.

MC²A-X Paul Revere

The Paul Revere 707 is a dedicated test aircraft for new command, control and communications, and airborne battle management technology and concepts, being developed for the planned Boeing E-10A Multi-Mission Command and Control aircraft (MC²A), which is the planned replacement for the USAF's AWACS, JSTARS and Rivet Joint aircraft. These aircraft may be replaced by a single MC²A, or by separate MC²A airframes.

The USAF's E-10s may be based on the Boeing 767-400ER airliner, or on the smaller, less expensive 767-200ER. The new aircraft may gain its capabilities in a progressive manner, with a next-generation air-to-ground radar by 2010, the air search radar and advanced battle management systems around 2015, and signals intelligence-gathering equipment in 2020.

The MC²A-X was converted by the Massachusetts Institute of Technology Lincoln Laboratory, and is owned by the US government, though it flies with a civil registration, and is flown and maintained by civilian personnel. The Paul Revere MC²A-X testbed made its first flight after conversion on 18 April 2002 at Hanscom AFB, Massachusetts. Systems tested during the first flight included the Tactical Common Data Link (TCDL), which will be the critical element in both sending and receiving data from airborne, ground and space sensors and the AOC.

Between 15 July and 5 August 2004, the Paul Revere 707 participated in the 2004 Joint Expeditionary Force Experiment (JEFX 04) at Nellis AFB, crewed by Air Force and Department of Defense personnel and government contractors. The aircraft operated as an airborne relay centre and command post, and used its multiple datalinks to exchange and fuse information from a variety of airborne and space sensors, and the Combined Air and Space Operations Centre (CAOC). The aircraft was also able to send and receive data to and from the Pentagon in real time.

The Paul Revere testbed is based on an airframe that has been in use with AFMC for some time (N2138T, now N404PA). In its current form it is outfitted with numerous datalinks and advanced airborne command and control equipment, and acts as a testbed for advanced battle management techniques in support of the USAF's E-10 programme.

RE-3A

One of the eight KE-3A tankers ordered by the Royal Saudi Air Force was subsequently converted as an Elint platform, perhaps before it had been completed as a tanker. It is not known whether the aircraft ever flew as a KE-3A. The aircraft (1901, original serial 1817) is now in 'RC-135 Lite' configuration, fitted with a Boeing supplied Tactical Airborne Surveillance System (TASS) and with cheek fairings like those of the RC-135V and a new fairing below the rear fuselage, larger and further forward than the fairing usually associated with the tanker's boom installation. The aircraft is reportedly designated as the RE-3A or KE-3A TAS.

Right: 1901 is the sole RE-3A in service with the Royal Saudi Air Force. Judging by its antenna fit, its capabilities are thought to be roughly comparable with those of the RC-135V/W Rivet Joint.

RE-3B

Once the decision was taken that USAF E-8s would be converted 707-320 airliners, the CFM56-engined YE-8B testbed became surplus to requirements. Initially sold to Omega, it was subsequently exported to Saudi Arabia as the sole RE-3B (serial 1902). The aircraft was equipped with the Improved Tactical Airborne Surveillance System (ITASS) and serves alongside the sole RE-3A with No. 19 Squadron.

Iranian EW conversion

At least one of Iran's surviving Boeing 707-3J9C tanker/transports (5-8305) has been converted for Sigint missions, though it is not known whether the conversion was undertaken by Iranian industry, in Israel, or locally with Israeli assistance. Though the aircraft does have 'cheek' fairings on the forward fuselage, these are much further aft than on the Israeli Barboor and Phalcon, or the Israeli-modified South African aircraft, and are 'longer' and 'thinner', while the aircraft has another box-like antenna fairing further forward on the lower fuselage. The conversion has included the removal of tanker capability, and a new bulged fairing has been added below the rear fuselage, ahead of the bulge normally associated with the refuelling boom installation.

No details have been released concerning the single Iranian 707 configured for Sigint missions. The side-looking antenna fairings would suggest an Elint (radar classification) capability.

Early IAI EW conversions

Israel is believed to have undertaken its own 707 tanker, Elint and EW conversions, and has had aircraft modified for the Elint and EW roles by E-Systems in the USA. Of the IDF/AF's eight, short-bodied 707-120s, at least one (001) was modified to Sigint configuration with prominent 'cheek' fairings (though this may have been to allow the aircraft to act as a trials and test aircraft for the definitive Barboor and another (008) reportedly served as a Command Post.

Of eleven turbojet-powered 707-320s used by the IDF/AF, at least eight are thought to have been modified as command posts (113, 115, 118 and 119), EW jammers (117 and 120) or Sigint platforms (113, 128 and 137). Some of these may have been the two aircraft modified to 'Spring Flower' Comint standards (or with 'Spring Flower' Comint equipment) by E-Systems during the late 1970s, while the code name 'Wide Fox' has also been associated with an Israeli Elint system or configuration.

By 2001, only 120, 128 and 137 remained in use. 128 was then re-engined with turbofans, and the other two aircraft have been withdrawn from use.

This Sigint conversion, seen derelict at Ben Gurion, was based on the short-body 707-120. The mission equipment was probably installed in the US by E-Systems.

In-service IAI EW conversions

A number of Israel's 707 tanker, Elint and AEW conversions still serve with the IDF/AF, though sources differ considerably as to the numbers of aircraft in service in each category.

Of four surviving JT4A-engined aircraft, two (RC-707 137 and KC-707 140) have recently been retired, while another (120) has been leased to IAI. The final aircraft (the remaining RC-707, 128) has been re-engined with turbofans. Of eleven fan-engined 707s delivered to the IDF/AF, 246 (probably a Phalcon testbed) and 258 have been withdrawn from use, leaving nine aircraft, plus the re-engined 128.

When 14 aircraft remained in use, one source indicated that there were five KC-707 Saknayee ('Pelican') tankers, two RC-707 Barboor ('Swan') Elint aircraft, two EC-707 Chasidah ('Stork') EW/ECM aircraft, and two AEW-configured Tavas ('Peacock') or Bdolach aircraft, as well as three aircraft in standard Re'em ('Unicorn') transport fit. Another source suggested that there were four tankers, nine assorted EW versions, and at least one dedicated medical aircraft.

After recent withdrawals, there should be four 'plain' tankers, one Barboor Elint aircraft, and five others, probably comprising two EW jammers and three airborne command posts (ABCP), at least one of each retaining a tanker capability. It seems unlikely that any aircraft remain in the original Re'em transport configuration.

EC-707 Chasidah

Though the EC-707 designation and Chasidah ('Stork') name are known to apply to at least two electronic combat IDF/AF Boeing 707s, the plethora of 'electronic 707' configurations in service makes it difficult to ascertain exactly which aircraft the designation and name describe. As far as can be ascertained, they apply to an EW jammer configuration, which has changed over time. It was originally characterised by four large, broad rectangular blade antennas above the rear fuselage, with a fifth below, as fitted to both 242 and 255 by 1996-97. These types of configuration may indicate that the aircraft are or were EW 'jammer' platforms.

This antenna fit seems to have been replaced with five smaller antennas above the rear fuselage, an array of 10 very small antennas below the rear fuselage, and arrays of swept blade and 'T'-shaped antennas above and below the forward fuselage. It is not known whether the changed antenna fit indicated a change in role from jamming to airborne command post.

The new antenna fit had been applied to 255 by late 2000, and is also fitted to 290, an aircraft which also carries a refuelling boom. Some reports suggest that 242 has also become a tanker, but it is not known whether tanker capability was added when the aircraft became redundant in their original role, or whether they have become dual-role tanker/EW or tanker/command post machines. The EC-707 designation is also associated with the name Shen'hav.

Characterised by an extensive antenna farm, this aircraft (242) is believed to be an EC-707 Chasidah (Stork) jammer.

The dorsal antennas and blocked out cabin windows on this aircraft would suggest some kind of electronic role.

Above: RC-707 Barboor 4X-JYL/128 is seen in 1980 during its delivery from E-Systems to the IDF/AF.

Below: 4X-JYH/264 has variously been described as a VIP transport or command post. It now has tanker capability.

IDF/AF in-service 707 versions

AF serial/reg	C/n	Built as	Engine	Variant	Notes
Current					
128/4X-JYL	18374	707-329	JT4A	RC-707	Sigint. Re-engined with JT3D
242/4X-JYQ	20110	707-344C	JT3D-7	EC-707	EW/ABCP, perhaps with tanker capability
248/4X-JYU	20429	707-331C	JT3D-3	KC-707	Tanker
250/4X-JYY	20428	707-331C	JT3D-3	KC-707	Tanker
255/4X-JYB	20629	707-3H7C	JT3D-3	EC-707	EW/ABCP
260/4X-JYN	20716	707-3J6B	JT3D-7	KC-707	Tanker
264/4X-JYH	20721	707-3J6C	JT3D-7	??-707?	VIP/ABCP; now with tanker capability
272/4X-JYV	21096	707-3L6C	JT3D-7	??-707?	possible VIP/ABCP aircraft with saddleback
275/4X-J??	21334	707-3P1C	JT3D-3	KC-707	Tanker
290/4X-J??	21956	707-3W6C	JT3D-3	??-707?	Tanker/EW/ABCP with antenna array
Recently withdrawn					
120/4X-JYP	17921	707-328	JT4A	??-707?	Configuration unknown. Leased to IAI
137/4X-JYM	18460	707-329	JT4A	RC-707	Sigint. Stored Tel Aviv
140/4X-JYT	17625	707-329	JT4A	KC-707	Tanker. Wfu, for sale.
246/4X-JYS	20230	707-344C	JT3D-7	Phalcon?	Stored at Lod since 1995
258/4X-JYC	19291	707-328B	JT3D-3	??-707?	ABCP? Stored at Lod since c.2003

RC-707 Barboor, Topaz

It has been reported that four IAF 707s were converted as dedicated Sigint aircraft by E-Systems of Greenville, Texas, in 1976. Two aircraft were supposed to have been fitted with equipment similar to that carried by the USAF RC-135U Combat Sent (though there is no evidence of any such conversions) and two were fitted with the same equipment as the USAF RC-135W Rivet Joint. The RC-707 Barboor ('Swan') is an Elint platform whose configuration certainly apes that of the RC-135V/W, with similar AEELS (Automatic Elint Emitter Locating System) 'cheek' antenna arrays in box-like fairings between the forward entrance door and the wing root. At least one of the earliest, short-fuselage, jet-powered Israeli 707s (4X-JYE/001) was operated in this Elint/Sigint configuration, though perhaps only as a testbed or trials aircraft, before it was augmented or replaced by at least two 'production' Barboors, 128 and 137 (and perhaps 116 as well), which may have served with No. 134 Squadron. These were based on JT4A-engined 707-329 airframes, and were modified for their new role by E-Systems. 137 has since been retired, but 128 was re-engined with JT3D turbofans and remains in use with No. 120 Squadron. There have been reports that another of the IDF/AF's 707s (264 in ABCP/VIP configuration) will be converted to RC-707 standards to replace 128.

'EC-707' Airborne Command Post

The names Bdolach and Tavas ('Peacock') have been used to describe a 707 AEW configuration, with some sources maintaining that two examples of this aircraft actually entered IDF/AF service. As far as can be ascertained, the only Israeli Boeing 707 AEW variant has been the Phalcon (described below) which was exported to Chile, but which has not entered service locally. There are, however, a number of IDF/AF Boeing 707s that are believed to operate in an airborne command post role, and it is possible that the Bdolach name, and/or perhaps the name Tavas, actually apply to these aircraft, and that there has been some confusion between the AEW and airborne command post roles.

Aircraft 264 and 272 have both been identified as being possible command posts, though both have a slightly different configuration. 264 has worn an extremely smart, VIP-type colour scheme, but has a series of five swept blade antennas along its spine, and more below the forward fuselage. 272, by contrast, is painted white overall, and has relatively few external antennas, but does have a prominent dorsal antenna fairing above the centre section.

Israel has certainly made use of 707s as command posts, using them to support long-range missions such as the attack on the PLO HQ in Tunis, the Iraqi nuclear facilities at Osirak, and the recovery of hostages from Entebbe.

IAI EW conversions for export

Israel has undertaken a number of Sigint and EW Boeing 707 conversions for export customers, including single aircraft for Angola, Argentina and Spain, two aircraft for India, and four for South Africa.

Angola

Angola purchased one 707-321B (c/n 19374, D2-MAY) from the Chilean Air Force, via IAI, who reportedly modified the aircraft in Israel before delivery. The Israelis reportedly installed Elint equipment and a long-range stabilised camera system.

Argentina

After the Falklands War of 1982, Argentina expanded its 707 fleet, eventually receiving four extra aircraft. The FAA also launched a study programme for a new Elint aircraft. In 1986 one of the original three Boeing 707-387Cs (TC-93) was converted to Elint configuration by IAI, gaining the new serial VR-21. The aircraft is presently believed to be in store, but has been operated by the Grupo Guerra Electrónica, part of Escuadrón V at El Palomar AFB near Buenos Aires.

The single Argentine electronic aircraft is fitted with the Elta EL/L-8300 strategic Sigint suite, as also fitted to the Spanish SCAPA aircraft.

India

India's pair of 707-337Cs, K2899 (c/n 19988) and K2900 (VT-DVB c/n 19248) originally served in the transport and trials role, but have reportedly been fitted out by IAI as Sigint platforms. They serve with the Air Research Centre's Research and Analysis Wing at Palam.

South Africa

During the long 'border war' the South African Air Force used its Boeing 707s to support long-range operations by Buccaneer, Canberra, Mirage and Cheetah fighter-bombers and attack aircraft. The aircraft were used in the tanker role, of course, but the Sigint and EW roles were equally important.

The first three 707s delivered to No. 60 Squadron at Waterkloof were designed to be capable of rapid conversion to the Sigint role, and had prominent attachment points on the sides of the forward fuselage for the associated cheek-antenna arrays. These are similar to the AEELS (Automatic ELINT Emitter Locating System) antenna fairings on the USAF's RC-135U/V/W or on Israel's Barboor Elint aircraft, but do not extend so far aft towards the wing roots. In their Elint fit they also have an additional antenna fairing above the forward fuselage. Though the aircraft were modified with Israeli assistance, the mission kit is understood to include the Boeing Airborne Communications Intelligence System (BACIS) and ELTA L-8330 Sigint equipment.

The fourth SAAF Boeing 707 (1421/621) was delivered in a different fit, after reportedly having been modified as a Stand-Off Communications Jammer (SOCJ) by Grintek, probably with Israeli assistance. The SOCJ aircraft has an 'antenna farm' with nine blade aerials above the forward fuselage, serving the Grintek GSY1501 communications jammer and Sysdel Airborne Analysis System. The system is similar to that used by the communications jammer variants of the Douglas C-47TP and the Denel Oryx helicopter.

The Elint and EW roles steadily declined in importance following South Africa's democratisation, and the aircraft are primarily used as tankers and transports today. After a period of funding shortfalls, during which the fleet reduced to a single operational aircraft, the SAAF's 707s are believed to be extremely active once more.

Spain TM.17 Santiago/SCAPA

Spain's fourth 707 (707-351C, c/n 19164) is locally designated as the TM.17 Santiago, and serves with Escuadrón 408 in the electronic warfare, Elint, Sigint and Comint roles. The squadron moved before gaining its 707 in March 1998, co-locating beside the Centro de Inteligencia Aérea (Air Intelligence Centre). The aircraft was modified for its new role by IAI, gaining Elta and Indra Group avionics and electronics systems, including an Elta EL/L-8300 Sigint system. This is served by an extensive antenna farm below the belly, and with prominent antenna fairings below the fuselage.

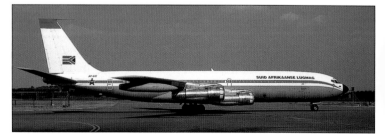

Above: 617 is a dual-role tanker/EW aircraft. the attachment points on the forward fuselage allow fitment of the BACIS Elint antenna fairing.

Below: Seen in clean transport configuration, 621 is the South African aircraft that can be converted for the SOCJ communications jammer role.

The aircraft (illustrated right) is also fitted with a Tamam Stabilised Long Range Observation System (SLOS) with a very high-resolution TV camera and video recording system. This has a range of over 62 miles (100 km) and has been used operationally to look into the western edge of North Africa, the Western Sahara and the Mediterranean. The acronym SCAPA is thought to be applied to the aircraft, with Santiago denoting the original programme.

IAI/Elta Phalcon/Condor

The Phalcon Airborne Early Warning, Command and Control System was developed by Israeli Aircraft Industries to meet Israeli Air Force (IAF) and export requirements. The system was first tested on a Boeing 707 airframe, with an Elta EL/M-2075 L-band conformally-mounted active phased-array 360° electronic-scanning antenna, consisting of 768 elements each individually controlled by a transmit/receive module. These elements were arranged in four groups, in nose, tail and cheek fairings, and together gave 360° coverage, without the complexity and expense of a conventional mechanically rotating radar antenna under a rotodome.

The Phalcon system incorporated considerable data fusion melding radar, IFF, ESM/Elint and Comint data, and with the ability to integrate data from off-board sensors to build a complete battlespace picture.

Reports suggest that an IDF/AF 707-344C (246/4X-JYS) was the first Phalcon, though this cannot be confirmed. The aircraft unveiled at Paris in 1993 was the first production conversion, and had been produced for the Chilean air force. Interestingly, Chile chose a three-antenna configuration, without the tail-mounted array, with about 260° coverage. The aircraft had a mission crew of two systems operators, two ESM/Elint operators, three radar operators, one Comint operator, two communications operators and a mission commander. The Chilean Phalcon-equipped 707 (serial 904, tested as 257/4X-JYI) is locally known as the Condor.

The Phalcon system has been marketed to a variety of countries, on a number of different platforms, with China being offered the radar system on a Russian-built Ilyushin/Beriev A-50 'Mainstay', for example, and India is expected to receive three Phalcon-equipped Il-76 aircraft. IAI is believed to

The sole Condor appeared at the 1993 Paris air show. As well as its IAI Elta EL/M-2075 Phalcon radar system, the aircraft also has EL/K-7031 Comint and EL/L-8312 Elint sensor suites.

have developed a miniaturised version of the Phalcon system for installation on three Gulfstream platforms for the IDF/AF, with S-band antennas in the nose and tail arrays and L-band antennas on the fuselage sides.

720 LRPA ASW test aircraft

Boeing used an ex-Eastern and ex-Trans Polar 720-025 (LN-TUW) as a prototype and proof-of-concept aircraft for the company's submission to meet Canada's Long Range Patrol Aircraft (LRPA) requirement. This outlined an ASW platform to replace Canada's ageing piston-engined Canadair Argus, and drew submissions from Breguet (Atlantic), Hawker Siddeley (Nimrod), Lockheed (P-3 Orion) and McDonnell Douglas (a version of the KC-10). With the 707 already in Canadian service in the transport and tanker roles, Boeing decided to offer a dedicated derivative of the 707-320 to meet the requirement, and hoped to offer the same aircraft to meet an RAAF requirement for a P-2H replacement.

Though the definitive 707 Long Range Patrol Aircraft would be based on the 707-320, Boeing installed representative mission equipment and systems (including an INS/Omega/Doppler-based navigation system, a Sperry SPZ-1-based dual-channel autopilot, MAD, sonobuoy launchers and tactical crew stations) in the 720-025, which it had repossessed in May 1971. The 720 (N3183B) gained prominent wingtip MAD fairings and 'Boeing ASW Test' titles on its fuselage sides. A small bulge below the aircraft's forward fuselage did not accommodate search radar (as has been suggested) but was actually just the Doppler antenna. The aircraft flew in its new guise on 6 April 1972.

During the flight test programme, the 720 was flown at low level (200 ft/61 m) over the water, and flew simulated ASW patterns that included aggressive 40° banked turns. With suitable reserves and high-speed dashes, the aircraft proved able to remain on station

Boeing's ASW test 720 is seen during trials. The main external feature was the wingtip-mounted MAD boom.

for between eight and 10 hours at a radius of 1,000 nm (1850 km) from base.

707-320B LRPA

The Canadian requirement demanded longer range and increased time on station compared to existing ASW aircraft, and incorporated some multi-mission (tanker and transport) capability. This dictated that the aircraft should be capable of being reconfigured for the long-range strategic transport role, or for tanker duties (carrying 20,000 lb/9072 kg more transferable fuel than the standard CC-137), though the latter requirement was subsequently quietly dropped. This demanded a larger airframe, and so the configuration proposed to Canada differed quite markedly from the ASW test aircraft, principally in that it was based on the turbofan-powered 707-320B. As such, the aircraft would have had a gross weight that was 104,500 lb (47400 kg) higher, with a 15-ft (4.57-m) increase in length and a similar increase in wingspan.

The aircraft was to have retained the prominent wingtip MAD fairings that had characterised the ASW test machine, but also had a longer, recontoured nose housing a 360° search radar. In plan view this was similar to the RC-135V/W nose, but was thicker in side elevation, with a slight 'droop'. There was to have been provision for a FLIR sensor forward of the nosewheel bay, with the Doppler and mapping and recce cameras behind, and an ESM antenna under the fuselage, roughly level with the trailing edge of the wing. The weapons bay was fitted below the rear fuselage, just ahead of the eight sonobuoy launch tubes.

Internally, the aircraft would have had four crew on the flight deck, with a galley and lavatory behind to starboard, opposite the entry door, and a 'dinette' to starboard, with four passenger seats to port behind that. The aircraft was to have been fitted with a forward freight door. The forward part of the tactical compartment was roughly in line with the wingroot, with a maintenance technician facing forward on the starboard side in front of the U-shaped tactical station. This accommodated two non-acoustic sensor operators facing forward on the starboard side, with the Navcom and Tacnav facing to starboard behind them, and the acoustic sensor operators facing aft behind that. Crew rest bunks were provided in front of the sonobuoy launchers, stowage and auxiliary lookout stations, and there was a second toilet in the extreme rear part of the cabin.

The Canadian requirement also dictated that the successful LRPA should be capable of detecting intrusions into the Arctic, and of flying environmental and fisheries protection, wildlife management, and ice reconnaissance sorties. These secondary roles required the installation of a number of specialised sensors, including FLIR and a removeable SLAR.

The Boeing LRPA was one of the two finalists in the Canadian competition (other contenders having been eliminated due to tight timescales), but the aircraft's long runway requirements, marginal performance characteristics at low level and worries about fatigue left it very much 'second-best' to the P-3. The CP-140 (a Canadianised P-3) was selected in November 1975, and work on ASW 707 variants was abandoned, since the chances of an Australian order (which were significantly smaller) had also slipped away. Had it entered service, the LRPA would almost certainly have used the designation CP-137.

This Boeing impression shows how the LRPA might have looked in Canadian service. Note the bulged nose profile.

Pioneers & Prototypes

VTOL flat-risers
Lockheed XV-4 and Ryan/GE XV-5

By the late 1950s, the global race to produce the first viable vertical take-off jet-powered warplane was well and truly on. While the British forged ahead with their Hawker P.1127 design, the US government funded research covering a wide range of propulsion systems and aircraft designs. Following the abandonment of the impressive but flawed vertical-rising Ryan X-13 design, attention shifted to new technologies, most notably that of engine augmentation, to produce flat-rising vehicles capable of producing more thrust than the engines alone could muster. Two companies, Lockheed-Georgia and General Electric (teamed with Ryan), were tasked with developing experimental aircraft capable of testing these radical new ideas. The following years would not only see the aircraft set many world firsts and achieve many of the programme goals, but also highlight the extreme dangers at the cutting edge of technology as all four aircraft would crash, claiming the lives of three of the country's finest test pilots.

Above: With its dorsal augmentation air intake doors fully open and the rear reaction nozzles working hard, the first of two XV-4As hovers close to the ground, having adopted its distinctive nose-high attitude caused by the rearwards inclination of the jet ejector manifolds.

The two Ryan/GE XV-5A Vertifans sit on the apron at Edwards AFB during a break from the rigorous US Army evaluation trials in 1965. Easily visible on the upper surface of the wings are the large doors covering the 5-ft diameter lift fans. The door opened upwards and inwards during vertical operations, allowing ambient air to be sucked through the fan system, increasing thrust by as much as 200 per cent.

VTOL flat-risers

As the Cold War intensified during the mid-1950s, the threat to NATO's airfields posed by nuclear attack – capable of knocking large proportions of a country's air assets out of a conflict – became a major concern for war planners. Rapidly, a technological race to develop a new class of jet combat aircraft capable of landing and taking-off vertically, thereby negating the need for the vulnerable runways and airfield infrastructure, gathered pace. In the UK Bristol was underway in developing its BE.53/2 (later Rolls-Royce Pegasus) engine that would power the eventual Kestrel and Harrier V/STOL fighters, concentrating on a light but powerful special-purpose engine that used variable thrust to provide both lift and forward power.

In the US, with jet engines too weighty and not yet powerful enough to provide sufficient thrust to overcome the weight of an aircraft for vertical flight, a different solution was sought. Lockheed-Georgia, General Electric and Ryan began investigations into augmented jets. Two methods, using the same physical laws, but different engineering solutions, were adopted – Lockheed electing to develop an augmented jet propulsion system and Ryan/GE opting for augmented lift fans.

Jet augmentation

Jet augmentation had first been described by von Karman in 1949 and was a remarkably simple theory combining the law of momentum and Bernouilli's theorum. The exhaust emanating from the jet engine entered a large chamber before exiting through an opposite aperture. As the hot and high-speed exhaust air exited the

Rising vertically from earth, the second XV-5A (62-4506) clearly depicts the open lift fan exit louvres in the wings and open pitch fan exit control doors on the nose that provided the pilot with control in this hazardous flight envelope. The relatively low rpm of the fans, combined with the mixing of cool ambient air during augmentation, allowed personnel to approach within a few feet of the aircraft in a low hover.

chamber it drew in ambient air to replace it. The momentum provided by the ambient air as it was sucked into the chamber provided the additional (augmented) thrust. In addition, the hot exhausted gas mixed with the cooler ambient air within the chamber, increasing the pressure of the gas and further increasing the energy, and therefore thrust, of the system.

The amount of additional thrust did, however, depend on the size ratio between the jet pipe entering the chamber and the chamber itself. Von Karman's theory predicted that the cross-sectional area of the chamber needed to be five times the size of the entrance tube to increase thrust by 20 per cent (32 per cent with the additional mixing and heating of air) and 10 times for a 36 per cent increase in thrust (75 per cent with mixing). This method of obtaining additional 'free' thrust, without the added penalty of exotic fuel or liquid injection systems, was an attractive proposition for designers looking to give aircraft the critical positive thrust-to-weight ratio needed for vertical flight.

Whereas General Electric and Ryan had elected to develop the augmentation process through the use of lift fans, Lockheed began initial experiments using the basic jet augmentation process.

Early laboratory tests of various types of augmentor jet had yielded promising results, as the exact geometry of the tubes and mixing chambers was perfected to give almost complete mixing of the jet efflux and ambient air – and therefore the most efficient results. Single-stage mixing gave some 50 per cent augmentation, multiple injectors brought an increase of thrust of almost 100 per cent and a three-stage system multiplied thrust by 2.3.

Lockheed's early experiments

By late 1957 the US Army was looking closely at developments in technology that could allow a feasible vertical take-off jet aircraft. Boosted by the raw data emanating from the laboratory tests, Lockheed's Georgia division established a privately-financed programme with the aim of developing an experimental aircraft to demonstrate the new jet ejector augmented thrust theory to meet US Army requirements.

In early 1959 Lockheed constructed and tested a wind-tunnel model enabling the engineers to develop the ideal aerodynamic qualities for the airframe. To proceed, however, the company needed to ensure that the augmentation theory, so promising in the laboratory, would translate to full-sized aircraft. To meet these needs Lockheed constructed and conducted a range of tests using a flying test rig (see box on page 121).

In mid-1959 the aircraft was allocated the company designation Model 330 Hummingbird, and in August the company presented the finalised design to the US Army Transport Research Command (Trecom) as an integrated VTOL aircraft system for battlefield surveillance and target acquisition.

After much deliberation the US Army, after recognising the potential, but tentative that the technology was not yet mature enough for the funding of fully-fledged service prototypes, elected in July 1961 to award Lockheed with an initial $2.5 million contract for two experimental research aircraft, omitting all equipment and systems related to its proposed combat role and

Pioneers & Prototypes

thus reducing the cost and the risk. The two aircraft received the US Army designation VZ-10 (serials 62-4503 and 62-4504).

With all major initial design work complete, Lockheed immediately set to work fulfilling the US Army's mission brief to test the feasibility of the untried augmented jet ejector concept. Little was known of the flight control systems needed for the complex transition from vertical engine-borne flight to horizontal wing-borne flight, just one of the many problems, the engineers would have to overcome.

Lift fan genesis

In 1957, as Lockheed was conducting its initial research in augmented jet power, Ryan thrust itself to the forefront of cutting-edge military fighter design by unveiling the X-13 Vertijet – the world's first jet-powered VTOL aircraft. A culmination of 10 years of VTOL research and design, the 'tailsitting' X-13 featured a delta wing and a 10,000-lb (44.5-kN) thrust Rolls-Royce engine. Operating from a ground trailer, the X-13 took off vertically, transitioned immediately into horizontal flight, then returned to a vertical attitude for let-down and landing. Although the aircraft proved impractical for front-line military use, the concept of producing a jet-powered VTOL had fired the imagination of Ryan's engineers and designers, and the company began work on new VTOL 'flat-riser' (an aircraft that can rise and descend vertically from a horizontal attitude) designs for military evaluation.

As the world marvelled at the outlandish looks of the X-13, Ryan was quietly working on the development of the augmented lift fan as an alternative VTOL powerplant. By coincidence, General Electric (GE) had also begun preliminary research into the lift fan theory, and it was the combination of these two companies that would attempt to meet the same US Army specification for which the Lockheed Model 330 and the Hawker P.1127 (designated XV-6A) were competing.

The basic ideas leading to the development of the lifting fan system were developed independently by GE's Peter Kappus and Ryan's John Peterson during the mid-1950s. A two-phase study by GE and Trecom in 1957-58 concluded that the augmented lift fan was the optimum system for jet VTOL operations, just as Ryan was working on its own lift fan contract for a USAF programme and, ironically, had intended to nominate GE as its subcontractor.

General Electric's Flight Propulsion Laboratory Department had commenced work on a lift fan propulsion concept in the mid-1950s, following the award of a preliminary study contract from the US Army. In 1959 GE was awarded a contract to develop and construct a working lift fan system and, by the end of the year, the first prototype fans were ready for testing at GE's outdoor testing facility at Evendale, Ohio.

The next stage of testing saw NASA build two full-scale lift fan aircraft models for testing the

General Electric and Ryan, buoyed by extremely positive laboratory results, were convinced that augmented lift fan power would power the VTOL military jet warplanes of the future. This initial artist's impression of the VZ-11 shows remarkable similarity to the eventual design, apart from the horizontal stabiliser that was moved to the top of the fin to reduce ground effects in the hover.

NASA/Ames wind tunnel at Moffett, California. A series of five tests was conducted on the first of the models, evaluating the fan in both fuselage and wing positions. After almost 160 hours of wind tunnel testing a second fan was delivered for the second full-scale model – this time more representative of the VZ-11, with a fan in each wing. A third full-scale model, matching the VZ-11's aerodynamic layout and fan arrangement, was added towards the end of the programme. The tests at both GE and NASA pointed to considerable potential – the fans had not suffered a single mechanical failure in over 300 hours of running time by mid-1962.

In December 1962 final full-scale tests were conducted, with the fans in the exact configuration as on the aircraft, to ensure the flightworthiness and integrity of the system before the first fan system was delivered to Ryan for installation. One of the most important tests, known as the 'torture test', involved the diverter valves switching at full military power to drive the fan. The ambient temperature around the fan rose from 75°F to 800°F (24°C to 427°C) in less than a second, however, the fan system proved it could

Lockheed's initial design studies

As Lockheed's detailed design work, headed by Al Mooney (of Mooney Aircraft fame) gathered pace from late 1957, the challenges of containing the large, heavy and cumbersome augmentation system within the fuselage became apparent. A large bulky fuselage would be necessary to contain the system, yet sufficient space for fuel and test equipment would also be required. In addition, the aircraft would need to possess adequate aerodynamic qualities to operate at speeds in excess of 450 kt. Lockheed's studies proceeded through a number of design iterations (under the temporary Design Designation GL-224) before reaching a close approximation of the resulting VZ-10 in the guise of GL-228-14.

One of the first designs, GL-223-3 (above right) shows the problems the designers had in reconciling aerodynamic efficiency with the need to contain the massive jet augmentation system. The engine intakes are seen here mounted high on the forward fuselage to help prevent hot gas ingestion in a low hover. As the design matured with sleeker lines (GL-228-8), the possibility of selecting a low-mounted swept wing behind integrated engine intakes was examined (below). The best compromise between performance and hover practicalities, however, emerged with GL-224-14, with pod-mounted engines high on the fuselage and a narrow chord straight wing.

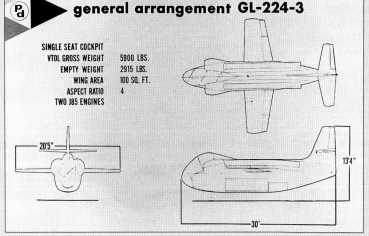

general arrangement GL-224-3

SINGLE SEAT COCKPIT
VTOL GROSS WEIGHT 5900 LBS.
EMPTY WEIGHT 2915 LBS.
WING AREA 100 SQ. FT.
ASPECT RATIO 4
TWO J85 ENGINES

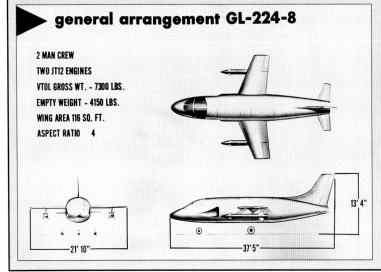

general arrangement GL-224-8

2 MAN CREW
TWO JT12 ENGINES
VTOL GROSS WT. - 7300 LBS.
EMPTY WEIGHT - 4150 LBS.
WING AREA 116 SQ. FT.
ASPECT RATIO 4

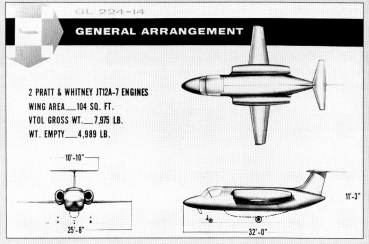

GENERAL ARRANGEMENT GL 224-14

2 PRATT & WHITNEY JT12A-7 ENGINES
WING AREA 104 SQ. FT.
VTOL GROSS WT. 7,975 LB.
WT. EMPTY 4,989 LB.

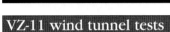

VZ-11 wind tunnel tests

As Ryan began to refine its VZ-11 design, configurations were tested in a number of wind tunnels including the high-speed David Taylor Model Basin facility in Maryland (above). These wind tunnel tests led to a number of modifications including the incorporation of a T-tail to improve airflow and stability and prevent the effects of fan-wash causing control problems in a low hover. NASA Langely also constructed a free-flight model incorporating different VZ-11 arrangements, providing valuable information. These provided a host of valuable data, most especially concerning the optimum design for the engine inlet and data on high- and low-speed performance, placing Ryan in an ideal position for selection for the US Army development contract.

NASA/Ames also built and tested a number of scale and full-sized models in its Moffett, California, wind tunnel facility. NASA was a key partner, working extensively in finalising the configuration and size of the wing-mounted lift fans and the nose-mounted control fan.

withstand these large temperature changes and continue to perform satisfactorily.

As GE and NASA worked on perfecting the fan design, the US Army opened a design competition to all US aircraft manufacturers, to build the two flight test aircraft in which the GE augmented fan design would be tested. Ryan was selected as the winner and set to work perfecting the aerodynamics of the fuselage and wings.

In 1959 Ryan gained valuable technological data using a purpose-built VTOL research aircraft, designated as the VZ-3RY Vertiplane. It used large flaps to deflect the propeller's slipstream, allowing the aircraft to take-off and land vertically. Both Ryan and NASA extensively utilised the Vertiplane to prove the feasibility of this V/STOL principle.

A full-scale VZ-11 mock-up was unveiled by the Army, General Electric and Ryan at Ryan's Lindbergh Field San Diego plant on 17 August 1961. Three months later the US Army Transportation Research Command ordered two VZ-11s from GE under a $10.5 million contract, of which a little more than half was to be allocated to the airframe supplier Ryan. Performance objectives required a five-minute hover, followed by a flight cruise at best endurance for a specified period, followed by a transition to a vertical landing with 10 per cent fuel reserves.

The aircraft was to be powered by General Electric's X353-5 Lift Fan system and would be tested by the US Army, along with the two Lockheed VZ-10s, under a programme to evaluate VTOL jet aircraft for future combat surveillance or target acquisition missions, blending battlefield mobility with high performance and allowing the aircraft to 'live with the troops'.

Before the flight test programme commenced, Ryan built an XV-5A flight simulator in a new purpose-built facility to give prospective pilots a feel for the unique characteristics of jet-borne vertical flight.

Hummingbird flight tests

While GE and Ryan continued to refine their VZ-11 design the XV-4A project continued apace, despite the inevitable problems, and in the spring of 1962 both aircraft were rolled-out within a few weeks of each other.

On 7 July 1962, only 53 weeks after the signing of the initial contract with the US Army, the XV-4A (the VZ-10 became the XV-4A and the VZ-11 became the XV-5A at this time during the revision of US military designations) made its maiden flight in conventional mode.

A number of minor problems were identified during the first flight, including the need for excessive trim changes during retraction of the landing gear and flaps. However, the Hummingbird was reported to fly well with good response from the hydraulically-driven elevators, ailerons and rudder.

The flight test programme was conducted at Dobbins AFB, Georgia, and, over the next few weeks, a further seven conventional flight tests were conducted by four different pilots, each reporting good flying characteristics as the envelope was slowly increased to 250 kt (463 km/h) and 10,000 ft (3048 m) altitude. These first conventional flights were conducted without the augmented jet ejector system fitted, and explored the standard flight envelope.

With initial conventional flight tests deemed successful, the programme progressed to the critical phase – vertical flight. To ease the transition from vertical to horizontal flight, the twin manifolds and associated ejectors were fixed at a 12° rearwards angle, adding a forward component to the thrust, however, this ensured that to remain static in the hover the XV-4A would have to lift-off with, and maintain, a 12° nose-high attitude, a situation that caused the engineers numerous problems. To investigate the flying characteristics at this attitude Lockheed completed construction of a large test rig, able to hold the aircraft in position at a height well out of ground effect. A series of successful

Lockheed's augmented jet test rig

To prove the jet ejector augmentation theory, Lockheed designed and built a flying test rig, initially powered by a pair of Fairchild J44 turbojets (right). By subtly changing the geometry of the jet tubes, mixing chamber and air intakes, the rig managed to lift a gross weight of 2,600 lb using engines with a combined nominal thrust of 2,000 lb. This augmentation of some 30 per cent convinced that the theory and aircraft was viable and design work on the real aircraft accelerated. Later the test rig was re-engined with two more powerful Continental J69 engines and was used extensively in flight tests to provide data for the complex control systems incorporated in the final design. In this configuration, and following further tweaks of the materials, and geometry of the augmentation chamber and exhaust system, an augmentation rate of 45 per cent was measured.

The custom-built engine test rig fulfilled a valuable role in not only testing and enhancing the performance of the jet augmentation powerplant, but also provided invaluable data for the eventual flight control system.

Pioneers & Prototypes

Inside the Lockheed XV-4A

Construction of the two XV-4As began almost immediately after the US Army contract was signed. The aircraft's shape and dimensions were largely governed by the need to accommodate the bulky mixing chamber within the fuselage. The 5-ft (1.52-m) diameter main fuselage was necessarily boxy, not only to provide sufficient room for the mixing chambers at the centre of gravity, three fuselage tanks with total capacity of 285 US gal (1079 litres) and aircraft electronics and hydraulics, but also to provide relatively flat dorsal and ventral surfaces for the incorporation of the large longitudinal doors that opened along the upper surface for ingestion of the ambient air into the mixing chamber. The rear fuselage ended with a bullet fairing, which housed a brake parachute.

The simple 26-ft (7.92-m) span trapezoidal-shaped wing was mounted mid-fuselage and was of thin-enough section to provide high-speed flight in the region of 500 mph (805 km/h). With slight dihedral and of narrow section, the wing incorporated single-slotted trailing-edge flaps (operated in only two positions – up or down) and ailerons. At the front of the fuselage Lockheed fitted a conventional side-by-side two-man cockpit, covered by a bulbous canopy providing good visibility forwards and laterally, however, rearwards vision was restricted. The pilot occupied the left-hand seat, positioned on a rocket-powered ejection seat taken from a Douglas A-4E Skyhawk and modified for zero-zero performance. The right-hand seat was reserved for testing and recording equipment.

Nacelles above the wing, on either side of the fuselage, housed the Pratt & Whitney JT12A-3(LH) turbojets selected as the best compromise of size, power and weight, providing both lift and cruise thrust. The nacelles also housed the retractable main landing gear that rotated rearwards to be housed in the lower rear section of the nacelle. The single-wheel nose gear similarly retracted rearwards into the lower forward fuselage. Lockheed, in its wisdom, ensured that the landing gear was tough and strong enough to sustain a large vertical rate of descent, an event highly likely in the testing environment of an untried new vertical landing concept. The gear was designed to absorb a nominal maximum sink rate of 16.6 ft (5.1 m) per second for vertical landings and 10 ft (3.05 m) per second for conventional touchdowns. Grateful test pilots would

Before the Lockheed XV-4A could attempt vertical lift-off and flight the aircraft's VTOL characteristics, engine response and vertical thrust needed to be extensively examined. To achieve this Lockheed constructed a static test rig equipped with sensors to record the powerplant's performance. The first XV-4A is seen here (above) attached to the rig with the jet ejectors and twin manifolds easily visible through the open ventral doors. Once Lockheed was satisfied that the aircraft had sufficient power for vertical flight the XV-4A conducted a series of tethered flights from 30 November 1962, lifting itself into the air vertically for the first time (above left).

soon lay testament to the strength of the gear as vertical descents approached and on one occasion exceeded 14 ft per second.

After initial iterations of the design featured a conventional fuselage-mounted tailplane, Lockheed elected to introduce a T-tail to the final design in an attempt to keep the horizontal stabiliser clear of disturbed air while the aircraft was in a low hover. This problem would, however, manifest itself during the flight test programme in the form of moderate to severe vibrations.

Key to the success or failure of the entire programme was, however, the incorporation and demonstration of the jet-ejector augmentation system. In normal wingborne flight the engines exhausted rearwards through short jet pipes in an entirely conventional manner. For vertical flight Lockheed settled on a mixing chamber/manifold system. A diverter valve situated in each jet pipe redirected the jet exhaust from the mixing chamber through a system of pipes into the manifolds (each engine fed both manifolds, preventing asymmetric lift in case of single engine failure). On each side of the upper fuselage a long narrow door was situated, which opened to allow the ambient air to be drawn into the mixing chamber. The doors could be adjusted in flight to help modulate the rate of ambient airflow into the manifolds to decrease or increase thrust. At the base of each manifold was a set of five ejector nozzles, again covered for conventional flight by narrow doors. As engine thrust and therefore jet exhaust velocity increased, the flow of ambient air into the manifolds through the open upper doors also increased. The air/gas mass flow would increase by some 40 per cent at maximum thrust, and the mixing of the hot exhaust with the cooler air within the mixing chamber caused an expansion and corresponding increase in pressure that added additional momentum to the flow. The efficiency of the augmentation system would be key to the success of the project. Lockheed engineers had predicted an increase approaching 1.5 times the thrust from the engines alone (50 per cent augmentation), proving more than enough power to vertically lift the aircraft, even while carrying a full fuel load (located in a tank between the two manifolds).

The 'free' additional thrust did not, however, come without penalty. The augmentation and exhaust systems and associated pipes were heavy (constructed largely of titanium and stainless steel) and took up a large proportion of the aircraft's internal capacity. Additionally, the complexity of the relationship between the size and geometric shape of the augmentation system and the exhaust ejectors tested the engineers to the limits as they strived to meet the theoretical potential of the system. As development progressed the man-hours and cost of producing each augmentation system was of increasing concern.

Above: XV-4A 62-4503 begins its maiden free-flight vertical take-off in May 1963. With the jet augmentor system yet to provide the theoretical power output, the early flights were conducted with very low fuel loads and at maximum power. Despite the low fuel load, hot gases re-ingested into the engines ensured that a hover could not be maintained for more than a minute before loss of thrust forced the aircraft to land. Note that the canopy covering the pilot has been removed, allowing for a rapid egress in an emergency.

Right: With responsive ailerons and good acceleration, the XV-4A was a pleasant and almost viceless aircraft to fly in conventional mode. The large canopy provided the pilot with an excellent field-of-view – especially important during hovering flight close to the ground.

VTOL flat-risers

Inside the Ryan/GE XV-5A

By late 1962 Ryan/GE had finalised the design of the XV-5A and construction commenced. The aircraft was powered by two standard GE J85-5 turbojets mounted high on the fuselage, two 5-ft (1.52-m) diameter tip turbine-driven fans submerged in the wings and a smaller fan in the nose of the fuselage. The 3-ft (91-cm) diameter nose fan was used to provide lift, trim and pitch control. Nose fan thrust modulator doors were pivoted to permit fan flow to be vectored for precise pitch control and thrust could be smoothly modulated over a range of 1,500 lb upwards and 200 lb downwards. The louvre-type nose inlet doors were fully open in VTOL flight, transforming when fully closed to a smooth aerodynamically clean surface for conventional flight. The position of the J85 turbojets high on the fuselage minimised the likelihood of foreign object ingestion and helped prevent re-circulation of fan exhaust through the engine (resulting in a loss of thrust).

The positioning of the engines also left a good percentage of the fuselage capacity for fuel, landing gear and test equipment. For vertical flight, diverter valves directed the jet exhaust to the tip turbines that drove the fans. Hinged doors above the fan opened during vertical flight to allow augmentation (sucking in of additional ambient air to increase thrust) through the open surfaces. Hinging on the fan centre support, the doors closed when in flight to provide an air-tight seal and form an aerodynamic fairing. The augmentation was predicted to increase thrust by some 300 per cent, but in reality only around 150 per cent was achieved. Crossover ducting between the engines and the fans insured that 60 per cent of the power was available with only a single engine operating. Thrust spoilers were located at the engine exhaust nozzles, allowing the engines to run at full rpm, and the fans to spin up quickly during the transition back to vertical flight without unwanted additional forward thrust. Integration of the engines and fan system was the most complex of any aircraft up to that time.

The main landing gear could be positioned aft for vertical take-off (to prevent the struts and wheels from interfering with the air exhausted through the fans and producing a subsequent loss of thrust) and forward for conventional take-off (close to the centre of gravity and therefore providing an easier rotation moment for lift-off). The gear was set in the correct position using a switch in the cockpit.

The aircraft's wings had a span of 29.83 ft (9.09 m) and had a noticeably wide chord to allow the fitment of the two lift fans, but were as small as possible to obtain the highest possible wing loading. The wings had 6° of anhedral to eliminate predicted yaw and roll problems. Flaps were fitted directly behind the fan and added to the augmentation during hovering. Ailerons were positioned on the outer wing.

The side-by-side two-man cockpit was fitted-out for pilot only (left-hand seat) seated on a new lightweight ejection seat developed for the programme by North American. The seat was designed to provide 'through-the-canopy' ejection from anywhere between 0-450 kt indicated air speed.

The cockpit layout was designed to be instantly recognisable to helicopter pilots, with a centred control column, rudder pedals and a stick to the pilot's left with a twist grip throttle. A pair of throttles, slaved to the twist grip throttle, were positioned on the centre console. The control column was mechanically connected to the wing fan exit louvres, nose fan exit doors, ailerons and elevators. During the hover sideways movement of the stick differentially adjusted the louvres on each wing fan, thereby controlling roll. Forward and back movement of the control column adjusted the amount of thrust to the nose fan and therefore controlled pitch. In conventional flight the column controlled the ailerons and elevators as on a normal aircraft. The rudder pedals, as well as controlling the rudder, also vectored the main fan exit louvres, therefore providing the necessary yaw control during slow speed or vertical flight. Pulling the twist grip stick upwards opened the sets of louvre doors on the main and nose fans and thus increased vertical thrust for take-off or vertical climb. The twist grip controlled the thrust of the engine and therefore the lift of all three fans. Also located on this stick was the conversion switch for the change to/from vertical and conventional flight. Additionally, the aircraft was fitted with a stability augmentation system comprising a series of gyros that could detect changes in roll, pitch and yaw, and adjust the nose and wing fan settings to prevent loss of control. The aircraft could be flown with or without the system turned on.

Power transmission was accomplished through pneumatic coupling, eliminating the need for gear boxes and drive shafts, and therefore reducing maintenance and parts problems. Jet exhaust passed through a turbine scroll before forcing the low-speed large-diameter axial-flow fans to rotate at around 2,600 rpm, resulting in a relatively low velocity and low temperature efflux, thus allowing the aircraft to operate from extremely confined spaces. The fan developed 100 per cent lift around 4 seconds after the valve was flicked to feed the fan.

The engine intakes were located high on the fuselage to minimise debris ingestion when operating from unprepared runways.

Left: The XV-5's two-man cockpit was fitted with controls and instrumentation for the pilot only. Positioned in front of the pilot was the nose lift fan, seen here with the louvre-type entry doors open.

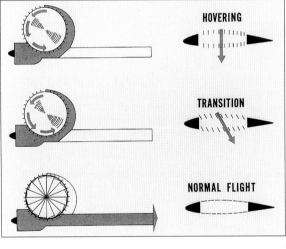

Left: This diagram demonstrates the transition from hovering flight to normal flight. In the hover, the engine exhaust is directed via diverter valves to the wing fans, with the louvres in a vertical position. The lift fan louvres then vector during the transition to provide a forward thrust component. Finally, the diverter valves open, the engine thrust is exhausted through the conventional jet pipe and the lift fan and nose fan exit and entry doors close.

Ryan/GE VZ-10 cutaway key

1. Nose fan inlet louvres
2. General Electric GE-X376 pitch control fan
3. Nose fan thrust reversers and nose fairings
4. Zero-zero ejection seat
5. Conventional control column
6. Rudder pedals
7. Throttles
8. Lift control stick
9. Instrument panel
10. Hydraulic compartment
11. Electrical compartment
12. Nosewheel landing gear
13. Fuel tank
14. Single split intake
15. Crossover duct
16. Nose fan supply duct
17. Main fan folding cover
18. Exit louvre actuators
19. Two-position main landing gear
20. Flap
21. Aileron
22. Engine jet pipe
23. Thrust spoilers
24. General Electric GE-X353-5 lift fan
25. All-moving horizontal stabiliser
26. Elevators
27. Vertical fin
28. Rudder
29. GE J85 turbojet
30. Diverter valve

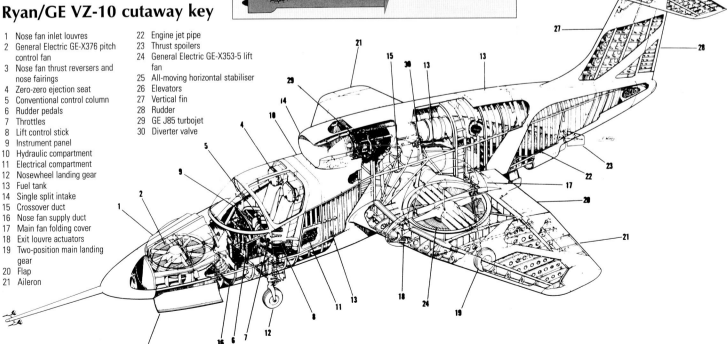

The second XV-5A (62-4506) begins the transition from vertical to conventional flight over Edwards AFB during 1965. The wing fan entry doors are fully open as the lift fans spool up to full rpm in around three seconds. As well as providing the test base for XV-5A flight trials, the USAF also supported the Vertifan project by providing the GE J85 engines and diverter valves. The latter had been developed for the earlier Bell XF-109 research project.

hovering test flights was conducted in the rig during October and November 1962 and, at the end of the month, the Hummingbird was deemed ready to lift itself vertically from the ground for the first time – albeit tethered with short restraining straps.

Even as the first tethered hovering flight was made on 30 November 1962, concerns that the augmented jet powerplant was not producing the predicted thrust were beginning to intensify. The initial tethered tests were hampered by inlet losses, unsatisfactory configuration of the mixing chamber and deficient roll control.

The XV-4A had been designed with an empty weight of around 5,000 lb (2268 kg) and a vertical take-off weight of some 7,200 lb (3266 kg). The twin Pratt & Whitney JT12A turbojets produced a total thrust of 6,600 lb (29.37 kN) and the success of the design depended on augmentation producing an additional 40 per cent of thrust, taking the theoretical power to 9,240 lb (41.12 kN). Loss of energy as the air flowed through the manifolds and out of the ejectors, combined with less than 100 per cent mixing of the air was predicted to reduce actual output to 8,375 lb (37.27 kN) – more than enough thrust reserve for safe vertical flight. In reality, however, the 6,600-lb thrust from the engines had reduced to 6,000 lb (26.7 kN) by the time the exhaust entered the manifolds. To further exacerbate the problem the augmentation achieved was only some 30 per cent, and further loss of efficiency as the air exited the ejectors ensured the attained thrust was only a little over 7,500 lb (33.37 kN) – resulting in an extremely marginal thrust-to-weight ratio of 1.05:1.

The problems achieving the desired level of augmentation caused a hiatus in the XV-4A flight test programme through the winter of 1962-63, as engineers attempted to improve the net augmentation achieved. With a slight increase in performance, the XV-4A achieved untethered vertical lift off and hover on 24 May 1963, albeit at low fuel weight to provide the necessary margin of performance. On 8 November 1963 the programme's next major milestone was achieved as XV-4A (62-4503) performed the first transition from vertical to horizontal flight and then reversed the process for landing.

In hovering flight the XV-4A was relatively easy to control. The pitch of the aircraft was controlled by the use of downward-pointing nozzles situated at the nose and tail and fed using exhaust air from the engines. Rated at 450-lb (2-kN) thrust, the nozzles could swivel in opposite directions for yaw control and each nozzle

XV-4A 62-2503 approaches the main runway at Dobbins AFB during initial Lockheed flight trials in August 1962. The augmented jet ejector system had yet to be fitted and the aircraft lacked the reactor thrusters on the wing tips, nose and tail.

was fed by a separate engine to allow a degree of redundancy in the case of engine failure. Roll control was achieved by the use of a reaction control thruster on each wing tip. Fed by engine compressor bleed air, the thrusters could quickly negate any tendency for the aircraft to drop a wing whilst in the hover. Once the transition to wingborne flight had been achieved, the pilot manually turned off the air flow to all four control nozzles.

To aid the pilot during the hover and low speed flight the XV-4A was fitted with an automatic stability control system adapted from that used on contemporary helicopter designs. Without the system turned on the aircraft was extremely difficult to control, however, if necessary, the system could be overridden by the pilot using large control column or rudder pedal inputs.

For the transition to horizontal flight the aircraft firstly had to pitch forwards to bring into effect the rearwards-facing jet ejectors. This was achieved by an automatic deflection of the elevator 30° downwards. High-speed air flowing over the surface of the elevator from a boundary layer control system forced the tail of the aircraft upwards and helped prevent the horizontal stabiliser from stalling. As the aircraft transitioned to a horizontal attitude (giving a 12° forward thrust component) the aircraft slowly accelerated, although the marginal thrust-to-weight ratio ensured the aircraft often settled back on the runway as it accelerated, especially at higher fuel weights. Re-ingestion of hot air gases (bouncing off the runway) into the engine at the hover and slow speeds also reduced the performance of the engine, preventing the aircraft from climbing to a higher altitude before beginning the transition.

Once the aircraft had accelerated to some 30 kt, the hot air ingestion ceased and the wings began to create lift. At this point the nose could be dropped further to 10° nose-down and the acceleration rapidly increased, allowing the aircraft to climb away. As the speed reached around 80 kt, one engine was diverted from the jet ejectors to normal thrust – the asymmetrical flight being aided by the aircraft's jet pipes being angled slightly away from the fuselage. The second engine transitioned to horizontal thrust at around 120 kt. Both sets of doors on the upper and lower fuselage were then closed.

To return to hovering flight the process was reversed. The pilot retarded the throttles on both engines to reduce speed before one and then the other engine transferred to lift thrust.

XV-5A takes to the air

In the autumn of 1962, as the XV-4A was achieving its first tethered vertical flight, and following a major design review that added a spin chute, fire seals between the engine compressors and turbines, and improved forward visibility, the manufacture of the two Ryan/GE flight test aircraft, by then redesignated XV-5A, was ready to commence.

The integration of the augmented fan system was complex and it was not until 26 February 1964 that the XV-5A was officially rolled out. A series of ground runs and taxi tests concluded on 23 May 1964, three days prior to the first flight.

Republic Aviation, which had provided financial support to the project, also provided its senior test pilot and, on 26 May 1964, the first flight was flown by Republic test pilot V. H. Schaeffer. The 19-minute conventional flight from Edwards AFB, reached 8,000 ft (2438 m), however, the aircraft had yet to receive the full fan augmentation system to allow vertical flight.

With the lift fans soon in place, hover tests began on 16 July 1964, the early trials revealing that the exit louvre door hydraulic actuators (providing roll control) were undersized.

Part of the US Army Phase II evaluation tests involved assessing the XV-5A's suitability in the combat rescue role. Tests performed at Edwards simulated air-sea rescue operations using two life rafts stationed on a pond, one anchored and one free. The free raft was rapidly blown to the edge of the pond (as it did during comparative trials with the CH-21), however, the raft with the sea anchor stayed within 20 ft of the aircraft. Visibility problems were encountered as the aircraft created a fine spray as it hovered (left). Further rescue tests were completed by fitting the aircraft with an electric winch for the retrieval of a 235-lb (107-kg) dummy (right). The dummy was instrumented to record strain and 'g' forces and had temperature-sensing paint (tempilac) on the helmet and shoulders to determine the effects of the heat from the fan downwash. No noticeable strain or prohibitive heat effects were registered on the dummy.

Phase I contractor flight tests continued until January 1965, when the two XV-5As were deemed ready for the official US Army evaluation programme (Phase II).

XV-4A testing takes its toll

Meanwhile, in the spring of 1964 the XV-4A programme was progressing satisfactorily as the transition from vertical to horizontal and back was perfected, and the design team struggled manfully to increase the augmentation rate and therefore thrust of the engines. By this time the combined test flight programme for both aircraft had encompassed 82 hovering flights (total 8 hours flying time) and 69 conventional flights (some 28 hours flight time). Seven complete transitions had been achieved and the flight envelope had expanded to include stall recovery procedure, in-flight single-engine shut-down and cruise, high-speed climb and limited aerobatic manoeuvres.

Flight tests revealed the XV-4A had an impressive initial sea level rate of climb of 18,000 ft (5486 m) per minute and could reach 12,000 ft (3658 m) in 54 seconds. Due to the limited internal space for fuel, however, the aircraft's endurance and range was minimal for conventional flight and extremely short for vertical flight.

As the initial stage of flight testing was drawing to a close, tragedy struck when, on 10 June 1964, the first XV-4A (62-4503) was destroyed in a crash that killed the civilian army test pilot. Reports from the chase aircraft stated the XV-4A had just completed a transition from conventional to vertical flight at 3,000 ft (914 m) when control was lost. The Hummingbird came down between the operating base at Dobbins AFB and the town of Woodstock, hitting the foundations of a house under construction and injuring a worker. All test flying was suspended, pending an inquiry into the crash.

With the sole surviving aircraft grounded, Lockheed took the opportunity to improve the marginal augmentor performance. A variety of manifold and ejector configurations was fitted to 62-4504 and evaluated on the static test rig. These modifications did improve the total thrust to a respectable 8,500 lb (37.82 kN), although this was at the expense of additional weight.

XV-5A Phase II – Army evaluation flight tests

With Phase I contractor flight tests satisfactorily achieved, the XV-5As were handed over to the US Army for official evaluation. Beginning in January 1965 the XV-5As logged almost 130 flight hours in 336 flights during the Phase II test programme conducted at Edwards AFB and supported by a resident Ryan/GE team. The evaluation was completed under the direction of the US Army Aviation Materiel Laboratories (AVLABS), Fort Eustis, Virginia, with the flight test element conducted by AVLABS and US Army Aviation Test Activity (ATA). The majority of flights were flown by Chief Engineering Test Pilot Val Schaeffer and Major William Welter of the ATA.

Initial Phase II tests investigated and evaluated basic conventional and hovering flight parameters and performance, the aircraft demonstrating good climb performance and a maximum speed of 526 mph (847 km/h). The XV-5A then flew a closed course in hover configuration, completing the course in 5 mins 30 secs. The pilot then flew the same course in a Bell UH-1B for comparison – taking 3 mins 50 secs.

The XV-5A also made hovering descents over a small pond, was air taxied from the Army hangar area (standard army helicopter procedure) and a 235-lb anthropomorphic dummy was raised up on a 50-ft cable to within 4-ft of the hovering XV-5A using an electric winch fitted on the outside of the airframe. Following the dummy tests, Ryan flight test director John Burhans walked directly underneath the XV-5A, dressed in full standard US military flight clothing and helmet, to evaluate problems encountered by a man on the ground reaching the rescue seat or sling. At 60 ft Burhans reported he could walk freely beneath the aircraft, reporting that "wind blast and noise were not objectionable. It would have been possible to climb into a rescue seat or sling without difficulty". At 35 ft, Burhans had difficulty in walking towards the pick-up spot but eventually made it by leaning into the downwash. Tests at 25 ft and 10 ft resulted in him being unable to reach the pick-up point as airstream velocity was in the order of 60-80 mph. Noise or heat were however, not a problem.

Complex transitions from hovering to conventional flight were then conducted, including emergency transitions and transitions at 30° of roll, at both high and low altitudes. By the end of Phase II the XV-5A had completed 293 airborne transitions to/from VTOL flight. Phase II flight tests concluded on 15 November 1965.

Phase II was followed by additional pilot training and austere environmental or erosion testing conducted at Edwards AFB and the Naval Ordnance Test Station, China Lake, California. Vertical take-offs were made from sod, alfalfa fields, ploughed dirt, a parachute drop zone, a standard Army T-17 membrane and on the unprepared desert floor (right).

The erosion tests were performed to determine the capability of the XV-5A to land on various surfaces and at no time was any foreign object damage reported. Before each test a USAF Piasecki CH-21 (of similar gross weight to the XV-5A) performed each manoeuvre at each site for safety reasons (below).

Pioneers & Prototypes

Above: XV-4B Hummingbird (62-4504) was rolled-out, resplendent in USAF markings, on 4 June 1968. Structural modifications included the fitment of extra horizontal surfaces on the tail fin and on the rear of the fuselage, and an extended tail cone housing the rear reactor thruster.

Right: Before the XV-4B's first flight numerous tests were conducted using a large truss to suspend the aircraft at differing attitudes and altitudes. The aircraft could either be locked in position during the tests, producing specific data relating to the forces generated by the engines, or tethered to the truss, allowing the pilot to rotate in a horizontal plane and move the aircraft vertically.

In late 1964, with conflict escalating in southeast Asia and placing more urgent demands on the US Army's time and funding, the XV-4A programme was quietly terminated. The Hummingbird, although demonstrating that the augmented jet ejector powerplant had definite prospects, had not met the requirements the US Army demanded and was nowhere near reaching a point where the Army could consider ordering pre-production prototypes. 62-5404 was subsequently transferred to NASA for wind tunnel testing at the Ames Research Center before being put into storage.

The XV-4A was the first US Army jet-powered aircraft to fly and was the world's first aircraft to take off vertically with the installed thrust being less than the operating weight of the vehicle. It was one of only two aircraft to employ the augmented jet ejector system, the other being the Rockwell XFV-12A, which never flew free.

High-profile Vertifan crash

The Phase II Army evaluation of the XV-5A ran from February to November 1965. Interspersed with the wide-ranging, performance, operational and handling trials the aircraft were called on to perform in front of assembled military observers, politicians and guests. It was at one of these events that tragedy struck.

On 27 April 1965 the first XV-5A crashed at Edwards AFB, during a demonstration for several hundred reporters, US military officials and civilian guests. Both aircraft were participating in the demonstration and 62-4505, flown by Ryan test pilot Lou Everett, had completed a vertical take-off and converted to conventional flight before conducting a low pass in front of the assembled guests at around 310 kt. After a sharp climbing turn to the left, the aircraft began decelerating prior to conversion back to fan-borne flight. At an altitude of around 800 ft (244 m) the pilot was progressing through the conversion check list when it was interrupted by words that sounded like "I've got to get out". The XV-5A had begun to nose over and dive towards the ground at an angle of 30-40° and Everett was fatally injured as he attempted an ejection at low altitude, probably outside the seat envelope. The other XV-5A, piloted by US Army pilot William Anderson, landed immediately and the aircraft was grounded pending the crash inquiry. The grounding was shortlived as investigations revealed that there was no basic defect with the design and no problems were found with the lift fan system.

The surviving XV-5A (62-4506) continued Phase II army testing and from November 1965-January 1966 conducted erosion trials to assess the suitability of fan lift aircraft for a variety of operational tasks.

By early 1966 Army evaluation had identified a number of areas of concern. The aircraft suffered severe buffet problems from the ground cushion effect at hover heights below 10 ft. The aircraft's ability to perform specific tasks in this flight regime were extremely limited and would have to be cured. In addition, the narrow stall margin during the transition to/from vertical flight was deemed unacceptable for service use by a standard military pilot and unworkable in a combat environment. Ground operations on the narrow track undercarriage were also unacceptable, and the aircraft was likely to swing into wind during crosswind conventional and vertical landings, placing the aircraft at risk. Also slow power response during vertical flight could lead to pilot-induced height oscillations, although these could normally be controlled through throttle manipulation.

With the Army concerns outlined Ryan/GE immediately grounded the aircraft for an extensive modification lay-up between January and August 1966. Among the changes introduced was a XV-4A-type sequential conversion of the engines during transition from vertical to horizontal flight, with one engine being directed to the conventional tailpipes followed by the other. This would result in improved rate of climb and a much greater stall margin. Drag-reducing modifications to the nose fan would allow conversion to happen at 110 kt rather than the previous 90 kt, and interconnection between the horizontal stabiliser and the engine diverter valves, coupled with improvements to the flight control system, would enhance the aircraft's stability in the low hover.

As the modification lay-up work was underway US Army interest in VTOL research began to wane. Conflict in Vietnam was diverting much of the funds and attention and the Army deemed none of the three research types (XV-4A, XV-5A and XV-6A) had shown sufficient maturity to merit ordering of pre-production prototypes of an operational variant.

Army out, USAF in

As Army interest in the XV-5A waned, however, possible salvation appeared as the USAF and US Navy expressed interest in follow-on tests to evaluate the aircraft as a combat search and rescue vehicle. With air crew losses in Vietnam of great concern, the plan involved a pair of Vertifans accompanying the strike fighter package to their target zone and plucking any downed aircrew from the ground using a rescue sling. The initial tests conducted during the Phase II evaluations would form the basis for the system development.

With the original Army funding having expired, the XV-5A project was handed to the USAF as the modified aircraft conducted its initial flight tests in August-September 1966. Happy that all was well, USAF test pilots were brought into the project to begin the initial rescue trials. Only days into the programme, however, an unfortunate incident occurred that would finally end military procurement interest in the XV-5A.

On 5 October 1966 the surviving XV-5A crashed at Edwards AFB during hover tests as part of the Strike Escort Rescue programme for the USAF, killing the USAF test pilot Maj. David Tittle. During a hovering manoeuvre the aircraft began an uncontrolled roll to the left. Tittle ejected at a height of 50 ft (15.24 m), but was killed when his parachute failed to open. Investigations revealed that the kapok rescue sling and associated metal hook assembly that

Piloted by Ryan Chief Test Pilot William A. Anderson, the extensively modified XV-5B begins conversion from conventional jet mode to fan mode under the watchful eye of NASA test pilot Ronald M. Gerdes in the accompanying Hiller UH-12. Following these initial test flights the aircraft was delivered to the NASA/Ames Research Center on 28 August 1968.

the XV-5A was carrying as part of the trials was ingested into the port wing lift fan, causing a loss of thrust. As the hook was thrown clear of the fan and penetrated the port side of the fuselage, and Tittle initiated the ejection sequence, the fan began to return to normal operating speed. The aircraft had returned to an almost wings-level attitude when it hit the ground and, although the landing gear collapsed on impact, the crash could have been survivable. The XV-5A project was over having accumulated 338 flights amassing 138 hours of flight time.

XV-4B Hummingbird II

Although the US Army's XV-4A programme was officially dead, Lockheed fully intended to utilise the surviving aircraft further for ongoing VTOL research. After over a year in storage the XV-4A was resurrected as Lockheed investigated other vertical lift systems. Augmented fans (similar to those fitted to the XV-5), direct lift jets and improved augmented jet ejector powerplants were all considered for installation. At this time the USAF was becoming increasingly interested in fielding VTOL jet aircraft and was actively participating in the development of the HS Kestrel/Harrier in the UK. The USAF established the VTOL Flight Control Technology Program under the control of the Air Force Flight Dynamics Laboratory to experiment with various operating procedures, techniques and handling qualities of jet-powered VTOL aircraft. Lockheed's ongoing work with the XV-4A was seen as an ideal match for this research and the US Army agreed to donate the Hummingbird to the Air Force for further research.

In September 1966 Lockheed-Georgia received a USAF contract to extensively modify the XV-4 to incorporate an entirely new propulsion system, variable-stability flight control systems and new all-weather instrumentation. Lockheed had favoured introducing lift fans, however, for expediency, and to meet USAF criteria, it was decided to opt for multiple lift jets to provide the vertical thrust component. Various types and combinations of engines were examined before Lockheed opted for a six-engine solution, four providing lift only and two providing forward thrust and, when necessary, lift through the use of deflector valves.

The engine selected was the General Electric YJ85-GE-19, each unit producing some 3,000 lb (13.35 kN) of thrust. The four lift engines occupied only the central part chamber previously occupied by the augmentation chamber, leaving room for additional fuel. The additional thrust and extra fuel allowed the maximum vertical take-off weight to be increased to 12,600 lb (5715 kg), although the significant increased thrust provided a healthy thrust-to-weight ratio. The added weight did, however, restrict the descent rate to 13 ft (4 m) per sec for vertical landings.

Each of the four lift engines was fitted with a nozzle capable of vectoring 10° fore and aft to provide control during hovering and aid acceleration/deceleration during transitions. To protect the main landing gear from the effects of heat from the engines, Goodrich developed special high-temperature resistant tyres.

High-end performance was also restricted to 260 kt (Mach 0.53) and a g loading of +3/-1.5. An innovative feature fitted to the XV-4B was the Douglas-developed Escapac 1D-3 ejection seat. The seat was designed to provide the pilot with safe egress from anywhere within the aircraft's operating envelope. A 'ballistic spreader device' was incorporated to insure rapid opening of the parachute at zero speeds and low altitude. Unlike in the XV-4A, both cockpit positions were fitted with seats and each pilot had his own set of cockpit controls

A new electro-mechanical fly-by-wire control system with dual redundancy was fitted. A major failure of the fly-by-wire control system would automatically trigger activation of a back-up mechanical system. The reaction jets on the wingtips, nose and tail of the Hummingbird II were modified to operate using engine bleed air only and, unlike on the XV-5A, only functioned when the aircraft's flight situation demanded.

Prior to the fitting of the new lift jets they were extensively trialled on a specially designed rig, enabling the engineers to determine the optimum inlet design. Wind tunnel tests of the airframe were conducted at NASA's Langley Flight Research Center and the Air Force Flight Dynamics Laboratory worked tirelessly to perfect the flight control system.

After almost two years of extensive re-engineering and reconstruction, on 4 June 1968 62-4504 was rolled-out for full-scale engine and taxiing tests prior to the first flight. After some minor alterations to the lift jet nozzles, the first flight, in conventional mode, was made on 28 September 1968 – the first aircraft in the western world to fly with fly-by-wire primary controls. Lockheed was scheduled to complete a range of initial test flights with the aircraft before it was handed over to the USAF for further flight research at Edwards AFB in June 1969.

The test flight programme progressed slowly and steadily until by March 1969 the XV-4B was approaching the phase where the first hovering test flight was to be made. This was never to happen. On a conventional test flight on 14 March 1969 the aircraft suddenly entered a rapid roll while climbing through 8,000 ft (2438 m), after departing Dobbins AFB. The pilot, Harlan J. Quamme, was unable to regain control and ejected, suffering minor injuries.

With both XV-4s now destroyed, the Hummingbird II project came to an end, with the XV-4B never having demonstrated vertical flight. Plans to build a second XV-4B were cancelled as the USAF lost interest in jet VTOL designs – a situation that would remain until the 1990s and the launch of the JSF project.

XV-5B NASA trials

The second XV-5A would, however, be resurrected, despite its crash having ended USAF interest in the programme. 62-4506 was deemed repairable, and in March 1967 Ryan/GE was awarded a $700,000 contract to restore the aircraft to flying condition, incorporating a number of modifications. The landing gear was moved outboard of the wings to improve ground handling and the fuel system, avionics, cockpit layout and ejection seat were all enhanced. Allocated the NASA aircraft number 705 and painted in a smart white livery, the aircraft took to the air again on 15 July 1968 as the XV-5B.

Over the next six years the XV-5B was involved in a wide range of NASA trials examining all aspects of VTOL operations, contributing much to VTOL transport and fighter research programmes, particularly in the fields of terminal area operations and stability and control during VTOL conversion.

Following the XV-5B's retirement from NASA service in 1974 the aircraft was placed into storage before allocation to the Army Aviation Museum at Fort Rucker, Alabama, where it remains today as the sole testament to the engineers, pilots and designers that risked so much in extending the boundaries of vertical flight.

Daniel J. March

All four XV-4s and XV-5s were involved in serious crashes, killing three test pilots. The XV-5A 62-4506 was rebuilt as the XV-5B for NASA and, in contrast to the earlier aircraft, had a good safety record providing valuable data during VTOL and boundary layer research programmes, until January 1974.

WARPLANE CLASSIC

Messerschmitt Bf 110
Underrated Warrior

At the start of World War II the growing Bf 110 *Zerstörer* (destroyer) force was the apple of Göring's eye, and its early achievements justified his faith. However, the mauling suffered by the type at the hands of the RAF in 1940 all but removed it from the Luftwaffe's plans. Nevertheless, through its capabilities as a ground attacker and night-fighter, it clawed its way back to prominence.

Top right: Maps are consulted in this posed mass briefing. The white dragon on a black shield worn by the Bf 110C behind denotes its allocation to I./ZG 52, one of nine Bf 110 Gruppen assigned to the assault on France and the Low Countries.

Below: This aircraft from the Gruppenstab of I./ZG 2 is seen during the French campaign, at a time when mottled grey camouflage began to replace the original dark green scheme.

During and since World War II, several historians have decried the Messerschmitt Bf 110 as a humiliating failure. That analysis is based mainly on the losses suffered while operating as a bomber escort during the Battle of Britain in 1940. Certainly the Bf 110 units took heavy losses from the RAF's Spitfires and Hurricanes, yet it is unfair to single-out the Bf 110, alone, for criticism on that account. No twin-engined fighter of the early-war period – not the French Potez 63, not the Dutch Fokker G.1, certainly not the fighter version of the Bristol Blenheim – could engage in turning fights with modern enemy single-engined fighters with much prospect of success. In truth, the Bf 110 was an extremely versatile fighting machine. Later, as a night-fighter, it would find its true *métier* and it proved a highly effective killer. It also operated successfully as a daylight bomber-destroyer, and as a high-speed reconnaissance aircraft and long-range fighter-bomber.

The Messerschmitt Bf 110 stemmed from a 1934 Luftwaffe requirement for a twin-engined heavy fighter, whose primary role was to clear a path through the enemy's fighter defences for the formations of bombers following it. As a separate role, it was also to be capable of fighting running actions against enemy bomber formations flying over or near Reich territory. For these roles the heavy fighter needed a longer range and a heavier armament, and to carry substantially more ammunition, than could a comparable single-seat fighter type.

The Messerschmitt Bf 110 prototype made its maiden flight on 12 May 1936, and two further prototypes joined the test programme before the end of that year. Those initial prototypes were powered by Daimler Benz DB 600A engines, which developed 986 hp (736 kW) for take-off. The motor suffered several shortcomings, however, and Daimler Benz phased it out of production in favour of the

newer DB 601, which developed 1,050 hp (783 kW) for take-off. That new engine was chosen to power the Bf 110.

Production of the DB 601 was slow, however, and most of those available went to power the Bf 109E fighter that had a higher priority. As a result, the Bf 110A-0, the B-0 and the B-1 pre-production aircraft were fitted with Junkers Jumo 210D engines that developed only 680 hp (507 kW) for take-off. Although this powerplant was less than ideal, it meant that this interim type could go into limited production, and be sent to test establishments and fighter training schools to prepare the way for production variants to enter front-line service as rapidly as possible.

Only at the end of 1938 did the DB 601A engine become available in quantity. The Luftwaffe received the pre-production batch of 10 Bf 110C-0 aircraft early in 1939, followed by the C-1 full production variant. By the end of August 1939 the Luftwaffe had accepted 159 examples of the C-1. These equipped the three Zerstörer units – I. (Zerstörer) Gruppe of Lehrgeschwader 1, I. Gruppe of Zerstörergeschwader 1 and I. Gruppe of Zerstörergeschwader 76. On 2 September 1939, the day after the invasion of Poland, the three Gruppen reported having a total of 95 Bf 110Cs on strength, of which 82 were serviceable. Seven other Gruppen had also been designated as Zerstörer units, but due to the shortage of Bf 110s these units went into action over Poland flying Bf 109s.

First combat

Over Poland the Bf 110 was initially employed in the bomber-escort role, in which it proved highly successful. The *Zerstörer* pilots found that, provided they maintained high speed and did not attempt to dogfight with the slower but more manoeuvrable Polish PZL 11 fighters, they could prevent the latter from engaging the bombers effectively. In

Above: Bf 110Cs of 3./ZG 2 patrol along the Channel coast – probably France – in 1940. At the time most of western Europe had fallen under German control with relative ease, and the Bf 110 had played a significant part in the lightning victories. However, a very painful lesson awaited the Zerstörerwaffe across the English Channel. The Luftwaffe began the Battle of Britain with 237 serviceable aircraft, and over 220 Bf 110s were lost during the campaign – most in air combat with the RAF.

Above: A schwarm of Bf 110Cs is seen on patrol. The aircraft was only really effective as a fighter at medium to high altitudes, and then only against poorly flown opponents in inferior equipment.

Right: Camaraderie remained high in the Zerstörerwaffe throughout its checkered history. The use of the famous Wespen (wasp) badge was begun by III./ZG 76 in Norway, and was continued by SKG 210 and ZG 1 on the Eastern front. This is a Bf 110D-3.

Warplane Classic

Right: The Bf 110 V1 was the first of the breed, undertaking its maiden flight on 12 May 1936 at the Messerschmitt company airfield at Augsburg-Haunstetten, with Rudolf Opitz at the controls. The main early problem was a swinging on take-off and landing, requiring undercarriage modifications.

Prototypes and pre-production

Messerschmitt built three prototypes and four pre-production Bf 110As.

Above: This is the first of four Bf 110A pre-production aircraft, sporting the Junkers Jumo 210Da engines which were substituted after the poor performance of the DB 600A had been revealed in the three prototype machines. Like the DB 600, the Jumo was characterised by a deep chin intake. Note the original nose profile, which was changed to a more pointed and less drooped shape from the fourth and final Bf 110A.

two weeks of intensive combat the Polish fighter force was virtually wiped out. That allowed the Bf 110 units to experiment with a new role, that of strafing Polish army units pulling back to regroup for the defence of Warsaw.

By then Great Britain and France had declared war on Germany, and it was important to guard against the potential threat in the west. So, from mid-September, several Luftwaffe units moved out of Poland to take up defensive positions in western Germany. The three Bf 110 Gruppen were included in those units re-deployed in that way.

During the Polish campaign the Luftwaffe had reason to be well pleased with the Bf 110. It had been effective in both the bomber-escort and ground-strafing roles, and during the short campaign losses amounted to just a dozen Bf 110s. On 27 September the remaining Polish forces laid down their arms.

Above and right: Sporting the aerodynamically cleaner nose section, these aircraft are Bf 110B-0s, which were delivered from Augsburg without armament in the early summer of 1938. The civil-registered aircraft in the line-up (D-AISY) was the first Bf 110B-0, which first flew on 19 April 1938. Armed B-1s were completed from July.

The Battle of the Bight

So began the so-called 'Phoney War' or 'Sitzkrieg', the period of relatively little activity as both sides strove to build their strength in readiness for the struggle ahead. On 18 December 1939 that calm ended when the RAF sent 24 Wellington bombers to conduct a daylight armed reconnaissance mission over the Heligoland Bight area. Two bombers returned early with technical troubles, but the remainder continued with the mission. At that time the RAF was prohibited from attacking land targets in Germany, since this might endanger the lives of civilians. For that reason, the bombers were permitted to attack only enemy naval units located at sea. None was found, and the bombers turned around and headed for home.

A Freya radar sited on the island of Wangerooge had observed the raiders' approach, and 16 Bf 110s of I./ZG 76, together with 34 Bf 109s, took off to engage. It was a clear day and the defending fighters quickly caught up with the retreating bombers. The Wellingtons held tight formation as their gunners tried to drive off the attackers with a defensive crossfire, but their rifle calibre machine-guns lacked the punch to do so. At that stage of the war RAF bombers lacked both armour protection for vital areas and self-sealing fuel tanks. As a result, the fighters' 20-mm cannon proved to the highly destructive, and bomber after bomber went down. Others suffered damage and had to leave the protection of the formation. That was the position when Leutnant Gustav Uellenbeck and three other Bf 110 pilots joined the fray. Uellenbeck later reported:

"I was with the second formation, flying on a heading of 120° about 50 km [31 miles] north of Ameland. Suddenly we came across two Wellingtons flying 300 metres [984 ft] below us on the opposite heading. I attacked the leading bomber from the side and it started to burn. I then attacked the second one from the left and above. When he didn't go down I sat off, 300 metres behind him, and fired all my weapons. Shortly afterwards he dived into the sea. At this time I received a bullet wound between the neck and the left shoulder. The radio operator was hit by the same bullet on his left wrist. I then returned to Jever."

Of the 22 Wellingtons taking part in the action only 10 regained the shores of Britain, and two of those crashed on landing. Bf 110Cs of I./ZG 76 were credited with the destruction of nine bombers, the rest went to the Bf 109s. Luftwaffe losses amounted to two Bf 109s shot down, and several fighters returned with minor damage. It had been a clear-cut victory for the defenders, and for the next three years no large enemy bomber formation would dare to violate German airspace in daylight.

New roles for the Bf 110

During the first six months of 1940, Bf 110 production averaged just over a hundred aircraft per month. That provided sufficient to bring the Zerstörer force up to its full strength at 10 Gruppen, and it allowed the type to be introduced into other roles.

In the late spring of 1940 the Luftwaffe established a fighter-bomber unit, Erprobungsgruppe 210, with two Staffeln of Bf 110s and one of Bf 109s. This unit operated mainly the C-4 variant of the Bf 110, with racks to carry two 250-kg (551-lb) bombs under the fuselage. The unit also received a few (about three) examples of the C-6 variant,

By the time of the invasion of Poland only one Zerstörergruppe (I.(Z)/LG 1) had fully converted to the Bf 110C. Both I./ZG 1 and I./ZG 76 continued to operate the older Jumo-powered Bf 110B-1 model. Here a group of 2./ZG 1 aircraft is seen while escorting a formation of Ju 87 Stukas on a raid against Warsaw on 8 September 1939.

fitted with a 30-mm Rheinmetall MK 101 cannon in a blister mounting under the fuselage. The new unit's role was to mount fighter-bomber operations and develop an effective pattern of tactics for these.

In addition, just over a score of Bf 110C-5s formed part of the equipment of four reconnaissance Gruppen. These aircraft had their 20-mm cannon removed, and had a vertically mounted Rb 50/30 reconnaissance camera installed mid-way along the cabin.

Scandinavian adventures

The 'Phoney War' came to an abrupt end on 9 April 1940. With improving weather, the Wehrmacht made its first move in what was intended to be a rapid conquest of western Europe, the occupation of Denmark. Led by Hauptmann Wolfgang Falck, I./ZG 1 and its 26 serviceable Bf 110Cs was tasked with escorting bombers in a show of force over Danish cities, covering paratroop drops from Ju 52s, and occupying several Danish airfields. The occupation ran smoothly, with one Danish Fokker D.XXI shot down by Falck as it tried to intercept, and numerous other aircraft being destroyed in strafing attacks.

I./ZG 1 landed at Aalborg, where it remained for some weeks, action being restricted mainly to a few early attempts at using the Bf 110 in the night-fighter role against RAF bombers that occasionally flew over Denmark.

On the same day as Denmark was occupied, the Wehrmacht began its operation to secure Norway. Leading the Zerstörers north was Hauptmann Günther Reinecke, Gruppenkommandeur of I./ZG 76, but the opening day of the Norway campaign ended in shambolic scenes of confusion. The Gruppe had 29 serviceable Bf 110Cs available, and 16 of them launched from Westerland-Sylt to escort troop-carrying Ju 52s. 1. Staffel headed towards the airfield at Oslo-Fornebu, while 3. Staffel made for Stavanger-Sola. The other elements of I./ZG 76 moved forward to newly occupied Aalborg to act as a reserve.

Range was critical to the Norwegian mission – despite its good range performance the Bf 110 had insufficient fuel to return to Germany or Denmark. As the two Staffeln launched they encountered thick cloud and fog, leading to great confusion. The 3. Staffel leader ordered his aircraft to cancel the mission before they reached the point of no return, but only four responded. Two others collided in mid-air, and two continued to Stavanger where they landed after the paratroops had dropped.

Meanwhile, 1. Staffel continued on to Oslo, but without the first wave of Ju 52s that was to drop paratroops. The Bf 110s received an abort message, but they had flown too far and had no choice but to continue. As they approached Oslo they were engaged by Norwegian Gladiators, and two from each side were shot down. The remaining Bf 110s strafed Fornebu airfield and awaited the Ju 52s. Of course, the first wave had turned back, and the second wave had been instructed to land rather than drop their troops. As they landed the defenders on the ground opened up. Short on fuel, the 1. Staffel Bf 110s elected to land. One ran off the end of the runway, but the others got down. They were deployed on the ground so that their guns could be used

With their brief front-line careers ended by deliveries of the definitive Bf 110C, the majority of B models were reworked to the Bf 110B-3 version. They were issued to Zerstörer training units, often with all armament removed and extra radio equipment installed. This aircraft was assigned to the Ergänzungs (Zerstörer) Staffel in autumn 1940. The B-3s stayed in service until at least 1941.

Above: A pre-production Bf 110C-0 is seen prior to delivery to the Luftwaffe for trials. The smoother lines of the DB 601 cowl are apparent, as are the blast troughs for the two MG FF 20-mm cannon.

Left: The Bf 110C was given production priority in 1939, with the result that Gotha and Focke-Wulf established production lines to augment Messerschmitt output. This is the 25th Bf 110C-1 built by Focke-Wulf at its Bremen works. Note the access ladder to the rear cockpit.

Warplane Classic

Assault on Poland

All of the Luftwaffe's 90 serviceable Bf 110s were thrown into the attack on Poland, including those assigned to the evaluation unit, which operated from the isolated German enclave of East Prussia.

Above: A 1./ZG 1 Bf 110C undergoes an engine change at the height of the Polish campaign – just visible are the two thin white bands used by the Luftwaffe as a campaign identification marking.

Below: The wolf's head badge on the nose and 'L1' codes identify this Bf 110 as belonging to I.(Z)/LG 1.

Below right: This Focke-Wulf-built Bf 110C-1 was tested with large chin radiators and the oil coolers moved to the top of the engine cowlings.

Below: After the end of the Polish campaign the rush was on to convert several more Gruppen to the Bf 110. Here I./ZG 52 crews practise formation flying over the Alps in their new mounts in February 1940.

against the defences. While this was going on, the defenders received the order to withdraw, and Norwegian resistance in the south subsequently ended.

I./ZG 76 regrouped at Stavanger in anticipation of the push north. A few RAF coastal patrol aircraft were intercepted and shot down before the Anglo-French landings in central Norway on 18/19 April. Supporting them were RAF Gladiators flown in from the aircraft-carrier HMS *Furious*. For the remainder of the month and into early May the landing forces were driven out until just a small enclave remained at Narvik, in the far north.

To cater for the enormous distances involved in flying bomber escort missions to Narvik, a long-range modification had been hastily prepared. This consisted of a 1175-litre (310-US gal) fuel tank mounted under the belly, covered by a wooden fairing. Dubbed 'Dackelbauch' ('Dachshund belly') by the crews and officially designated Bf 110D-1/R1, the long-range fighter was issued to a special Staffel organised within I./ZG 76, which began operations from a forward base at Trondheim-Vaernes on 18 May.

Disliked by its crews, the 'Dackelbauch' nevertheless scored well against the RAF's Gladiators and Hurricanes during the remainder of May, up to the evacuation of British forces from Narvik in early June. I./ZG 76 remained in Norway to defend the newly occupied territory, participating in the infamous 13 June slaughter of an attack by Blackburn Skuas on *Scharnhorst*, which effectively ended the career of the dive-bomber in Royal Navy service.

Despite the bungled start, the Norwegian campaign highlighted the Bf 110's effectiveness in operations against thinly deployed defences over great ranges. As described later, I./ZG 76 would soon have a rude awakening, though, during its one attempt to influence the course of the battle over Britain.

France and the Low Countries

A month after the two Zerstörergruppen had headed north, a much larger campaign was launched into France, Belgium and the Netherlands. Initially equipped with Bf 109s, the remainder of the Zerstörerwaffe had converted to the Bf 110C. With the recall of I./ZG 1 from Denmark, nine Gruppen were available for the new campaign, which began on 10 May.

Bf 110s were employed in the roles for which they had already become highly regarded. They escorted bombers on deep strikes, and they set out to cripple the opposition's airpower during strafing raids on airfields. In the Netherlands they strafed airfield defences in advance of paradrops at Rotterdam and Hamstede. Victories began to mount rapidly, but so did losses, especially to RAF Hurricanes – a harbinger of the mauling the Zerstörerwaffen would face over England.

Tactics became an increasingly important part of the air battle, the Germans' well-drilled manoeuvres often carrying the day against less disciplined opposition. When threatened, Bf 110 formations would quickly adopt a defensive 'wagon circle', in which the forward- and aft-firing armament could protect the formation from attack on all sides. As the fighting neared the Channel coast, Bf 110s increasingly encountered RAF Spitfires, but these engagements were far from one-sided. Intense air battles continued until the night of 2/3 June, when British forces completed their evacuation from Dunkirk.

Although the Bf 110s had proved generally successful in the Luftwaffe's operations over western Europe – especially against the Dutch and Belgians – they had suffered over 60 losses and the air of invincibility gained in the Polish and Scandinavian campaigns had been threatened. Serious suspicions had been raised as to the survivability of the Bf 110 in the face of well-flown (if poorly organised) French and British single-seat opposition. Yet, for now, the reputation of the Zerstörerwaffe remained intact, if a little shaken.

For the remaining days of the campaign the Bf 110s chased occasional French aircraft until the final capitulation. During this period Bf 110s also found themselves

Above: This Bf 110 served with 1./ZG 76 and is seen at its base in Norway in August 1940. After the eviction of Anglo-French forces from Norway had been completed in June, I. Gruppe remained to provide coastal patrols. It also flew one ill-advised mission during the Battle of Britain.

Denmark and Norway

Left: This staged propaganda photo shows I./ZG 1 commander Wolfgang Falck's eighth kill – a Danish Fokker D.XXI – being 'painted' on the tail of his Bf 110C. Close inspection reveals the other victory bars to represent three Polish and four British victims.

fighting an unusual opponent – the Swiss air force. Operations over France had resulted in a number of incursions into neutral Swiss airspace by German aircraft, and a few had been shot down by Swiss Bf 109s. On 4 June an irate Hermann Göring launched a retaliatory mission in which a large number of Bf 110s deliberately flew over Switzerland to entice up the defences. One Bf 110 and a Swiss Bf 109 were lost in the ensuing fight. Four days later a similar mission resulted in the loss of another Swiss Bf 109 – and four Bf 110s! No such further missions were launched.

Battle of Britain

Despite the strong hints emanating from the latter stages of the battle for France, the Bf 110 was thrown headlong into the fray over Britain, and during August 1940 the Bf 110 suffered its first major setback in action. In the course of operations over England in the bomber-escort role, Bf 110 units faced a head-on confrontation with RAF Fighter Command. The latter was a well-established force, flying modern fighter types. Whenever the German twin-engined fighters engaged in stand-up fights with the Spitfires and Hurricanes, they usually got the worst of the encounter.

At that time Leutnant Joachim Köpsell flew Bf 110s with I. Gruppe of ZG 26; he told this writer of some of the problems he faced when operating the aircraft in the bomber escort role: "When, during our fighter training we had flown other fighters – biplanes – we knew how a fighter should handle. Then the 110 came as a disappointment, it was so heavy it was difficult to manoeuvre into a firing position.

"Usually flying close formation with bombers meant flying about 100 metres [328 ft] away, throttled back to maintain position on the bombers. Once enemy fighters started to attack we had to accelerate to fighting speed. This took time, because the 110 had poor acceleration."

Although the heavy fighters had a difficult time when they tried to engage RAF fighters, their pilots had one escape tactic they felt they could rely on if hard-pressed. The Bf 110's Daimler Benz engines employed fuel injection, which meant that they continued to run if the pilot pushed forwards on the control column and applied negative g. The Merlin engines fitted to RAF fighters, on the other hand, were fitted with float carburettors and these engines cut out whenever the pilot applied negative g.

Joachim Köpsell described one occasion when these factors saved him and his aircraft from almost certain destruction during a bomber escort mission over Essex:

"My radio operator reported a Hurricane diving on us from the left. I dived away, and each time he opened fire I jerked the control column forwards. When the Hurricane pilot tried to follow, his engine cut out and he dropped back. This happened three or four times until in the end we were both going down almost vertically towards the sea. The Hurricane broke off the chase, and I had to haul on the stick to pull out of the dive close to the surface."

Köpsell levelled off close to the water, and the Hurricane was nowhere to be seen. His aircraft had taken some hits, and the port main wheel tyre had been shot through. On touching down at his base at Clairmarais near St Omer, the aircraft swung violently through three-quarters of a circle, before coming to rest with no further damage.

For Erprobungsgruppe 210, operating against targets in southern England in the fighter-bomber role from Calais-Marck airfield, things went well at first. Initially the unit delivered attacks on pinpoint targets, the aircraft descend-

Hastily developed to meet the needs of the Norwegian campaign, the 'Dackelbauch' plywood and fabric belly tank was first tested by the Bf 110D-0 (illustrated), before fielding on the Bf 110D-1/R1. The tank was intended to be jettisoned when empty or when fighters were encountered, but in practice it frequently 'hung up'. The effect on handling and performance was significant, the aircraft being very difficult to control on take-off with full fuel. As the tank emptied, the sloshing fuel inside also had an adverse effect on handling. When the tank was 'empty', it remained full of highly explosive petrol fumes, the ignition of which was believed to be the cause of a number of mysterious disappearances.

Left: A Bf 110D-1 of 2./ZG 76 taxis at its Norwegian base. Following I./ZG 76's disastrous raid during the Battle of Britain, the 'Dackelbauch' tank was swiftly abandoned in favour of a simpler arrangement in which a standard drop tank was hung from the centreline. Later Bf 110D langstrecken Zerstörer (long-range destroyer) variants introduced drop tanks under the wings.

A pair of Bf 110Cs from 1. Staffel/ZG 2 flies over occupied France. Led by Hauptmann (later Major) Johannes Gentzen – the most successful pilot of the Polish campaign, ZG 2's I. Gruppe was based at Darmstadt prior to the May campaign, moving forward into Belgium and ending up at Caen-Carpiquet in time for the assault on Britain. Gentzen was killed in Belgium while attempting to intercept two RAF Blenheims that had just bombed 'his' airfield.

ing in shallow dives from 10,000 to 3,000 ft (3048 to 914 m) to deliver their bombs. Then, on the morning of 12 August, the pace of operations was stepped up when the unit's Bf 110s launched attacks on the Chain Home radar stations at Dunkirk (in Kent), Pevensey and Rye. All three radar stations suffered damage and all except Dunkirk had to go off the air for a few hours while repairs were affected. All the raiders returned from the mission.

After landing, the aircraft were quickly refuelled and rearmed. Then 14 Bf 110s and seven Bf 109s from the Gruppe took off to attack the Fighter Command airfield at Manston. The raiders cratered the landing area in several places rendering it temporarily unserviceable, and inflicted damage on two hangars and some other buildings. Again, the raiders all returned safely.

Later that afternoon, Erprobungsgruppe 210 was in action yet again, this time against the Fighter Command airfield at Hawkinge. The attack wrecked one hangar and the station workshops, but the airfield continued in use. Yet again, all the raiders returned safely. By any standards, 12 August was a successful day for Erprobungsgruppe 210.

Black day for the Zerstörerwaffe

Three days later, it was a different story. For the sort of tactics the unit used, the margin between success and failure can sometimes be narrow indeed. Late on the afternoon of the 15th, 15 Bf 110s and eight Bf 109s set out to attack the important Fighter Command sector airfield at Kenley. One Bf 110 returned early with technical trouble, the others continuing with the mission. Probably due to a bank of haze, the unit commander and attack leader, Hauptmann Walter Rubensdörffer, missed Kenley and instead led the attackers to the nearby airfield at Croydon. As the Messerschmitts commenced their attack, two squadrons of Hurricanes caught up with them. In the ensuing *melée* six Bf 110s, including that flown by Rubensdörffer, and a Bf 109, were destroyed and two further Bf 110s suffered battle damage. Of those aboard the aircraft six were killed and seven were taken prisoner.

Earlier on the 15th, other Bf 110 units had also taken a

RAF combat appraisal

On 21 July 1940 a Bf 110C-4 of 4. Staffel of Fernaufklärungsgruppe 14 (5F+CM), conducting a photographic reconnaissance mission over southern England, came under attack from Hurricanes of No. 238 Squadron and force-landed near Chichester. The aircraft was almost intact and RAF technicians restored it to flying condition using parts salvaged from another Bf 110. After the camera installation was removed, the aircraft was ballasted to represent the weight of the two cannon and ammunition for all weapons carried by the fighter version.

In February 1941 the aircraft was taken on RAF charge as AX772, and test-flown against a representative selection of RAF aircraft. Excerpts from the report on the Air Fighting Development Unit trial are given below. Note that the RAF report always referred to the aircraft as an Me 110; the Bf designation for Messerschmitt fighters did not become generally known in Britain until some years after the war.

FLYING CHARACTERISTICS
General: The aircraft is very pleasant to fly and easy to take off and land. It handles more like a single-engined fighter than a twin, the controls being comparatively light at all speeds. It was found that when dived at an indicated air speed of 340 mph [547 km/h] the controls do not stiffen appreciably and the aircraft is still fully manoeuvrable.
Performance: Comparative speed trials were carried out with a Hurricane Mk I and a Spitfire Mk VB. Trials were not carried out above 20,000 ft [6096 m], as the performance of the engines of the Me 110 was not reliable above this height. At 5,000 ft [1524 m] the Hurricane was 15 to 20 mph [24 to 32 km/h] faster than the Me 110 but at 20,000 ft the Me 110 was 2 or 3 miles faster than the Hurricane. The Spitfire was considerably faster at all altitudes, the difference being about 40 mph [64 km/h] at

20,000 ft. The rated altitude of the Me 110 was about 17,000 ft [5182 m].
Instrument flying: The aircraft can be trimmed to fly 'hands-off' and this, combined with the well-balanced controls and good view of the instrument panel, makes instrument flying easy.
Single-engined flying: The aircraft flies well on either engine. It is able to climb and maintain height easily without feathering the propeller. With the propeller [of the stopped engine] feathered, the single engine performance is improved considerably, and the aircraft may be turned comfortably both with and against the live engine.
Low flying: This aircraft is very suitable for low flying due to its good manoeuvrability, handling qualities and the pilot's view.
Formation flying: The aircraft is easy to handle in formation due to the pilot's good view, good throttle response and decelerations.
Search: The all-round view from the pilot's cockpit is exceptional, an unusual feature being that when looking to the rear it is possible to see the whole tail

The 4.(F)/14 Bf 110C-4 (Werk Nummer 2177) which force-landed at Goodwood in July 1940 is seen after repair using parts from another crashed aircraft.

unit through the Perspex canopy. The usual German practice of fitting flat panels gives the crew an undistorted view. The view from the rear cockpit is also excellent.
Slipstream: The slipstream at 300 ft [91 m] is strong and drops about 5° below the aircraft. At 600 ft [183 m] it is dead astern and at its strongest. It then gradually diminishes until at 900 ft [274 m] it is not likely to upset the aim of a fighter pilot.
Manoeuvrability: The Me 110 is very manoeuvrable and no difficulty was found in placing and holding a bomber target in the sight. Results of ciné films taken show very steady shooting. Its manoeuvrability was compared with that of a Hurricane Mk I and a Spitfire Mk VB at various altitudes.
Turning: The Me 110 was positioned immediately behind each of the other aircraft in turn. The leading aircraft than turned as quickly as it could with the Me 110 attempting to follow. The number of turns taken for the leading aircraft to position itself on the tail of the Me 110 was noted. It was found that each of the single-seater aircraft could easily out-turn the Me 110 at low altitude, but the advantage was less marked as height was increased up to 25,000 ft [7620 m] when the average number of turns required was four.
Diving: The controls do not stiffen appreciably during a dive and with direct injection fuel systems the engines do not cut under application of negative g. To test the tactical aspect of this, the Me 110, Hurricane

Werk-Nr 2177 was tested by the RAE before passing to the Air Fighting Development Unit (No. 1426 Flight) at Duxford. It later joined the Enemy Aircraft Flight at Tangmere. Traces of the original Luftwaffe code (5F+CM) still show through the RAF camouflage.

severe drubbing from RAF Fighter Command. Shortly after mid-day a force of 63 Heinkel He 111 bombers of KG 26, escorted by 21 Bf 110D-1s of I. Gruppe of ZG 76, headed across the North Sea to attack the RAF bomber airfields at Dishforth and Linton-on-Ouse. Planners at Luftflotte 5 headquarters at Oslo expected that, by this stage of the battle, RAF Fighter Command would have pulled most of its units to southern England, leaving the north almost devoid of fighter protection.

How wrong they were. As the raiders neared the coast, two squadrons of Spitfires and parts of two squadrons of Hurricanes pounced on them. Each Bf 110D-1 carried a big 1175-litre (310-US gal) fixed tank in a streamlined fairing under the forward fuselage. This fixture reduced the twin-engined fighter's speed and manoeuvrability. During the ensuing action the first Bf 110 to go down was that flown by Hauptmann Restemayer, the Gruppenkommandeur. Probably his aircraft had taken hits on the auxiliary tank, which by then was empty of gas but full of explosive vapour; the fighter blew up violently and fell into the sea, taking both crewmen with it. In the minutes that followed seven more Bf 110s followed it down, as did eight of the He 111 bombers they were trying to escort. For their part the Bf 110s made several claims, but RAF records confirm the loss of only one Hurricane during the engagement.

Also on that day, Bf 110Cs of the other two Gruppen of ZG 76, and the Stab flight, engaged in operations over southern England. They, too, suffered heavy losses: 10 were shot down, and two more made it back to France damaged beyond repair. ZG 2 also lost a Bf 110 that day, to bring the total loss of this type to 28 aircraft.

Above: A schwarm of V.(Z)/LG 1 aircraft flies over Paris, the Arc de Triomphe being visible in the lower left corner. One of the aircraft is still in the old dark green scheme.

Above left: A ZG 26 Bf 110C wheels over the French countryside. The Geschwader provided three Gruppen and the Geschwaderstab for the May 1940 campaign.

Mk I and Spitfire Mk VB were flown in line abreast. The control column of the Me 110 was then pushed forward very quickly. The Hurricane and Spitfire lost a few lengths as their engines cut, but were able to make up the lost distance rapidly and hold the Me 110 throughout the remainder of the dive.

During the tactical trials it was found impossible for the pilot of the Me 110 to maintain maximum possible output from the engines and keep the rpm within the limits when engaged in air combat. The pitch of the propellers is infinitely variable and is changed manually by the pilot who has to hold two electric switches in an 'up' or 'down' position while the pitch is changing. The rate of change of pitch appears very slow compared with the rapid variation in rpm, when the aircraft is suddenly put into a dive or climb. This is particularly noticeable in the dive and during dogfighting when it is very easy to exceed the maximum permissible rpm.

ATTACKS

Astern attacks: A fighter attacking the Me 110 from astern can avoid the return fire from the single MG 15 machine-gun providing it keeps below the tailplane. While some protection is given to the crew by armour plate, the engines are unprotected.

Quarter and beam attacks: Return fire can be avoided if the attack is made from below. From level or above up to 60°, return fire from the rear gun can only be effective when the fighter is between 30° and astern. If the Me 110 is taking evasive action it has been found that, due to the limitations of the free gun mounting and the effect of g, accurate shooting is extremely difficult. The pilot and crew have no armour protection from these attacks.

Head-on attacks: The Me 110 pilot and crew are well protected with armour plate from frontal attacks and an attacking fighter would of course be met with the return fire of the fixed armament of four machine-guns and two cannon.

EVASIVE ACTION

Comparative climbs were carried out with Hurricane Mk I and Spitfire Mk VB at heights up to 20,000 ft. It found that there is little difference between the rates of climb of the Me 110 and Hurricane, but the Hurricane fell away when it attempted to follow and shoot in the climb, owing to the much steeper [climbing] angle of the Me 110. The Spitfire can easily outclimb the Me 110, but does so at a faster airspeed and less steeply. If it attempted to follow the Me 110 it sacrificed its superior rate of climb. Trials showed that at heights below 5,000 ft [1524 m] it could follow and shoot with reasonable accuracy, but it fell behind as height was gained up to about 15,000 ft [4572 m]. Above this height, however, due to the state of the Me 110's engines and the fact that the Spitfire was reaching its rated altitude, it was able to close range on the climb. It was thought that, had the Spitfire attempted to fire its cannon between 5,000 and 15,000 ft, it would have stalled as its airspeed was very low. Generally, therefore, if the Me 110 attempts to evade in a long climb the fighter should not follow but should gain height at its own best speed in order to avoid being at a tactical disadvantage.

As the trials have shown, the Me 110 can be easily outmanoeuvred, especially at low altitudes by single-engined fighters, but, having good deceleration and a steep climbing angle, it can temporarily break off from the attacking fighter by pulling its nose up to its steepest climbing position, thus causing the attacker to overshoot. This does not give the Me 110 any tactical advantage if the fighter comes in to attack again quickly.

If a fighter is positioning to attack from astern and is seen early by the crew of the Me 110, a quick turn through 180° when the fighter is about 3,000 ft [914 m] away will give the fighter a difficult head-on attack and allow the Me 110 a chance of using its front armament. Although this manoeuvre can be repeated several times, sooner or later the fighter will be able to close range and obtain the advantage.

Right: As well as the Bf 110C-4 that was captured in July 1940, the RAF acquired various other Bf 110s. This aircraft force-landed near Mosul in Iraq during the brief Luftwaffe intervention in April 1941. It was towed to Habbaniyah where it was repaired and given the serial HK846 and the name The Belle of Berlin. *Having flown at Heliopolis in Egypt for some while, it set off for South Africa but crashed in Sudan.*

Below: Two Bf 110G-4/R3 night-fighters that came to the UK were 730037 (right) and 730301 (left). The latter was a I./NJG 3 aircraft captured intact at Grove in Jutland, and is now on display in the RAF Museum at Hendon. It is the most authentic Bf 110 to have survived.

This Bf 110C wears the well-known 'Bernburger Jäger' badge of I./ZG 2. The Gruppe was heavily involved in the Battle of Britain, flying bomber-escort missions in which it suffered severe losses. Gruppenkommandeur Ernst Ott was killed on 11 August.

Battle of Britain

This campaign was to be the last time that the much-vaunted Zerstörerwaffe were deployed *en masse*. Operations began in July 1940 aimed at denying the enemy the use of the Channel, before Adlertag (Eagle Day) – the campaign against the RAF – began on 13 August. Losses were so heavy over the next six days that Bf 110s appeared much less frequently over Britain after 18 August, but when they did losses remained high. The reduced operational rate was due entirely to a lack of aircraft, with raids being mounted only when sufficient aircraft could be assembled.

This formation of 1./ZG 26 aircraft is seen at around the time of the Battle of Britain. The parent I. Gruppe was based at St Omer and Abbeville, with the White Cliffs of Dover just minutes' flying time away. The Geschwader suffered terribly at the hands of the RAF, losing 15 aircraft in a single day. The much-vaunted defensive circle tactic – occasionally involving up to 60 Bf 110s – had the positive effect of attracting RAF fighters away from the German bombers, but only so they could line up to decimate the cumbersome Bf 110s.

The slaughter continued on two of the three days that followed. On the 16th the Bf 110s were committed in smaller numbers, and eight were destroyed. On the 17th, poor weather prevented large-scale air operations. Then, on the 18th, it was the turn of ZG 26 to suffer a drubbing: it lost 15 Bf 110s. A reconnaissance Bf 110 of Lehrgeschwader 2 also failed to return, to bring the day's loss to 16 aircraft. Thus, during those four days the Bf 110 force lost 52 aircraft destroyed, nearly a quarter of the number in front-line service at the beginning of the period.

In the weeks to follow there were several days when the Bf 110 fighters and fighter-bombers did not operate at all, though when they did they continued to suffer serious losses. As late as 27 September these units were still being committed to battle, and on that day they lost 20 Bf 110s in action.

First night-fighter operations

In May 1940 the RAF began attacking targets in Germany at night, in retaliation for the bombing of Rotterdam. At the time the Luftwaffe had no effective night-fighter arm, night air defence being the responsibility of the Flak units. The latter proved unequal to this task and on 20 July Reichsmarschall Göring ordered Oberst Josef Kammhuber to set up a night-fighter force. The first unit formed, Nachtjagdgeschwader 1, included one Gruppe and part of a Gruppe equipped with Bf 110C-4s.

From these small beginnings the Luftwaffe night-fighter force expanded rapidly. By the end of 1940 it had a strength of 165 aircraft, mainly Bf 110s. At this time Luftwaffe night-fighters carried no special equipment for the new role, and the day- and night-fighter versions of the Bf 110C were interchangeable.

Initially the night-fighters delivered most of their attacks on RAF bombers that had been illuminated by searchlights. It soon became clear that this method was a slave to the weather, however. Anything more than a touch of cloud cover gave problems for the defenders.

To overcome this limitation, Kammhuber – now promoted to Generalmajor – introduced a system of ground radar control code-named Himmelbett (four-poster bed). By the end of 1941 a line of fighter control radar stations had been erected, with one station every 20 miles or so to form a barrier through which the raiders had to pass. The barrier was shaped like a giant inverted sickle: the 'handle' ran through Denmark from north to south, and the 'blade' curved through northern Germany, Holland, Belgium and eastern France to the Swiss frontier.

Each control station employed a Freya radar to provide 360° area surveillance out to a range of about 100 miles (160 km). The Freya operators directed into position two Giant Würzburg pencil-beam radars, each with an effective range of about 40 miles (64 km). One of the latter tracked the movements of the fighter, while the other tracked the bomber. From the plotting table at each ground station, the fighter controller radioed instructions to direct the night-fighter on to its target.

Under the Himmelbett system the night-fighter units' operational areas were clearly defined, and generally within 30 miles (48 km) of their base. In this role the rela-

Bf 110 operational flying units, 17 August 1940

On 17 August 1940, shortly after the Battle of Britain had begun in earnest, the Luftwaffe had a total of 415 Bf 110s serving with front-line units, of which 277 were serviceable. The lion's share of these aircraft – 300 – served with the long-range fighter/escort fighter units. The remaining 115 were divided between the newly established fighter-bomber, night-fighter and reconnaissance units. The Lehrgeschwader was a tactical development unit which operated as a regular front-line unit.

LONG-RANGE FIGHTER/ESCORT UNITS	Total	Serviceable
Lehrgeschwader 1		
V. Gruppe	36	24
Zerstörergeschwader 2		
Stab	4	2
I. Gruppe	33	19
II. Gruppe	36	23
Zerstörergeschwader 26		
Stab	2	2
I. Gruppe	31	21
II. Gruppe	36	29
III. Gruppe	33	21
Zerstörergeschwader 76		
Stab	2	2
I. Gruppe	34	20
II. Gruppe	30	24
III. Gruppe	23	12
FIGHTER-BOMBER UNIT		
Erprobungsgruppe 210[1]	20	8

NIGHT-FIGHTER UNITS	Total	Serviceable
Nachtjagdgeschwader 1		
Stab	3	2
I. Gruppe	33	24
III. Gruppe[1]	33	20
RECONNAISSANCE UNITS		
Lehrgeschwader 2		
7. Staffel[2]	9	8
Aufklärungsgruppe 11		
2. Staffel[2]	5	4
Aufklärungsgruppe 14		
4. Staffel[2]	6	6
Aufklärungsgruppe 22		
1. Staffel[2]	6	6

Notes
1. unit also operated Messerschmitt Bf 109s
2. unit also operated Dornier Do 17s

II./ZG 76 adopted sharkmouth ('Haifisch') markings, as well as the flags of the countries in which it had operated, as worn on this Bf 110C-4 (above). Below is a group of Erprobungsgruppe 210 fighter-bombers.

Messerschmitt Bf 110

tively nimble Bf 110 gained the edge in popularity over the less handy Ju 88s and Do 17s that were also employed.

Bf 110 in the Balkans

After the near decimation of the Bf 110 force during the Battle of Britain, the Zerstörerwaffe was systematically dismantled in the months that followed and the units were re-roled. Many were given night-fighting roles to challenge the growing frequency and strength of RAF night raids, while three Gruppen were given coastal patrol roles around the German Bight (II./ZG 76), Norway (III./ZG 76) and Sicily (III./ZG 26). However, events in the south and east would see the restoration of the Zerstörerwaffe as a fighting force in areas where the opposition was generally weak, and operational ranges were great.

In early 1941 plans were being drawn up for the invasion of the Soviet Union. To underpin the intended success of this massive venture, the southern flank had to be secure, a task entrusted to Germany's Italian allies. A rushed and ill-planned Italian invasion of Albania, plus uncertainties concerning some Balkan states, led Hitler to intervene directly. While the ultimately successful outcome of this campaign allowed the invasion of Russia to proceed unhindered by resistance from the south, the necessity to intervene in the Balkans certainly delayed the launch of Operation Barbarossa by a few weeks, a delay which contributed significantly to its ultimate failure.

Tasked with hastily assembling an air force to simultaneously attack Yugoslavia and Greece, Luftflotte 4 found itself needing a long-range fighter. Two Bf 110 Gruppen – I. and II./NJG 4 – had recently adopted the night-fighter role, but reverted to their former incarnations as I. and II./ZG 26 for the new campaign. The units flew to Bulgaria, from where the opening raids on Yugoslavia were launched on 6 April. III./ZG 26 also joined in, flying its aircraft from mainland Italy across the Adriatic Sea.

Yugoslavia fell quickly. During the brief but hectic air fighting, the Bf 110s (a mix of C and E models) came up

Above: Bf 110D-1 night-fighters of I./NJG 3 are seen in October 1940, the month of the unit's reformation as a night-fighter unit from the remnants of V.(Z)/LG 1. The aircraft in the background still wears the codes of 7./LG 1.

Above left: This Bf 110 wears the codes of the Geschwaderstab KG 55. The bomber unit, equipped with He 111s, used the Bf 110 for reconnaissance purposes.

Deploying south

Above: These Bf 110s are from 7./ZG 26. Aircraft from this Sicily-based unit flew to Taranto to fly missions in support of the Balkans campaign.

Above right: ZG 26 Bf 110s fly over Greece. During the brief Balkans campaign the Bf 110 recaptured some of the 'glory' of its earlier exploits, proving more than a match for the Greek and RAF defenders.

Right: This III./ZG 26 aircraft at rest (possibly in Crete or southern Italy) combines the white Mediterranean theatre band (for commonality with Italian aircraft) with yellow Balkans campaign markings.

Warplane Classic

Above: *This scene shows 8./ZG 26's dispersal area at a desert airstrip. Facilities in North Africa were often rudimentary, at best, especially as the aircraft moved forward and back across the desert to keep up with the see-saw progress of the ground war. The aircraft in the foreground is a newly arrived attrition replacement, and has yet to receive the Staffel's codes.*

Above right: *A state-of-the-art fighting machine meets one of the desert's most valuable and most ancient commodities. The war in North Africa introduced many troops – from both sides – to very different cultures. The Bf 110E carries the letter 'N' on its cowling to signify that it was powered by the DB 601N engine and should be filled with 96-octane C3 fuel.*

against Yugoslav Bf 109s, I./ZG 26 losing five of its aircraft on the opening day. However, with aerial opposition swiftly neutralised by strafing and aerial combat, the German ground forces overcame effective opposition swiftly and swung south into Greece, joined by a second line of advance that struck out from Bulgaria.

Despite the intervention of the RAF, the Bf 110s performed well over Greece, destroying many aircraft on the ground. There were also some hectic air battles, none more so than the fight over Piraeus on 20 April, during which 5./ZG 26 shot down five RAF Hurricanes for the loss of two Bf 110s. Among the victims that day was South African 'Pat' Pattle, the RAF's highest-scoring ace of the war with at least 50 kills to his credit.

With mainland Greece under German control, the island of Crete was the next target. The two Bf 110 Gruppen were joined by II./ZG 76 for the Crete campaign, which involved the Bf 110s mainly in strafing and bombing attacks on Commonwealth ground defences and naval ships.

Iraqi interlude

With Crete secure, the bulk of II./ZG 76 returned to the German Bight and was subsequently retasked with night-fighting. However, one Staffel was retained in the theatre for an unusual mission.

In April 1941 a military *coup* in Iraq had brought to power a pro-Axis government. In response, the British landed a large troop force to protect their interests, which was met by an Iraqi siege of the RAF airfield at Habbaniyah. Eager to capitalise on this unrest, Germany hastily put together a force to be dispatched to Iraq to aid the government in its fight with the British. Denied the large force originally requested because of the demands of the upcoming Soviet invasion, the Iraqi expeditionary force, known as Sonderkommando Junck, comprised a single Staffel of He 111 bombers and another of Bf 110s (4./ZG 76), supported by some Ju 52 and Ju 90 transports.

The Luftwaffe's adventure in Iraq was brief. The aircraft of Sonderkommando Junck flew to Iraq via Vichy-held Syria and arrived at the Iraqi airfield at Mosul in mid-May. Operations chiefly involved attacks on Habbaniyah airfield and British ground positions. Two RAF Gladiators were shot down, but 4./ZG 76 suffered its fair share of attrition, notably during RAF counter-attacks on Mosul. On 25 May neither of the last two serviceable Bf 110s returned from a mission, and 4./ZG 76's contribution was over. The Staffel's personnel were evacuated by Ju 90.

Mediterranean Zerstörers

As mentioned earlier, III./ZG 26 had been dispatched to the Mediterranean theatre after the end of the Battle of Britain, arriving in Sicily in December 1940. They ranged over the Mediterranean on a number of tasks, including anti-shipping patrols, bomber escorts, protecting the air and sea lanes to Axis forces in North Africa, and attacks against the British redoubt on Malta. The Gruppe was usually dispersed between bases in Sicily, Crete and Africa, although was often brought together for key operations.

The fiercest fighting was in North Africa, where the Bf 110s were embroiled in the see-saw war as Axis forces advanced and retreated a number of times. Many of the missions were fighter-bomber sorties against Commonwealth airstrips and ground forces. III./ZG 76 was involved as a complete entity in the ground attack role during the German advance that led up the stalemate at El Alamein, but thereafter only one Staffel remained in North Africa. Bf 110s based on Crete subsequently encountered USAAF day bombers flying from Palestine.

After the Commonwealth break-out from El Alamein, and the Allied Torch landings in northwest Africa in November 1942, German forces were swiftly pushed back into Tunisia, and most of III./ZG 26 withdrew to Trapani in Sicily to fly increasingly desperate protection missions for

Above: *One of the main roles in the Mediterranean theatre was escorting transport aircraft. These Bf 110s are from 8./ZG 26.*

Right: *These 9./ZG 26 aircraft are Bf 110D-3s, able to carry large drop tanks underwing and equipped with a liferaft in an extended tail. The cockerel badge was that of the Staffel while the ladybird was that of III. Gruppe/ZG 26.*

Above: 2./Aufklärungsgruppe 14 flew Bf 110s in the reconnaissance role, with cameras replacing the cannon. The Staffel was Rommel's 'eyes' from March 1941 to June 1942.

the Tunisian air bridge. A few Bf 110s remained in the shrinking Axis enclave, and they achieved some notable victories over Allied aircraft, including P-38 Lightnings.

When Tunis fell to the Allies in May 1943 III./ZG 26 pulled back to near Rome, charged with protecting the Italian capital and surrounding area from daylight attacks by USAAF bombers. However, the Allied landings on Sicily saw the Gruppe adopt the close support missions again. Before the island fell to the Allies, III./ZG 26 was recalled to Germany to join the daylight Reich defence force.

In the latter few weeks of the North African campaign, III./ZG 26's Bf 110s had been joined by those of II./ZG 1, which was withdrawn from the Eastern front. In July they were dispatched from Italy to the Brittany coast for a brief period of flying patrols over the Bay of Biscay, before joining the Reich defence units in October.

Zerstörers in the East

Fresh from their triumphs over the Balkans and Crete, I. and II./ZG 26 transited via Germany to Suwalki in Poland, their start-point for the invasion of the Soviet Union in June 1941. Just 51 serviceable Bf 110s were available for the heavy fighter role in Operation Barbarossa out of a force of 2,500 front-line Luftwaffe aircraft.

While the single-seat fighter pilots scored heavily in the opening days, the Zerstörer crews encountered few opponents. This was mainly because the Red air force flew low, where the Bf 110 could not engage successfully. There were, however, rows and rows of parked aircraft on Soviet airfields, and the Bf 110s flew numerous strafing sorties. From the outset of the campaign the Bf 110 was being viewed as primarily a ground attack tool, although the aircraft generally came out on top when aerial opposition was encountered at sufficient altitude.

Both Bf 110 Gruppen supported the Wehrmacht's northern push towards Leningrad, the advance halting close to the city in late August. I. and II./ZG 26 spent some time in the Leningrad area, chiefly employed on enforcing the blockade by attacking road and rail traffic, before being rotated back to Germany. Some Staffeln returned to the central sector and saw some action before being recalled in April 1942 to Germany, where they transitioned – as I. and II./NJG 4 – back to the night-fighter role they had been performing before their Balkans call-up.

This was not to be the end of the Bf 110 in the attack role on the Russian front. In the initial attack on the Soviet Union two further Gruppen were involved, both tasked exclusively with ground attack. They were I. and II./Schnellkampfgeschwader 210, which had been formed from Erprobungsgruppe 210 (of Battle of Britain fame) and III./ZG 76, respectively.

SKG 210 Bf 110s operated almost universally with bombs, usually making dive attacks before following up with strafing runs. While aerial combat was normally avoided, SKG 210 crews did score some victories. Assigned to the central sector, the SKG 210 Gruppen supported the

Above: An 8./ZG 26 Bf 110 cruises alongside a factory-coded aircraft during a patrol over the Aegean Sea in 1942.

Below: This 7./ZG 26 Bf 110E provides a welcome respite from the desert sun for German troops. This model introduced wing bomb racks in addition to the ability to carry drop tanks.

One of a number of trials programmes undertaken by the Bf 110 concerned the 'Dobbas', a collapsible freight container which was carried under the belly of a Bf 110E.

Bf 110 reinstated in production

From the summer of 1941 deliveries of the Bf 110F started to taper off as the type's intended replacement, the Messerschmitt Me 210 multi-role combat aircraft, began to come off the production lines at Regensburg and Augsburg. In August 45 Bf 110s were delivered, falling to 15 in October, eight in November, one in December and none in January 1942.

However, by the end of 1941 it was clear that, as it stood, the Me 210 was unsuitable for general service use. In their attempts to rectify its failings Messerschmitt engineers applied various incremental modifications to the aircraft, but with little success. Finally, in January 1942, the Luftwaffe ordered the termination of the Me 210. At the same time it ordered the Bf 110 to be reinstated in production immediately.

Thus, instead of being phased out of service in most of its roles during 1942, the Bf 110 was to continue in service until the end of the conflict. In February 1942 the Luftwaffe received four aircraft, followed by 51 in March and 61 in April. During 1942 the Luftwaffe took delivery of 581.

Messerschmitt Bf 110 production by year

1939	1940	1941	1942	1943	1944	1945	Total
315	1,231	786	581	1,509	1,518	45	5,985

Bf 110s in Russia

Above: One of II./SKG 210's fighter-bombers flies over the central sector in the winter of 1941. Although the rear of the aircraft has soluble white distemper applied neatly, there are only traces on the forward half, and the 'Wespen' badge is untouched.

Below: A Bf 110D-3 crew consults the map before delivering its aircraft to a unit in Russia. Note the extended tail of this variant.

Receiving maintenance at a forward landing strip in Russia is a camera-equipped Bf 110 of 3. Staffel, Aufklärungsgruppe (Fern) 11.

Far right: A II./ZG 76 Zerstörer lands at a waterlogged airstrip on the way to Leningrad in the autumn of 1941. The airfield was also home to the Bf 109Fs of II./JG 54.

Below: A gaggle of fighter-bombers from II./SKG 210 flies over Russia in the summer of 1941. The aircraft in the foreground carries conventional bombs under the fuselage and 2-kg (4.4-lb) SD 2 'butterfly' anti-personnel bombs under the wings, as denoted by the forest of dangling arming cables.

main advance on Moscow, ending 1941 to the southwest of the Soviet capital. Early in 1942 the 'fast bomber' SKG designation was dropped, the two Gruppen becoming I. and II./ZG 1.

In the 1942 offensive the Bf 110s supported Army Group South as it headed for the Caucasus and Stalingrad, and by August were based close to the city in support of 6. Armee. Operations tuned increasingly to anti-armour missions as Soviet ground resistance stiffened. Despite suffering heavy losses during numerous close support missions, Bf 110s remained in action over Stalingrad until the end of January 1943, when the few survivors were withdrawn.

In March II. Gruppe was sent to the Mediterranean in the final days of the North African campaign, while I. Gruppe stayed in Russia long enough to participate in the massive tank battle at Kursk. It returned to Germany at the end of July to be redesignated as I./ZG 26. Only one small element of ZG 1 remained in Russia: 10.(Nachtjagd) Staffel. This small night-fighting unit had been established in September 1942, and was conspicuously successful. In August 1943 it was rechristened Nachtjagdschwarm/ Luftflotte 4 and flew on into 1944.

Mention should also be made of another Eastern front Bf 110 unit that started life as a three-aircraft Zerstörerkette at Kirkenes in Norway, within the Bf 109-operating JG 77. By the time of Barbarossa it had expanded to Staffel strength as 1.(Z)/JG 77, and flew bomber-escort missions to Murmansk. It moved to Finland, and increasingly became involved in ground attack missions, as well as scoring regularly over aerial opposition. The unit became part of JG 5 as 6., then 10., and finally 13. (Zerstörer) Staffel, withdrawing from Finland back to Norway in February 1944 and becoming 10./ZG 26.

Planned replacement

During the second half of 1941 production of the Bf 110 was scaled down, as the Messerschmitt factories tooled up to mass-produce the Messerschmitt Me 210 scheduled to be

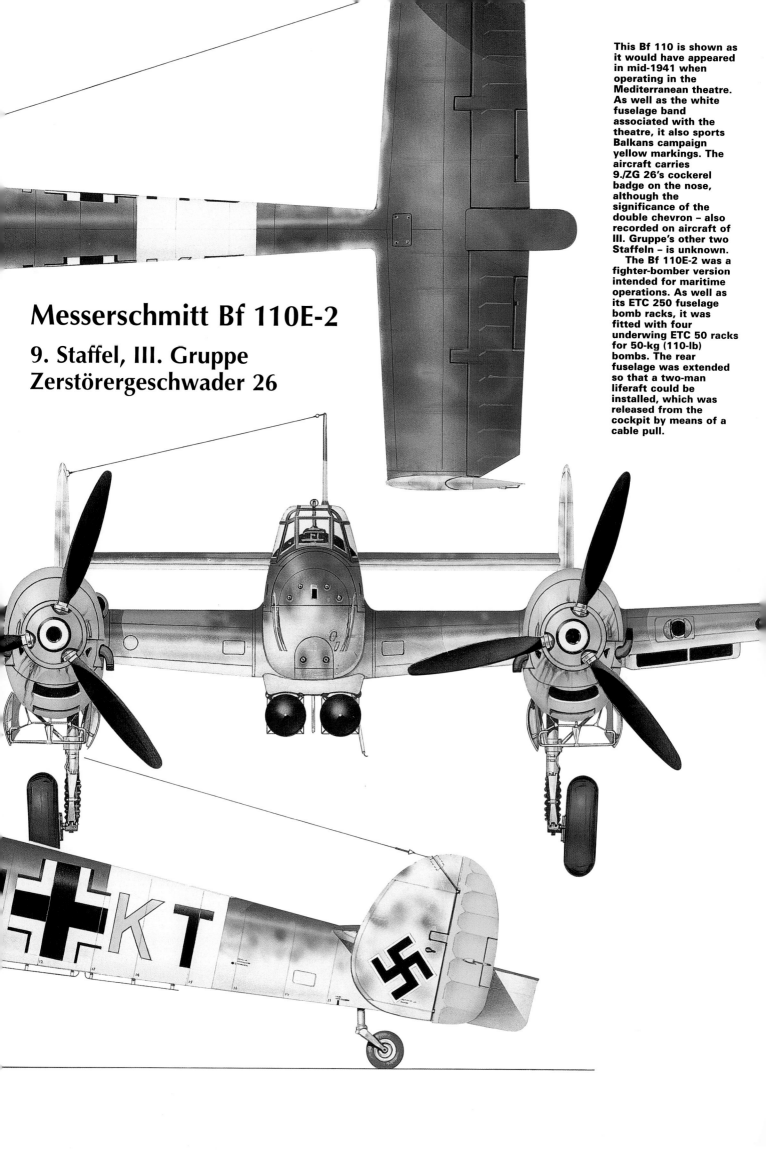

Messerschmitt Bf 110E-2
9. Staffel, III. Gruppe Zerstörergeschwader 26

This Bf 110 is shown as it would have appeared in mid-1941 when operating in the Mediterranean theatre. As well as the white fuselage band associated with the theatre, it also sports Balkans campaign yellow markings. The aircraft carries 9./ZG 26's cockerel badge on the nose, although the significance of the double chevron – also recorded on aircraft of III. Gruppe's other two Staffeln – is unknown.

The Bf 110E-2 was a fighter-bomber version intended for maritime operations. As well as its ETC 250 fuselage bomb racks, it was fitted with four underwing ETC 50 racks for 50-kg (110-lb) bombs. The rear fuselage was extended so that a two-man liferaft could be installed, which was released from the cockpit by means of a cable pull.

Crew accommodation
The crew comprised a pilot and a radar operator, housed under the long glazed cockpit canopy. The crewman could either face forwards to operate the radar and radio, or rearwards to operate the MG 81Z. During the final months of the conflict it was usual to carry an additional crewman, a gunner, to keep a full-time watch for enemy night-fighters approaching from the rear.

Drop tanks
For the night-fighter role, in which loiter time was often an important factor, Bf 110G-4s usually carried 900-litre (238-US gal) drop tanks.

Messerschmitt Bf 110G-4d/R3
Geschwaderstab Nachtjagdgeschwader 6

Armament
This varied, but typically that fitted to night fighters comprised two 30-mm Rheinmetall MK 108 cannon with 135 rounds per gun in the upper half of the nose and two 20-mm Mauser MG 151/20 cannon with 300 and 350 rpg respectively in the lower half of the nose. A Mauser MG 81Z paired installation of 7.9-mm machine-guns provided for rear defence. Several Bf 110s carried the *schräge Musik* installation instead of the MG 81Z, with two 20-mm MG FF or MG 151 cannon arranged to fire upwards and forwards at an angle of about 60° from the rear cockpit bulkhead. A maximum bomb load of 700 kg (1,543 lb) could also be carried during night ground attack operations.

FuG 220 Lichtenstein SN-2d
This radar weighed around 70 kg (154 lb) and had a maximum range of about 4 km (2.5 miles). It was effective over search angles of +/-60° from the centreline in azimuth, and +/-50° in elevation.

Flame-dampers
A necessary evil for the wartime night-fighter, flame-dampers over the exhaust stubs created significant drag. On the Bf 110 the problem was exacerbated by the need to duct the exhaust gases from the outboard exhaust stubs to the top of the wing to avoid ingestion by the underslung radiators. A few aircraft were fitted with the much neater Eberspacher system, which had a single long pipe under the wing, reaching past the radiator intake.

Messerschmitt Bf 110G-4c/R3
Dimensions: wing span 53 ft 4 in (16.26 m); length 42 ft 10 in (13.05 m); height 13 ft 8½ in (41.18 m); wing area 413 sq ft (38.37 m²)
Weights: empty 11,220 lb (5089 kg); normal loaded 20,700 lb (9390 kg); maximum loaded 21,800 lb (9888 kg)
Performance: maximum speed 342 mph (550 km/h) at 22,900 ft (6980 m), 311 mph (500 km/h) at sea level; range 560 miles (900 km) or 1,305 miles (2100 km) with two 238-US gal (900-litre) drop tanks; service ceiling 26,250 ft (8000 m); absolute ceiling 36,090 ft (11000 m); maximum climb rate 2,165 ft (660 m) per minute

Warplane Classic

Gun armament

A number of ventral gun installations were developed for the Bf 110 in place of the lower fuselage cannon. Above is an MK 101 30-mm cannon, intended for strafing, on a Bf 110C-6, while below is a Bf 110G-2/R5 'bomber-destroyer' sporting a BK 3,7 (Flak 18) 37-mm cannon housed in a wood and fabric fairing, and fed with up to 72 rounds from tanks in the rear cockpit.

Above: The nose armament of four staggered 7.9-mm MG 17s remained standard throughout production, although many Bf 110Gs were instead modified with either two 20-mm MG 151s or the R3 installation of two 30-mm MK 108 cannon with 135 rounds each. Under the nose were two MG FF 20-mm cannon, each with 180 rounds (replaced by MG 151 from the Bf 110F-4).

Above right: 'Black men' load the ammunition magazines for the four MG 17s of a Bf 110C. The magazines loaded through the drop-down doors. Each gun had 1,000 rounds.

Right: This Bf 110 was tested with a remotely-controlled barbette in place of the standard MG 15 or MG 81Z machine-gun installation in the rear cockpit.

Bombs and rocket carriage

Bf 110 fighter-bombers carried bombs on ventral racks. This installation on a Bf 110C-4/B has paired ETC 250 racks (for 250-kg) bombs).

In late 1942 Bf 110F-2 (V19) tested the 73-mm RZ 65 rocket in a battery of 12 tubes for the anti-bomber role. RZ 65s were also tested for sideways firing.

A 210-mm rocket is loaded into the Werfergranate 21 launcher beneath a bomber-destroyer Bf 110G. The weapon was tested at the Tarnewitz Waffenprüfplatz.

Night-fighting equipment

Above: Flame-dampers were fitted over the exhaust stubs on both inner and outer faces of the engine nacelles, producing considerable drag.

Above and right: From early 1943 many Bf 110 night-fighters had schräge Musik upward-firing cannon installed, often using the MG FF cannon made redundant when the lower fuselage weapons were upgraded to MG 151s. The diagram shows the MG FF installation, with its drum feed, reserve drums and firing mechanism. The first schräge Musik kill reportedly went to a Bf 110 of II./NJG 5 in May 1943, flown by the concept's chief proponent, Hauptmann Rudolf Schönert.

Left: Spanner-Anlage, an infra-red detection set, was first fitted to the Bf 110D-1/U1, which had a glazed sensor in the nose. The equipment was not successful in combat, although refined telescopic versions – like this installation in a Bf 110G – continued in use.

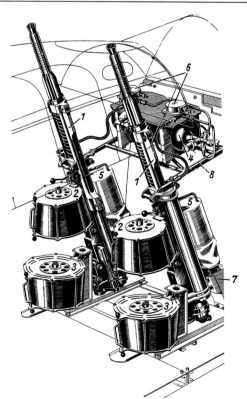

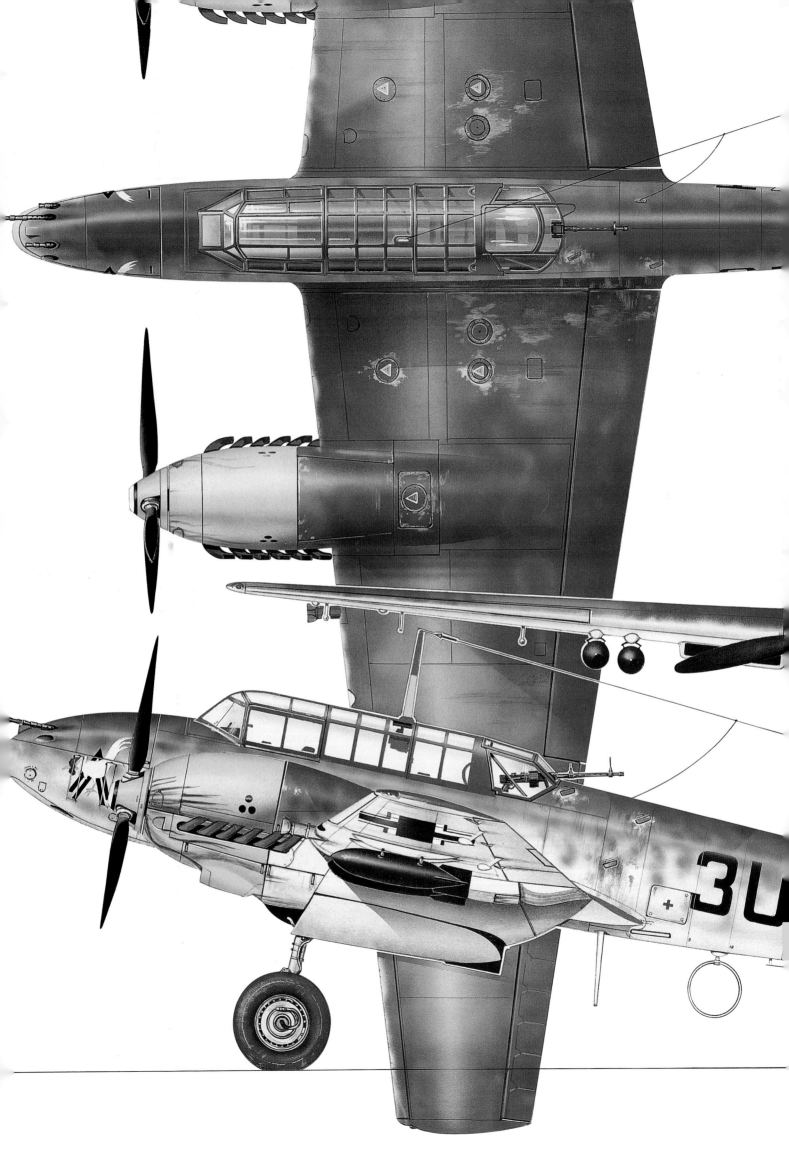

Messerschmitt Bf 110 variants

Bf 110 prototypes: the first prototype, V1, made its maiden flight on 12 May 1936. It was powered by two Daimler Benz DB 600A engines each developing 910 hp (679 kW). The V2 second prototype was essentially similar to the V1 with only minor changes. V3, the third prototype, again was similar but was fitted with the nose-mounted armament of four 7.9-mm MG 17 machine-guns. V3 first flew on 24 December 1936.

Bf 110A: intended as the pre-production model powered by DB 600Aa engines. That engine suffered from several shortcomings, however, and it was phased out of production in favour of the DB 601 with direct fuel injection. Production of the latter was slow, so to allow service testing and training to begin a small batch of **Bf 110A-0** aircraft was fitted with Junkers Jumo 210D engines developing 680 hp (507 kW) for take-off. Armament comprised four 7.9-mm MG 17 machine-guns in the nose and a flexibly mounted MG 15 of the same calibre in the rear cockpit. The rear of the engine nacelles was extended aft to cure a buffet problem, and the mainwheels were left partially exposed when retracted. Four were built.

Bf 110B: this became the first production variant of the fighter, with revised nose contours that were first tested on Bf 110A-04. The Bf 110B was to have been powered by two DB 601As, but there were delays in getting the engine into full production. The aircraft were each fitted with a pair of Jumo 210G engines developing 700 hp (522 kW) for take-off. Armament as for the A-0, but in addition there was provision to carry two 20-mm Oerlikon MG FF cannon in the lower part of the fuselage, hand-loaded by the radio operator/navigator. About 45 examples were produced, and were mostly employed on trials or as pilot conversion trainers. The first of 10 **Bf 110B-0** gunless pre-production aircraft flew on 19 April 1938. Production **Bf 110B-1**s had armament installed, whereas the **Bf 110B-2** had a camera installation in place of the cannon. Most were modified to **Bf 110B-3** standard, without cannon, to serve as Zerstörer trainers.

Unarmed Bf 110B-0 outside the Messerschmitt works

Bf 110C: generally similar to the B variant, the C was fitted with the long awaited DB 601A-1 engines. A pre-production batch of 10 **Bf 110C-0** fighters was delivered to the Luftwaffe early in 1939, followed by the first production variant, the **Bf 110C-1**. The Luftwaffe received 315 examples of this variant during 1939. Other sub-variants followed: the **Bf 110C-2** had FuG 10 Lorenz radio in place of FuG IIIa; the **Bf 110C-3** with minor changes to the MG FF cannon installation; and the **Bf 110C-4** which introduced armour protection for the crew. The **Bf 110C-4/B** was a fighter-bomber version with twin ETC 250 bomb racks and uprated DB 601N engines offering 1,200 hp (895 kW) for take-off. The **Bf 110C-5** was a DB 601A-1-powered reconnaissance variant with the cannon removed and a single Rb 50/30 camera mounted vertically mid-way along the cabin. A few had DB 601N engines and were known as **Bf 110C-5/N**. The **Bf 110C-6** fighter-bomber variant featured a single 30-mm MK 101 cannon in place of the two MG FF weapons, while the **Bf 110C-7** was a development of the C-4/B with stronger undercarriage and twin ETC 500 bomb racks. A number of early aircraft were converted for glider-towing as the **Bf 110C-1/U1**.

Bf 110D: this began life as a long-range modification of the C-3 fighter, with a 1175-litre (310-US gal) jettisonable tank under the forward fuselage covered by a streamlined fairing. Prototype conversions were designated **Bf 110D-0**, and were followed by production **Bf 110D-1/R1**s, which were interspersed with Bf 110C-3s on the line. Problems with the belly tank led to the **Bf 110D-1/R2**, which had 'wet' hardpoints under the wings for 900-litre (238-US gal) drop tanks. The **Bf 110D-1/U1** designation covered some night-fighters fitted with Spanner IR equipment. Development of long-range Bf 110s continued with the **Bf 110D-2** fighter-bomber based on the D-1/R2, and the **Bf 110D-3**. The latter had provision to carry drop tanks under the wings and an underfuselage oil tank for long-range maritime patrols. The tail was extended to house a two-man liferaft.

Bf 110D-3s of 8. (foreground) and 9. Staffel/ZG 26 fly along the North African coast

Bf 110E: essentially similar to the D variant but with four ETC 50 underwing bomb racks. Pre-production **Bf 110E-0**s and early **Bf 110E-1**s were powered by DB 601A-1s, but the DB 601N subsequently became standard. The **Bf 110E-1/U1** sub-variant carried flame-dampers and other modifications such as Spanner to suit it for the night-fighter role, while the **Bf 110E-1/U2** had a third crew seat. The **Bf 110E-1/R2** deleted the underwing racks but had two ETC 1000 racks under the fuselage. The **Bf 110E-2** was a fighter-bomber variant with extended liferaft tail, while the **Bf 110E-3** was optimised for reconnaissance.

Bf 110E-1/U2 of ZG 1 with four ETC 50 underwing racks and two fuselage ETC 250s

Bf 110F: generally similar to the E variant and produced in parallel, the F was fitted with more powerful DB 601F engines developing 1,350 hp (1007 kW) for take-off. The **Bf 110F-1** was a fighter-bomber variant featuring improved armour protection for the crew, such as a bullet-resistant windscreen. The **Bf 110F-2** was the heavy fighter version and the **Bf 110F-3** was for reconnaissance. The **Bf 110F-4** was a night-fighter version with revised tail surfaces, many of which were retrofitted with *schräge Musik* upward-firing armament in place of the rear cabin MG 15. Those produced by factory kits were designated **Bf 110F-4/U1**. The **Bf 110F-4a** was fitted with the FuG 202 Lichtenstein BC airborne interception radar, and there was also provision to carry two 30-mm MK 108 cannon in a tray under the fuselage. Later production Fs and subsequent variants had the 20-mm MG FF cannon replaced by MK 151/20 belt-fed weapons of the same calibre but with an increased rate of fire. The starboard weapon had 350 rounds, while that to port had 300.

Bf 110F-1 displaying the enlarged oil coolers that distinguished the variant

Bf 110G: following the demise of the Me 210, Messerschmitt engineers strove to improve the performance of the Bf 110 with the new G variant. This was generally similar to the F, but was fitted with the more powerful DB 605B-1 engine that developed 1,475 hp (1100 kW) for take-off. Defensive armament was changed to the MG 81Z, comprising two 7.9-mm machine-guns joined and firing rearwards on a flexible mounting. Most aircraft were built with the enlarged fins introduced on the F-4, but a number of night-fighters fitted with FuG 202 had the original fins. In accordance with the previous pattern of designations, the **Bf 110G-0** was the pre-production aircraft, the **Bf 110G-1** was the fighter-bomber version, the **Bf 110G-2** was the heavy fighter version, the **Bf 110G-3** was the reconnaissance version and the **Bf 110G-4** was the night-fighter. In the event, the G-1 was not produced as the G-2 could adequately cover both roles. To enable it to engage the US heavy bomber formations more effectively, the G-2 carried additional firepower. The **Bf 110G-2/R1** sub-variant carried a modified 37-mm anti-aircraft gun in a streamlined blister under the fuselage, while the **Bf 110G-2/R3** sub-variant had its forward-firing 7.9-mm machine-guns removed and two 30-mm MK 108 cannon mounted in their place, and two further 20-mm MK 151/20 weapons installed in a tray under the fuselage. Under the outer wing sections this variant carried tube launchers for four 210-mm unguided rockets. Some had the R2 modification that added GM 1 nitrous oxide boosting but removed the MG 81Z. G-4 night-fighters served in a number of sub-variants, depending on radar fit and armament. The baseline model had four MG 17s in the upper nose, but these were progressively replaced by two MK 108 cannon (R3 modification).

Bf 110H: the final production variant of the Bf 110 was generally similar to the G variant and was to have been built in parallel. It was to have featured several major changes compared with the G model, but in the event most of them were incorporated in the earlier variant. As a result, a late-production G model was externally little different from an H model. Chief differences were the use of the DB 605E engine and strengthened structure for an increase in maximum take-off weight to 18,930 lb (8587 kg). The pattern of sub-designations for the various roles followed that established earlier for the Bf 110F and G models.

Warplane Classic

Arctic Zerstörers

Right: This Bf 110 wears the codes 'LN+KR', indicating assignment to 13. Staffel (Zerstörer) of JG 5. The latter was the Arctic fighter Geschwader flying Bf 109s and then Fw 190s. The single Staffel of Bf 110s was assigned primarily for ground attack duties.

Above: This Bf 110E – a different 'LN+KR' from the one above – is seen patrolling the Arctic coastline in the summer of 1942. The Arctic Zerstörer Staffel had several different designations as its parent Geschwader reorganised and ultimately changed (to JG 77). Under the unit's unusual coding system all aircraft carried 'LN+xR', with only the individual aircraft (third) letter changing.

Above right: This 13.(Z)/JG 5 aircraft is a Bf 110F-2, as denoted by the enlarged oil coolers. The Staffel badge on the nose consisted of a Dachshund eating a Polikarpov I-16. From its base in Finland the Staffel maintained a continuous harassment against Soviet forces in the far north, especially around the ports of the Kola peninsula and along the roads south to Leningrad.

its replacement. Production of the Bf 110 ended in January 1942, but by then it was clear that the Me 210 suffered from dangerous handling characteristics. In particular it had a propensity to enter flat spins from which there was little chance of recovery. Several test crews had been lost in this way, and the type was judged unsuitable for combat operations. In January 1942, after several attempts to solve its problems, Me 210 production was terminated.

That left the Luftwaffe in a difficult position, with no aircraft to replace losses in several important roles. The only answer was to reinstate the Bf 110 in production. By the end of May 1942 monthly production of the Bf 110 had risen to 61 aircraft, and it remained at about that level for the remainder of the year.

Following the demise of the Me 210, the Messerschmitt company instigated a crash programme to improve the performance of the Bf 110 and extend its effective service life. The outcome was the G variant fitted with a pair of Daimler Benz DB 605B-1 engines, each of which developed 1,475 hp (1100 kW) for take-off. Otherwise, the G version was similar to the F version, which it replaced in production with the same pattern of sub-variants.

Night-fighter developments

The first variant of the Bf 110 built specially for the night-fighter role was the F-4. This was fitted with two 1,300-hp (970-kW) DB 601F motors in place of the 1,100-hp (820-kW) DB 601As in the C model. To supplement their already powerful armament, some F-4s carried two 30-mm MK 108 cannon in place of the rifle-calibre machine-guns. Production of the F-4 began in 1941, and by March 1942 it was running at 36 aircraft per month.

The tactic of directing ground-controlled night-fighters into visual contact with the bombers worked well enough on moonlit nights, but General Kammhuber saw that, in time, the bombers would also attack on dark nights. In anticipation of this he asked the Telefunken Company to build a lightweight radar set suitable for installation in a night-fighter. In February 1942 the set, the Lichtenstein BC with a maximum range of about two and a half miles (4 km), was introduced into service. By the following autumn the majority of German night-fighters carried the new radar.

The system of ground-controlled night interceptions, with a single line of defensive 'boxes' in front of the targets, worked well against RAF bombers when the latter passed through the defensive line at many separate points over a period of several hours.

As the RAF learned more about the Himmelbett system, however, it became clear that the system suffered a major weakness. Each defensive radar station could control only one interception at a time, and on average each interception lasted about 10 minutes. Thus, while a ground control station was engaged in controlling an interception, there existed an unguarded gap in the line through which any number of other bombers could pass unmolested. If the raiders flew in a relatively tight mass, concentrated in time and space, all except for the unfortunates being engaged would have free rides through the defensive line.

The RAF tried out the new tactic for the first time on the night of 30 May 1942, during the first Thousand-Bomber Raid on Cologne. The bombers all flew along the same route, and delivered their attacks during a two and a half-

Early night-fighters

The *Englandblitz* badge worn by the early night-fighters was originally designed by Victor Mölders, brother of the more famous Werner. It depicted a diving falcon and lightning bolt striking at southern England, and was first devised for Wolfgang Falck's I./ZG 1, which became I./NJG 1 in June 1940 with Falck as Geschwaderkommodore.

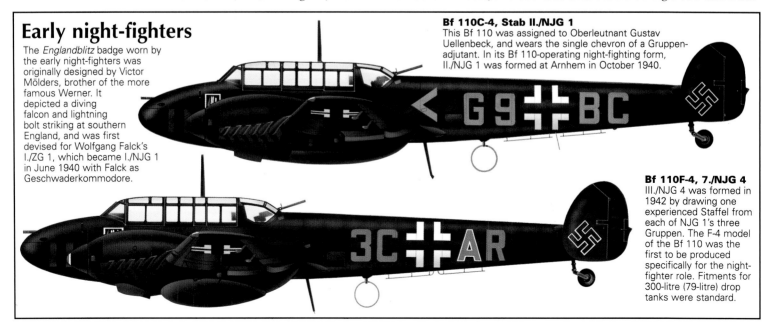

Bf 110C-4, Stab II./NJG 1
This Bf 110 was assigned to Oberleutnant Gustav Uellenbeck, and wears the single chevron of a Gruppen-adjutant. In its Bf 110-operating night-fighting form, II./NJG 1 was formed at Arnhem in October 1940.

Bf 110F-4, 7./NJG 4
III./NJG 4 was formed in 1942 by drawing one experienced Staffel from each of NJG 1's three Gruppen. The F-4 model of the Bf 110 was the first to be produced specifically for the night-fighter role. Fitments for 300-litre (79-litre) drop tanks were standard.

hour period, giving an average of seven bombers passing through the target area per minute (compared with two to three aircraft per minute during earlier raids). Forty-one aircraft failed to return, representing 3.8 per cent of the 1,046-strong attacking force. That was a significantly lower loss rate than hitherto.

During 1942 the various modifications to suit the Bf 110 for the night-fighting role caused a serious deterioration in its performance. The weight of the Lichtenstein radar, and the drag from its aerials and the large exhaust flame-dampers, clipped about 35 mph (56 km/h) off the fighter's maximum speed. Against the slower RAF bombers – the twin-engined Whitleys and Hampdens, and the four-engined Stirlings – the Bf 110F was fast enough, but against the higher performance Lancasters and Halifaxes the night-fighter had little speed advantage. It often occurred that the bomber passed outside the 40-mile (64-km) effective range of the Himmelbett ground station before the night-fighter completed its interception. At that point, unless its crew had good visual or radar contact with the bomber, the night-fighter had to break off the pursuit and return to the waiting area.

Himmelbett interception

At the end of 1942 the German night-fighter force comprised 389 aircraft, of which the vast majority – about 300 – were Bf 110s. During that year RAF Bomber Command lost 1,291 bombers in night attacks on targets in German-occupied Europe. An estimated two-thirds of those, some 850 aircraft, fell to night-fighters.

At this point it is of interest to look in detail at the working of the Himmelbett system, as it stood in the spring of 1943. Soon after midnight on the morning of 22 June 1943, Leutnant Heinz-Wolfgang Schnaufer took off in his Bf 110G-4 from St Trond in Belgium. He was ordered to orbit over Himmelbett station 'Meise', 15 miles (24 km) northeast of Brussels, to await 'trade'. A large force of RAF

Two views show III./NJG 4 aircraft, comprising Bf 110F-4s (top) and Bf 110C-4s (above). Attempts to establish NJG 4 in early 1941 were interrupted when the first two Gruppen were siphoned off to the war in the Balkans. When they returned to the night role in 1942 they were joined by a newly formed III. Gruppe.

NJG 1 aircraft, like this Bf 110C, usually carried the Gruppe number (in this case 'III') next to the Englandblitz badge.

The original caption for this photo states that these are ZG 76 Bf 110s providing cover for the 1942 'Channel Dash'. While the vessel in the photo is indeed Scharnhorst, the scene may easily have been recorded in another location, such as the Baltic or Norway. Furthermore, the Bf 110s involved in the Channel operation came from western-based night-fighter units, such as II./NJG 1.

Bf 110 operational flying units, 17 May 1943

On 17 May 1943 the Luftwaffe had a total of 473 Bf 110s serving with front-line units, of which 351 were serviceable. The lion's share of these aircraft (304) served with night-fighter units, while the remaining 170 served with long-range fighter and reconnaissance units. By now the Bf 110 had passed out of service in the fighter-bomber role. Two units require further explanation: 10. Staffel of ZG 1 was a night-fighter unit operating on the Eastern front and attached to ZG 1; and the Bf 110s attached to Stab of the Stuka dive-bomber unit StG 1 provided pre-strike and post-strike photography of targets.

	Total	Serviceable
LONG-RANGE FIGHTER UNITS		
Zerstörergeschwader 1		
Stab	2	2
I. Gruppe	38	30
II. Gruppe	32	8
Zerstörergeschwader 26		
III. Gruppe	29	20
Jagdgeschwader 5		
13. Staffel	12	8
NIGHT-FIGHTER UNITS		
Nachtjagdgeschwader 1		
Stab	4	4
I. Gruppe[1]	27	20
II. Gruppe[1]	26	17
III. Gruppe	23	20
IV. Gruppe[2]	22	16
Nachtjagdgeschwader 3		
Stab	2	2
I. Gruppe[1]	11	11[2]
III. Gruppe	23	18
Nachtjagdgeschwader 4		
Stab	1	1
I. Gruppe[1]	22	19
II. Gruppe	22	20
III. Gruppe[1]	24	22
IV. Gruppe[1]	23	23
Nachtjagdgeschwader 5		
Stab	2	1
I. Gruppe	26	26
II. Gruppe[1]	19	17
IV. Gruppe[3]	18	18
Zerstörergeschwader 1		
10. Staffel[4]	12	7
TACTICAL RECONNAISSANCE UNITS		
Nahaufklärungsgruppe 2[5]	9	9
Nahaufklärungsgruppe 4[5]	6	4
Nahaufklärungsgruppe 5[6]	8	7
Nahaufklärungsgruppe 6[7]	10	3
Nahaufklärungsgruppe 8[6]	10	7
Nahaufklärungsgruppe 14[8]	7	4
Sturzkampfgeschwader 1		
Stab[9]	6	5

Notes
1. unit also operated Dornier Do 217s
2. unit also operated Dornier Do 215s
3. unit also operated Junkers Ju 88s
4. Night-fighter unit
5. unit also operated Messerschmitt Bf 109s
6. unit also operated Henschel Hs 126s
7. unit also operated Hs 126s and Fw 189s
8. unit also operated Focke Wulf Fw 189s
9. Reconnaissance Bf 110s provided pre-strike and post-strike reconnaissance of targets for the dive-bomber unit

Warplane Classic

Right: Mottled light grey camouflage began to replace all-over black on night-fighters as early as the summer of 1941, although it was a long process. Radar was introduced from February 1942, and it was relatively rare to encounter a grey aircraft without the new equipment.

A black-painted Bf 110F-4a of 4./NJG 1 is jacked up for gun and sight harmonisation. F-4s were all powered by the high-octane DB 601F engine, hence the 'N' on the cowling to avoid any refuelling mishaps. The aircraft is fitted with the 'Matratzen' array for the Telefunken FuG 202 Lichtenstein BC radar.

This aircraft – almost certainly coded 'G9+AA' – was one of the last Bf 110Cs flown by the charismatic Wolfgang Falck, Kommodore of NJG 1 from its inception in 1940 until the summer of 1943. It was while commanding I./ZG 1 at Aalborg during the Danish occupation that Falck began experiments with the Bf 110 in the night-fighter role. Following the French campaign he and most of his Gruppe returned to Germany to establish NJG 1. Although widely credited as being the 'father' of the Nachtjagdwaffe, Falck himself scored no night victories.

bombers was coming in over the North Sea. The bombers' briefed route took most of them well to the east of 'Meise', but at 01.20 hours Schnaufer was informed of a lone bomber approaching the ground radar station from the west. On the ground the 13th Company, 211th Signals Regiment, manning the radar station 'Meise', worked as a well-knit team. One Giant Würzburg tracked Schnaufer's fighter, while the other swung to the west and began searching for the intruder.

The bomber was detected almost as soon as it came within range of the second Giant Würzburg. The fighter control officer attached to 13/211, Leutnant Kühnel, passed a constant stream of instructions, which placed Schnaufer in position behind the bomber. In the rear of the Messerschmitt Leutnant Baro, the radar operator, observed a small hump of blue light rise from the flickering base line of his screen: an aircraft, range 2,500 yards. Baro took over the commentary until, in Schnaufer's words:

"I recognised, 500 yards above and to the right, a Short

Stirling and succeeded in getting in an attack on the violently evading enemy aircraft. It caught fire in the fuselage and the wings and continued on, burning. Then it went into a dive and crashed two miles northeast of Aerschot." At first light Kühnel drove out to look at the wreckage of the bomber, and verify the claim. It was Heinz-Wolfgang Schnaufer's 13th victory.

The rigidly controlled Himmelbett system had another weakness that did not pass unnoticed in England. Just over a month later, during the attack on Hamburg on 24 July 1943, a new RAF tactic effectively removed the linchpin of the night air defence system. The answer was as simple as it was effective: one thousand strips of Window aluminum foil, each measuring 30 x 1.5 cm (12 x 0.6 in), held together by an elastic band. Each aircraft in the attacking force dropped one such bundle per minute. Dropped into the aircraft's slipstream, the foil strips formed a cloud that gave a radar echo similar to that from a large bomber. The three types of precision radar used to engage the bombers – the Lichtenstein equipment carried by night-fighters, and the Würzburg and Giant Würzburg ground radars, all worked close together in the frequency spectrum. That meant the bundle of Window described was effective against all three German radars. With a mass of Window clouds covering the screens of these precision radars, it was extremely difficult to mount interceptions at night.

Through the Window

One Luftwaffe pilot airborne that night later wrote (Wilhelm Johnen in *Duel Under the Stars*, William Kimber Ltd): "At 5000 metres my radio operator announced the first enemy on his Lichtenstein. I was delighted. I swung round on to the bearing, in the direction of the Ruhr, for in this way I was bound to approach the stream. Facius proceeded to report three or four targets on his screens. I hoped that I should have enough ammunition to deal with them all.

"Then Facius shouted: 'Tommy flying towards us at a great speed. Distance decreasing . . . 2,000 metres . . . 1,500 . . . 1,000 . . . 500 . . .' I was speechless. Facius already had a new target. 'Perhaps it was a German night-fighter on a westerly course,' I said to myself, and made for the next bomber. It was not long before Facius shouted again: 'Bomber coming for us at a hell of a speed. 2,000 ... 1,000 .. . 500 ... he's gone!' 'You're crackers, Facius,' I said jokingly. But soon I lost my sense of humour, for this crazy performance was repeated a score of times."

In response to the Window threat, the Luftwaffe made radical changes to its night-fighting tactics. Previously, single night-fighters had been vectored on to individual bombers passing through the line of Himmelbett boxes. Now all that changed. The fighters were scrambled by Gruppen, and vectored into the bomber stream where they

Reich defender

Bf 110G-4, Gruppenstab II./NJG 5
This G-4 was flown from Gütersloh by Oberleutnant Leopold 'Poldi' Fellerer, Kommandeur of II./NJG 5, and credited with 39 night kills (plus two by day). Some G-4s operated with just the early FuG 212 radar until the end of the war.

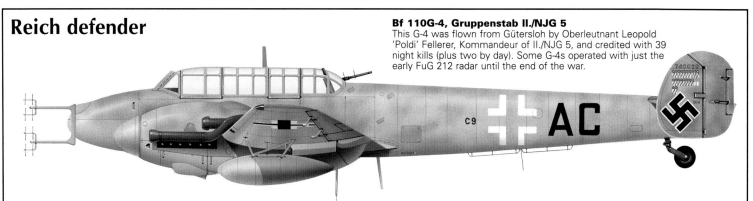

Messerschmitt Bf 110

Above: The standard early night-fighter radar was the FuG 202 Lichtenstein BC, which used a four-post 'Matratzen' (mattress) antenna array. The improved FuG 212 Lichtenstein C-1 used a similar array.

Night-fighter radars

Below: This Bf 110G-4/U6 carries outward-looking antennas for the FuG 221a Rosendaal-Halbe system – which homed on the Monica tail-warning radar of bombers – with a central Weitwinkel (wide-angle) array for Lichtenstein C-1 radar.

Below: The final radar installation to see widespread use in the Bf 110 was the FuG 220 Lichtenstein SN-2d, which dispensed with the central Weitwinkel antenna and featured angled antennas.

The radar installation in the Bf 110G-4b was known as Lichtenstein SN-2b, comprising four 'Hirschgeweih' (stag's antlers) antennas for the FuG 220 Lichtenstein SN-2 radar with a single Weitwinkel FuG 212 Lichtenstein C-1 antenna in the centre. The aircraft carries the Waffenwanne 151Z, a ventral pannier housing two MG 151 20-mm cannon and their ammunition.

were to engage the raiders using Lichtenstein radar or visually. The new tactic bore the code-named *zahme Sau* (Tame Boar), and the intention was to set up running battles with the raiders. Night-fighter crews were to engage the enemy planes until they ran out of ammunition or their fuel ran low, then they would put down at the nearest airfield. Losses of night-fighters due to their running out of fuel were risked and accepted, that was part and parcel of the *zahme Sau* tactics.

If in their course of their hot pursuit of bombers a night-fighter crew arrived over the target, it was to continue the engagement. There they would exploit the light from fires on the ground, searchlights and target-markers, to silhouette the bombers so the latter could be attacked visually. The tactic of engaging bombers over the target bore the code-name *wilde Sau* (Wild Boar), and was employed by both single-engined and twin-engined night-fighters.

The Luftwaffe night-fighter force first employed the *zahme Sau* tactics *en masse* on the evening of 17 August 1943, when the Royal Air Force dispatched 597 heavy bombers to smash the V-weapons research establishment at Peenemünde. It was a brilliant moonlit night, ideal for the new German methods.

Oberleutnant Hans Meihsner, a Bf 110 pilot with II./NJG 3, was one of those in action that night. He was scrambled from Jagel at 02.15 hours on the 18th and ordered to southeastern Denmark. He was lucky – his controller had placed him right in the path of the bomber stream as it headed for home. He recalled:

"Unfortunately the R/T was so badly jammed that we could make no contact with [Himmelbett station] Ameise,

Fighters in the East

Left: Although normally based at Wels in Austria, II./NJG 1's Bf 110s were briefly dispatched to Mamaia in Romania in April 1944 to counter the USAAF's daylight bombing raids. Here a Bf 110G-2 lands over an Me 323 transport.

Below: These Bf 110Gs served with 5./NJG 200, which operated with some success on visual helle Nachtjagd (light night-fighting) duties on the southern sector of the Eastern front in 1943/44.

Pulk-Zerstörer

Heavily armed and with sufficient range to stalk bomber formations over long distances, the Bf 110 was ideal for the Pulk-Zerstörer (literally, formation-destroyer) role – providing, of course, there were no escort fighters to deal with. New weapons were added to enable the Bf 110s to stay outside the withering fire put up by the USAAF 'Viermots' (four-engined bombers).

Above: This scene was recorded on 13 January 1944 and shows II./ZG 76 Bf 110G-2s flying from their base at Neubiberg. The lead aircraft is armed with a BK 3,7 cannon. At this stage of the war the Bf 110 day-fighter force was suffering increasingly at the hands of USAAF escort fighters, and operations were largely restricted to outside the range of England-based P-47s. The appearance of the P-51 finally ended the Bf 110's effectiveness in the day role.

This aircraft tested the Bf 110G-2/R4 configuration, combining the BK 3,7 ventral cannon and two 30-mm MK 108s in the nose. Carriage of the BK 3,7 made the Bf 110 even more cumbersome than before.

This Bf 110G-2 – from either 5./ZG 26 or 5./ZG 76 – lands after a day-fighter sortie. The long barrels protruding from the nose indicate fitment of 20-mm MG 151/20 cannon, and another pair is carried in the Waffenwanne 151Z ventral tray, beneath the internal fuselage weapons.

and could get no information. Meanwhile we were at 11,000 ft [3353 m] as we approached the Apenrader Bight. My radar operator [Unteroffizier Josef Krinner] picked up several contacts on the Lichtenstein which passed across the tubes very quickly, so at first we took it to be Window. As the contacts were below us I went into a dive and picked up speed.

"At 2.54 a.m. I saw the first Lancaster at about 10,000 ft, flying directly in front of me on a westerly heading. I closed in and opened fire from about 150 yards, somewhat to the right and 150 ft below. Its No. 3 engine caught fire. As I broke away below him, return fire from the rear gunner passed to my left. From the beginning of the engagement both aircraft were caught up by our own searchlights. When the Lancaster pilot attempted to escape in a left diving turn he came into my sights again and I was able to give him a short burst. He went down crashing at 02.56 hours, a few hundred yards from Ufer.

"I immediately set off eastwards, obtained another contact from my radar operator, dived again, and saw a Lancaster flying directly above me on a westerly heading. On account of the good visibility I was able to keep the Lancaster in sight as I turned sharply round to a westerly heading. I fired from more or less the same position, again at No. 3 engine. He went into a dive and crashed at 03.01 hours, on the shore of the Apenrader Bight.

"After that I headed north, and my radar operator soon picked up yet another contact. I was able to make out the aircraft about 1,200 yards away. My first attack was the same as the others, from 150 yards range, a little to the right and 150 ft below. As the No. 3 engine caught fire we were held by a searchlight, and despite the moonlight it was dazzling. The Lancaster pilot pulled his aircraft up (perhaps he was also dazzled, or wanted to reduce his speed to make it easier for his crew to abandon). The enemy aircraft now filled my horizon. I pulled up to within 20 yards and with a few rounds set the No. 2 engine and the fuselage on fire. The aircraft broke up and crashed at 03.11 a.m., one and half miles to the west of Ustrup."

As the Lancaster dived away Meihsner's windscreen was suddenly covered in opaque oil from his victim's tanks, forcing him to break off the action. That night the RAF lost 41 bombers.

Differences should be noted between Heinz-Wolfgang Schnaufer's interception described earlier, which throughout was carefully controlled from the ground, and the lone-wolf nature of Meihsner's engagement. The latter was fortunate to find a dense part of the bomber stream, but once he was there his attacks were devastating.

Overcoming Window

The visual search tactics were successful during the light summer nights, but they were no more than a stopgap pending the introduction of new equipment to work through Window jamming. In the final months of 1943 three new devices entered service in the German night-fighter force, to assist the crews to find the bombers: SN-2, Flensburg and Naxos.

SN-2 was a new night-fighter radar working in the 90-MHz band, a part of the frequency spectrum where the types of Window then in use were ineffective. Given the vulnerability of the earlier Lichtenstein radar to this countermeasure, that attribute was of crucial importance. Initially SN-2 suffered from a serious disadvantage, however. Its minimum range – the minimum range at which it could detect targets – was about 1,200 ft (366 m). The radar could not show targets inside that distance. Yet 1,200 ft was somewhat further than a night-fighter crew could expect to see the prey visually, except on the clearest of nights. To bridge the gap between the radar's minimum range, and the likely maximum visual range at night, some night-fighters carried both SN-2 and the earlier Lichtenstein equipment device. This makeshift solution resulted in a forest of drag-producing aerials on the nose of the aircraft, but it was the best that could be done quickly until the radar's minimum range problems were solved.

Messerschmitt Bf 110

Flensburg and Naxos were radar homing systems to exploit radiations from the bombers. Flensburg homed on emissions of the British Monica tail-warning radar, while Naxos homed on emissions from the H₂S bombing radar.

With the introduction of these new electronic devices the Luftwaffe night-fighter force was once again able to work effectively on dark nights. Now the *zahme Sau* tactic could be developed to its logical conclusion, setting up long-running battles with the raiders which might extend for hundreds of miles.

In the autumn of 1943 the majority of the Luftwaffe night-fighter units still operated the Messerschmitt Bf 110, though by now the type was at the end of its development life. With the new electronic devices and their drag-producing aerials, heavier armament and the external fuel tanks to provide the extra range necessary for the *zahme Sau* operations, the fighter had very little performance margin over the heavy bombers it sought to engage.

Leutnant Günther Wolf, a Bf 110 pilot of III./NJG 5, told the writer of the problems he had flying this type on night operations from the autumn of 1943. "It was handicapped by the drag of the radar aerials, and the weight of the radar and other equipment for night use. I remember on one occasion I was following a Halifax on its bombing run over Berlin, trying to get into a firing position. He released his bombs and then somehow the crew must have detected me. He went into a diving turn, which I followed, then he pulled up his nose and climbed and climbed. When I tried to follow him, I stalled and fell out of the sky."

Daylight bomber-destroyer

In the late summer of 1943 the Bf 110 took on a new role, a task demanded by the increasing appearance of the USAAF's daylight bombers over the Reich. On 17 August 1943 the US Army Air Force launched its deepest daylight penetration missions into Germany so far, with a twin attack on the ball bearing factories at Schweinfurt and the Messerschmitt plant at Regensburg. The raiders inflicted heavy damage on both targets but at great cost to themselves: 60 B-17s failed to return. Just over a week later, Generalfeldmarschall Erhard Milch, Director General for Luftwaffe equipment, told a meeting of senior officers in Berlin that for the next one to two years the Luftwaffe would have to fight a defensive war, as it waited for the new jet aircraft types to become ready for action.

He continued: "We must definitely decide on priorities. That means the Bf 109 and the Fw 190, and the Bf 110 which bears the brunt of the night-fighting . . . everything must be staked on the 110. Only the 110 in sufficient numbers can give us the necessary relief at night. Moreover, the 110 can also be used by day. Compared with the other fighter types it has the great advantage of considerably longer range. After the raid on Regensburg, for example, the enemy bombers headed south for Africa. Our twin-engined fighters pursued them beyond Innsbruck and inflicted quite serious losses. That could not have been achieved with the 109 or the 190 because their limited endurance would have compelled them to land for refuelling and re-arming long before that. Thus the 110 is particularly important for both purposes."

The Bf 110s Milch referred to had been night-fighters flying behind the US formations, to pick off wounded bombers that were flying alone. Probably Milch had judged their performance on the strength of the crews' exaggerated initial claims, for in fact the Bf 110 night-fighters probably accounted for only about five B-17s that day.

Nevertheless, Milch's point was well made: the Bf 110 optimised for the daylight bomber-destroyer role would be a useful addition to the Reich air defence force. Accordingly, in the weeks that followed, Bf 110 crews were pulled back from the Eastern and Mediterranean fronts. On return to the homeland they joined Zerstörergeschwader 26 equipped with the new Bf 110G-2 variant.

To provide the firepower to engage US heavy bombers effectively, these Bf 110s were fitted with three important

Above: This 9./ZG 26 aircraft carries four 'Dödels' (rocket tubes) for 210-mm rockets.

Breaking up the formations

The strength of the USAAF 'heavies' lay in their mutual defensive firepower. If the tight formations could be broken up, individual 'dicke Autos' (Luftwaffe slang for the bombers, literally 'fast cars') could be attacked at will. Many Bf 110G-2s were fitted with the WGr 210 system, which fired spin-stabilised rockets with pre-set timed fuzes. Although less accurate than cannon, they were intended to break up the formation and did achieve some kills. Below is an in-cockpit view of a rocket being launched, while at right rocket explosions can be seen amid a B-17 formation, one of which (centre left) is trailing smoke.

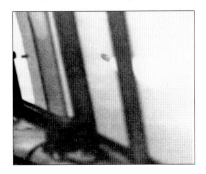

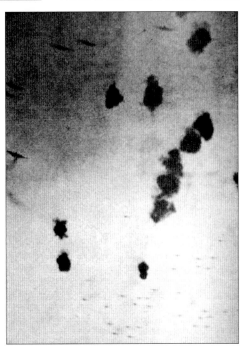

new weapons. The first of these was the 30-mm MK 108 cannon already carried by the night-fighter variants. This weapon fired 11-oz (312-g) high-explosive rounds at a rate of over six hundred per minute; on average three hits were sufficient to bring down a large bomber. Due to its relatively low muzzle velocity, 1,700 ft (518 m) per second, the weapon was ineffective at ranges much beyond 750 ft (230 m) however. That meant the defending fighters had to close to a point inside the bombers' defensive crossfire, if they were to use the weapon effectively.

The second of the new weapons was the 210-mm (8.27-in) rocket, adapted from the bombardment weapon in large-scale use by the German army. The tube-launched spin-stabilised rocket weighed 248 lb (112 kg). A time fuse

The hunter hunted – an FuG 220-equipped Bf 110G-4 falls prey to the P-51 of Captain Raymond Care of the 334th FS on 11 April 1944. Night-fighters were used in desperation against the daylight raids, but the extra drag of the radar arrays and flame-dampers meant they stood even less chance against the escort fighters than the G-2s flown by day-fighter units.

Warplane Classic

Luftwaffe Nachtexperten

Major Wilhelm Herget flew with NJGs 1, 3 and 4, and scored 57 night kills (plus 15 by day, many from the Battle of Britain). He was one of 22 Nachtjagdwaffe pilots who achieved 50 kills or more by night.

Oberst Helmut Lent successfully transitioned from Zerstörer to Nachtjagd roles, scoring 102 victories at night to add to eight by day and becoming the second highest-scoring night-fighter pilot.

The world's leading night-fighter ace, Heinz-Wolfgang Schnauffer survived the war with 121 victories to his credit. On 21 February 1945 he dispatched seven Lancasters in just 17 minutes.

This 9./NJG 3 Bf 110G-4 has ETC 50 bomb racks under the wings in place of the more normal drop tanks. Night-fighters were used for nocturnal ground attack duties in the latter half of 1944, following the Allied landings in Normandy.

detonated the 90-lb (41-kg) warhead at a preset range of between 1,800 and 3,600 ft (550 and 1100 m) from the launch point. The 210-mm rocket was a rather inaccurate weapon, however, and its main value was to inflict damage on bombers and force them to leave their formation. They could then be finished off at leisure. The Bf 110G-2 bomber-destroyer carried a hefty forward-firing armament comprising four 20-mm cannon, two 30-mm cannon and launchers for four 210-mm rockets.

As an alternative to the 20-mm and 30-mm cannon and the rockets, a few Bf 110G-2s were fitted with the R1 modification kit which included the third of the new air weapons: a Rheinmetall Bordkannon 3,7 heavy cannon fitted in a streamlined blister under the fuselage. This weapon was a modified version of the 37-mm anti-aircraft gun, and it fired 1.4-lb (0.63-kg) shells with a muzzle velocity of 2,800 ft (853 m) per second. With this weapon the Bf 110s could engage US heavy bombers from a distance of 3,000 ft (914 m), beyond the reach of their defensive fire, and a couple of hits would be sufficient to inflict lethal damage on a heavy bomber.

Generalfeldmarschall Erhard Milch's words in praise of the Bf 110 were true enough when he uttered them, at the end of August 1943. But the 'window of opportunity' for the Bf 110 to achieve much in the day bomber-destroyer role would prove to be narrow indeed.

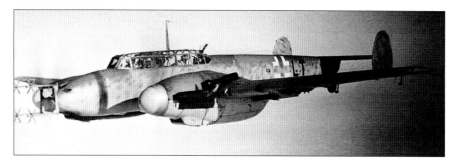

Below: This Bf 110G-2 of 7./NJG 1 – with black undersides and grey top surfaces – was probably used in the 'Beleuchter' role, in which it would track bomber streams and drop flares to mark their progress for other night-fighters.

Left: This aircraft, 'C9+EN', was a Bf 110G-4b/R3 flown by 34-kill ace Oberleutnant Wilhelm Johnen, Staffelkapitän of 5./NJG 5. On the night of 28 April 1944 Johnen was chasing an RAF aircraft when it entered Swiss airspace. Johnen pursued but was hit by flak and made an emergency landing at Dübendorf. To prevent the sensitive FuG 220 Lichtenstein SN-2b radar installation from falling into enemy hands, the Luftwaffe provided the Swiss with 12 Bf 109G-6s in return for the guaranteed destruction of the Bf 110.

On 4 October 1943 a force of 155 bombers made a deep-penetration attack on Frankfurt am Main, escorted by Thunderbolts fighters to a point just past the German frontier. Near Cologne Thunderbolts of the 56th Fighter Group, operating close to the limit of their radius of action, caught a Gruppe of Messerschmitt Bf 110G-2s moving into position to launch 210-mm rockets into the rear combat wing. The escorts shot down more than 10 of the heavily laden bomber-destroyers, without loss to themselves. The arrival of the escort held losses to just eight bombers.

The Luftwaffe reaction to the one-sided action on 4 October was to restrict the twin-engined fighters operating by day to the south and east of the line Bremen-Kassel-Frankfurt, supposedly to keep them out of reach of the US escorts.

Schweinfurt 2: day-fighters strike back

Ten days later, on 14 October, US heavy bombers set out to complete the destruction of the ball bearing plants at Schweinfurt. As during the previous attack on this target, it took the bombers far beyond the reach of their escorting Thunderbolts. Over München-Gladbach the Kommandeur of III./ZG 26, Hauptmann Fritz Schulz-Dickow, led 26 Bf 110s into action. The heavy fighters moved into position about 3,300 ft (1000 m) behind the bombers and launched salvoes of 210-mm rockets at them. Then the Bf 110s, together with Bf 109s and Fw 190s also engaging the raiders, closed in to finish off the damaged machines with cannon fire. During this action III./ZG 26 claimed the destruction of seven B-17s, for the loss of four Bf 110s.

That would turn out to be one of the major successes of

the Bf 110s in the role of daylight bomber-destroyers. In the weeks to follow the US escort fighters extended their reach, penetrating progressively further into Germany. As a result, the areas where the German heavy fighters could operate in relative safety shrank steadily. During the attack on Emden on 11 December 1943, for example, Thunderbolts of the 56th Fighter Group caught part of a force of 40 Bf 110s of I. and III./ZG 26 just after they had launched their rockets into one of the bomber formations. The Bf 110s claimed the destruction of two B-17s; but at a cost of seven Bf 110s shot down and two more which had to make forced landings.

From then on things got progressively worse for the Bf 110 units. On 20 February 1944 III./ZG 26 put up 16

The 'Haifisch' nose markings – originally used by II./ZG 76 – are an unusual feature of this Bf 110G-4c/R3 assigned to Stab I./NJG 4. The angled wingtip antenna arrays served the Siemens FuG 227 Flensburg system, which homed on the emissions of the RAF's Monica tail-warning radar, and various jammers such as Mandrel.

War diary of NJG 6

In March 1944 the hard-fought battle between Bomber Command and the Luftwaffe's night defenders was nearing its climax. The war diary of NJG 6 for a 12-day period during March 1944 gives some idea of the intensity of the battle being fought, as seen from the viewpoint of this unit. NJG 6 operated as part of Jagddivision 7, responsible for the defence of southern Germany. The Geschwader comprised two Gruppen of Bf 110s, I. Gruppe based at Mainz/Finthen and II. Gruppe at Stuttgart/Echterdingen. The unit also operated a few Ju 88s to seek out and report on the position of the bomber streams, and drop flares to mark them.

15 March: Target Stuttgart. Own take-off too early. Consequently there was a lack of fuel. Twenty-six Bf 110s and three Ju 88s took off. Three four-engined bombers shot down for certain, and two probable victories. Five Bf 110s crashed due to their running out of fuel, one made a belly landing and one force-landed at Zürich/Dübendorf [in Switzerland].
RAF losses 36 out of 863

18 March: British penetration into the area Frankfurt-Mannheim-Darmstadt [the target was Frankfurt]. Twenty-four Bf 110s and two Ju 88s took off. One bomber was shot down for certain, and three probable victories. One Bf 110 was shot down and one was rammed by an enemy night-fighter and crashed.
RAF losses 22 out of 846

22 March: Target Frankfurt. Twenty-one Bf 110s and two Ju 88s set out. Oberleutnant Becker scored six victories [Martin Becker was to end the war credited with 47 confirmed victories]. The air situation was not at all clear. The enemy turned when to the north of Terschelling, towards the south-east in the direction of Osnabrück, but this was not recognised. From Osnabrück on no contact was made. The enemy main force was not recognised until it was to the north of Frankfurt.
RAF losses 33 out of 816

23 March: Received false reports of an enemy force moving in an easterly direction. The target was Paris [in fact it was Laon]. Twenty Bf 110s and one Ju 88 took off, but in vain. *RAF losses 2 out of 143*

24 March: The enemy approached over the North Sea and Jutland, to Berlin. The return flight touched the northern tip of our own divisional area. Radio Beacon 12 was subjected to music-type interference. Crews encountered severe icing when breaking through the overcast. Vain attempts were made to make contact with the bomber stream during its return flight. Our own flares over Berlin were too high (20,000 ft [6096 m]). Very disciplined firing by the flak over Berlin. [i.e. the gunners restricted their fire to the lower altitudes to allow the night-fighters to engage]. Corps communication channels could be heard well, in spite of the enemy interference. In action were 11 Bf 110s to Berlin, five Bf 110s against the returning stream, three Bf 110s engaged in Himmelbett operations, and one Ju 88 reconnaissance aircraft. One victory to Oberleutnant Becker. *RAF losses 72 out of 811.*
(Other Luftwaffe units had obviously fared rather better)

26 March: About 500 bombers approached over the Zuider Zee on an easterly course towards the Rhine. They then turned south towards Essen-Oberhausen-Duisberg [the target was Essen]. Our radar and ground observers recognised the turn too late. Our own reconnaissance aircraft, a Ju 88 flown by Hauptmann Wallner, reported enemy activity only over the Ruhr area as a whole. The direction of the [enemy's] approach and return flights could not be recognised from the running commentary. Therefore it was not possible to get into the bomber stream. Due to the devious approach and the strong headwind, II./NJG. 6 did not arrive at the target before the end of the attack. Severe icing was reported. Twenty-one Bf 110s engaged on *zahme Sau* operations, three Bf 110s on Himmelbett missions, one Ju 88 on reconnaissance. Three Bf 110s ran out of fuel and crashed, and one made a belly landing. *RAF losses 9 out of 705*

A few night-fighters served in the far north with what was established in July 1944 as the Nachtjagdstaffel Finnland, becoming the NJSt Norwegen when the Luftwaffe retreated into Norway in November. Although this unit was principally equipped with Ju 88G-6s, it also had a few Bf 110s on strength, like this Bf 110G-4c/R3 seen at the final operating base at Oslo-Gardemoen. In March 1945 the Staffel became 4./NJG 3.

Although most Bf 110G night-fighters were fitted with the short-barrelled 30-mm MK 108 in the upper nose position, this aircraft has the longer-barrelled MG 151/20. The aircraft has the SN-2c version of FuG 220, a version of the original with decreased minimum detection range, in turn allowing the deletion of the single FuG 212 Weitwinkel array. A subsequent improvement involved angling the array at 45° to produce the SN-2d.
By the end of the war a very few Bf 110G-4s – including that flown by leading ace Schnaufer – had received the FuG 218 Neptun VR radar with a similar but smaller array mounted on a single support pole.

Final developments in the night war

Returning to the night battle, let us look at the *zahme Sau* tactic as it had developed by the beginning of 1944. When the order came to take off, crews boarded their aircraft and taxied to the take-off point. As they did so they received the radioed broadcasts giving the latest position, height, course and strength of the bomber stream. Each unit also received the code-names of the radio beacons it was to make for after take-off. Once airborne, the night-fighters climbed to the bombers' altitude and headed from beacon to beacon until they reached the force assembly beacon. The latter was chosen to be as near as possible to the anticipated flight path of the bombers. There might be as many as 50 fighters circling above the beacon in the darkness, constituting a flight safety hazard that would be unthinkable in peacetime. Yet, remarkably, few night-fighters were lost there in collisions. When the order came to leave the beacon, the night-fighters headed towards the latest position they had been given on the whereabouts of the bomber stream and began searching there.

Often the first positive indication that a night-fighter crew received of the proximity of the bomber stream was their aircraft juddering as it passed through the wake of churned air that trailed behind each bomber. At other times, the initial pick-up was on SN-2 radar or with a Flensburg or a Naxos homer. Once the crew had radar or visual contact with a bomber it was not to attack immediately. Instead it was to pass to its divisional control centre the bomber's location and heading, so that other night-fighters could be guided to that area. Only then could the crew deliver its attack.

Using these tactics the defenders inflicted serious losses on raiding forces on several occasions during the early months of 1944. On 21/22 January, 55 bombers were lost out of 648 attacking Magdeburg, and a week later 43 were shot down out of 683 attacking Berlin. But even these huge scores were eclipsed on 19/20 February when 78 bombers were shot down out of a force of 823 attacking Leipzig.

The greatest night battle of all time took place on 30/31 March, when 795 bombers set out to attack Nuremberg. Elements of 20 Gruppen of twin-engined fighters, with

Bf 110s to engage one of the raiding formations. Thunderbolts of the 56th Fighter Group again hit the German unit hard, as they 'bounced' part of the force from out of the sun and shot down 11 Bf 110s.

The defenders suffered another bad day on 6 March 1944, when the US 8th Air Force delivered its first large-scale attack on Berlin. Bf 110s of III./ZG 26, I. and II./ZG 76, as well as night-fighters from I., II. and III./NJG 5, engaged the raiding force in front of the capital with a total of 52 aircraft. With the bombers came the P-51s of the 4th, 354th and 357th Fighter Groups, however. In the *melée* that followed the latter shot down 17 Bf 110 bomber-destroyers and night-fighters, nearly one-third of those committed. In a similar action 10 days later, ZG 76 put up 43 Bf 110s to engage a US raiding formation near Augsburg, and lost 26 aircraft with 10 more making forced landings.

Such loss rates could not be sustained and within a few weeks both ZG 26 and ZG 76 had given up their Bf 110s in favour of the higher performance twin-engined Me 410 destroyer. In the face of the overwhelming forces of US escorts the Me 410s did little better, however, and in the summer of 1944 those units were disbanded.

Destruction of a Bf 110

During the final months of the World War II, Luftwaffe night-fighter crews faced the unwelcome attentions of Mosquito night-fighters from the RAF's No. 100 Group. The latter flew patrols deep into Germany, acting as escorts for the bombers. Typical of these operations was that during the early morning darkness of 19 March 1945, when bombers attacked Hanau and Witten. Flight Lieutenant Winn, flying a Mosquito of No. 85 Squadron, later reported:

"On patrol in the Hanau target area obtained several Perfectos contacts [Perfectos: device carried in RAF night-fighters to home on emissions from the German IFF identification equipment.], one of which was selected. After a 15-minute chase this was converted to AI [radar contact], range six miles. Continued in an easterly direction for a further five minutes and obtained a visual, range 1,500 ft [457 m], height 5,000 ft [1524 m]. Target was followed in and out of cloud. Eventually he commenced a gentle descent, followed by the Mosquito, which led to a fully lighted airfield (Kitsingen). After several chases round the airfield the enemy aircraft decided to land; closing to 400 ft [122 m] the target was recognised as an Me 110 with its undercarriage down. The first burst, fired from 100 ft [30 m], scored no strikes. The Mosquito had great difficulty in keeping behind without stalling. A second burst was fired as the enemy aircraft was on the approach, causing the port engine to explode. The enemy aircraft dived into the ground and was scattered over a wide area on the downward end of the flare path. 0521 hours."

The airfield was not Kitsingen but Gerolzhofen, situated a few miles to the north-east. Oberfahnrich Erichsen and his crew survived the crash, though all suffered wounds.

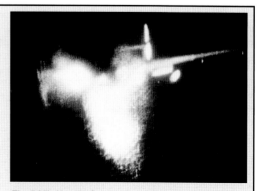

The RAF's No. 100 Group created 'Moskitopanik' in the Luftwaffe night-fighter units. Here a II./NJG 1 Bf 110 falls to a No. 85 Squadron 'Mossie' in November 1944.

about 200 aircraft, engaged the force and 94 bombers were shot down.

At this, the high-water mark of its existence, the Luftwaffe night-fighter force had 565 night-fighters on strength: Ju 88s, Bf 110s, Do 217s and He 219s. But still nearly 60 per cent of the force, about 320 aircraft, comprised Bf 110s.

Between 18 November 1943 and 31 March 1944 RAF Bomber Command lost 1,047 aircraft, of which three-quarters probably fell to night-fighters. With so many victories to be shared out it, is no wonder that the leading night-fighter aces amassed high scores. In October 1944 the then top scorer, Major Helmut Lent, Kommodore of NJG 3, was killed in a flying accident. At the time of his death his victory score stood at 110 (102 at night), all of them while flying the Bf 110.

By the end of the war ended Helmut Lent's huge score had been topped by another Bf 110 pilot, Heinz-Wolfgang Schnaufer, one of whose actions we observed earlier in this account. He ended the war as a Major and Kommodore of NJG 4, with 121 confirmed victories.

From the invasion of France in June 1944 until the war's end, the story of the Luftwaffe night-fighter force was one of unremitting decline. First the capture of France punched a great hole in the Luftwaffe's radar cover, seriously reducing the warning time available to the defenders. Then the RAF started jamming the SN-2 radar, rendering it almost useless. From the autumn of 1944 German fuel reserves fell drastically, as one by one the refineries and oil storage centres were bombed to ashes.

At the end 1944 the Luftwaffe night-fighter force had 913 aircraft on strength, but many of these remained on the ground for want of fuel. During this period, in addition to their accustomed tasks, Luftwaffe night-fighter units had also to operate in the ground attack role. For this purpose the Bf 110 night-fighter carried a rack for two 250-kg (551-lb) bombs under the fuselage and, if no drop tanks were carried, two 50-kg (110-lb) bombs under each wing. The use of the Bf 110 in this role was a measure of desperation: it was a hazardous business, and losses were high with few successes.

Production of the Bf 110 finally ended in March 1945, after nearly six thousand of these aircraft had been delivered to the Luftwaffe

The Bf 110 in retrospect

When it was introduced into service in 1939, the Bf 110 was the most effective long-range fighter in the world. It soon established a formidable reputation as a bomber-destroyer, and halted in its tracks the RAF campaign to attack warships operating in German coastal waters.

During the Battle of Britain in the summer of 1940 the Luftwaffe employed the Bf 110 in the bomber escort, fighter-bomber and photographic reconnaissance roles. In the escort role the type took a heavy beating at the hands of the far more agile RAF Spitfires and Hurricanes. Yet the type operated effectively in the roles of fighter-bomber and reconnaissance aircraft in the years that followed.

In the summer of 1943 the Bf 110 took on a new lease of life as a home defence day-fighter. So long as the US heavy bombers made deep penetration attacks that took them beyond the reach of their escorting fighters, the heavily armed Bf 110G-2s could operate against them with considerable effect. Yet by the end of 1943 the increased penetration range of the P-47 Thunderbolt, followed by the introduction of the superb P-51 Mustang, imposed heavy attrition on the Luftwaffe twin-engined fighter units. In the spring of 1944 the Bf 110 passed out of service in this role.

Yet it was as a night-fighter that the Bf 110 made its greatest impact on the conflict. From the summer of 1940 until the final year of the war the Bf 110 was the backbone of the Luftwaffe night-fighter force, and it proved a ferocious opponent for the RAF night raiders. Not only did it achieve more night victories than any other fighter type in World War II, but it probably achieved more night kills than all the other night-fighter types put together, during that conflict. Finally, but not least of its claims to fame, the Bf 110 was one of the very few World War II combat aircraft to operate in front-line service from the very first day of that conflict, to the very last.

Dr Alfred Price; additional material by David Donald

Above: *A trio of Bf 110G-4c/R3s from NJG 1 is seen on a daylight patrol, a risky practice that was largely abandoned in the spring of 1944. Although in the final months of the conflict crippling fuel shortages reduced sortie rates, the Nachtjagdwaffe remained a potent force until the end of hostilities, one of the few areas in which the Luftwaffe had not been completely outclassed by the Allies.*

Above left: *The Bf 110Gs of 12./NJG 3 are seen lined up – possibly in Norway – after they had surrendered. The rudders have been removed as a hasty means of immobilising the aircraft. The Staffel was based in Denmark and northern Germany in the later stages of the war.*

Foreign use of the Bf 110 was restricted to a few G-4 night-fighters being operated by the Italians and Romanians. However, the type served briefly in French markings both before and after the end of the war. The Groupement Patrie had been created spontaneously in North Africa to fly Bostons and Marylands, and one Bf 110 (below). This unit was subsequently formalised as GB I/34 'Béarn'.

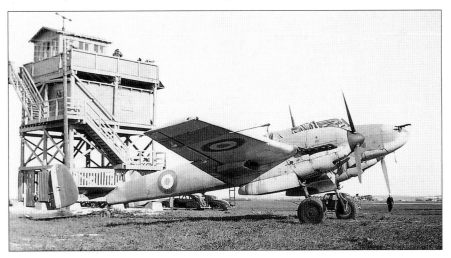

AIR COMBAT

RAF Liberators at war

Part 2: Service abroad, variants and operators

If the RAF's Liberators have been largely forgotten in many modern histories of the type, the part played by Coastal Command's aircraft in the Battle of the Atlantic has received greater attention in recent years, and even the transport aircraft that kept the Atlantic air bridge open have had a mention or two. By contrast, probably the least well-known RAF Liberators were those that were based overseas, although ironically these were the only RAF aircraft used in the type's originally intended heavy bomber role. This is a tragic omission, since RAF Liberators based in the Mediterranean, Middle East, West Africa and the Far East played a vital, war-winning role, amassing a record that was no less distinguished than that of their better known and more widely publicised USAAF counterparts.

The Liberator's relatively modest bomb load and its white-hot glowing supercharger discs (which acted like a locator beacon for enemy night fighters) ensured the type was soon rejected by Bomber Command, who had little use for an aircraft which could not tote a 4,000-lb (1814-kg) 'Cookie' bomb on night raids over the Reich. Plans to equip an entire home-based Bomber Group with Liberators were cancelled. The modern, high-performance, long-range Liberators thus became available to Commands who would otherwise have had to rely on Bomber Command's worn-out and obsolete cast-offs, including Armstrong Whitworth Whitleys and Vickers Wellingtons, and in these Commands the Liberators were destined to make a massive difference. In the Middle East, the Liberator gave the RAF an aircraft with the range to cover the entire theatre, while in the Far East, Liberator formations made attacks on targets that could be reached by no other Allied bomber until the advent of the B-29.

The first RAF Liberators based overseas were a quartet of Liberator IIs delivered to the Wellington-equipped No. 108 Squadron at Kabrit in Egypt during December 1941. Initially

Main picture: Finally replacing the obsolete Vickers Wellington from 1944, the Liberator B.Mk VIs and VIIIs provided RAF squadrons in Asia with a modern long-range bomber with which to strike installations deep within Japanese-held territory. Typical of the strategic targets selected, here No. 356 Squadron B.Mk VIs set Japanese oil installations ablaze as the Allies attempted to starve the opposition of fuel.

Above: RAF groundcrew prepare to load 500-lb (227-kg) bombs on a Liberator B.Mk VI at an Indian base in late 1944. The RAF's Liberators in Asia not only conducted day and night bombing raids, but also flew anti-shipping, supply, transport, maritime patrol and Elint missions.

RAF Liberators: Part 2

The Liberator II was thrust into combat in the Middle East and North Africa in early 1942. Able to operate from rough, sandy strips in Egypt, the aircraft proved far superior to the resident Vickers Wellingtons (seen in background; above) and Nos 159 and 160 Squadron aircraft flew day and night bombing missions in the Suez Canal zone. Most aircraft retained the mid-upper gun turret for defence against enemy fighters (right).

unarmed, the aircraft were fitted with turrets and plans were put in place for the new type to replace the unit's ageing Wellingtons. These plans were cancelled however, and only two of the aircraft were used for bombing missions, the rest used for supply-dropping before being withdrawn in June 1942. A small rump of No. 108 Squadron used the Liberator again during November and December 1942, before disbanding.

Combat in the desert

The next RAF Liberators in the Middle East were actually aircraft that had originally been intended for service in the Far East. Two squadrons, Nos 159 and 160, were formed at Molesworth and Thurleigh, respectively, in January 1942. No. 159 Squadron was held in the Middle East en route to India, flying long-range bombing missions from Fayid until May, when the squadron began to continue its journey east in dribs and drabs. No. 160 arrived in the Middle East in June 1942, and settled into operational flying while the Middle East Air Command's own bomber squadrons converted to the new type.

No. 257 Wing in the Suez Canal Zone had flown day- and night-bombing missions with its Wellingtons, before being redesignated as No. 205 Group in September 1941, by which time it was clear that a more modern, longer range bomber was required. Although an early decision was taken to re-equip the group with Liberators, the process was extremely slow. On 15 January 1943, a new unit, No. 178 Squadron, formed at Shandur (a Canal Zone airfield) from a cadre provided by No. 160 Squadron, and a 'Special Liberator Flight' formed at Gambut in Libya on 14 March 1943, later becoming part of No. 148 Squadron.

No. 178 Squadron did not become fully-equipped with the Liberator for almost a year, since its Liberator IIs were augmented by Halifaxes between May and September 1943.

No. 148 Squadron's Liberator IIs and Mk IIIs were phased out in January 1944, while No. 178's were replaced by Liberator B.Mk VIs from December 1943.

In the Middle East, Liberator Mk VIs usually flew without nose, belly or mid-upper guns fitted, relying on the tail turret alone. This was because it had been found that there was relatively little enemy night-fighter opposition in the areas in which the Liberators operated, and that such opposition never involved head-on attacks. This allowed the removal of the nose and mid-upper turret guns, and sometimes of the turrets themselves. The belly turret was dispensed with because the glow from the turbo-supercharger discs effectively robbed the turret's occupant of any night vision.

The Liberator Mk VIs of No. 178 Squadron were augmented by four further RAF Liberator bomber squadrons during 1944, while the Balkan Air Force SOE gained a further Liberator squadron, finally replacing the Liberator IIs that No. 148 Squadron had lost in January.

Before the new RAF Liberator units could become operational, however, two Dominion squadrons arrived in-theatre. These were the South African Nos 31 and 34 Squadrons, which formed No. 2 Wing, SAAF. This, in turn, formed one-quarter of the RAF's No. 205 Group at Foggia, alongside three wings of RAF Vickers Wellingtons, one of which included a single Liberator Squadron (No. 178). The group was commanded by Maj. Gen. J. T. Durrant, SAAF.

No. 31 Squadron arrived in North Africa on 30 January 1944, and converted to the Liberator with No. 1675 Conversion Unit at Lydda. Whereas Liberator aircrew destined for Coastal and Bomber Commands converted to type at Nassau and while SEAC Liberator aircrew converted in Canada, aircrew for the Middle East squadrons did so in theatre.

On 27 April 1944 No. 31 Squadron moved to a new base, Kilo 40, about 25 miles (40 km) north of Cairo, where it was soon joined by No. 34 Squadron. The unit flew its first operational sorties on 27 May 1944 against targets on Crete, and moved to Foggia in southern Italy on 16 June 1944. Early operations by the SAAF Liberators included mining operations in the Danube and attacks on oil targets, including Ploesti, from 26 July 1944.

The first of the new RAF units was the Halifax-equipped No. 614 Squadron at Amendola, which began receiving Liberator B.Mk VIIIs in August 1944. These were among some 36 Mk VIIIs delivered to the RAF in the Mediterranean with centimetric radar in place of the usual belly turret, allowing them to be used in the Pathfinder role.

No. 205 Group was tasked with long-range, strategic night saturation bombing, usually operating as a group (or in wings) using the maximum number of serviceable aircraft to attack a single target – perhaps a railway marshalling yard or an oil refinery – in exactly the same way as the home-based Bomber Command groups.

Just like Bomber Command, No. 205 Group used a 'Pathfinder Force' of Blind Illuminators, which dropped parachute flares to light up the ground for the Target Marker bombers, which flew three minutes behind. The Target Markers identified the target and marked it with brightly coloured marker flares. Six minutes behind the Blind Illuminators came the Main Bomber Force, which began dropping at the selected 'Blitz' time. Even the Main Force bombers flowed over the target in a particular order, with aircraft carrying 500-lb (227-kg) and 1,000-lb (454-kg) bombs preceding aircraft carrying 2,000-lb (907-kg) 'Cookies', which typically bombed at 'Blitz' plus three and with the incendiary bombers following these at 'Blitz' plus four.

The Liberator II and B.Mk III versions, used by the first RAF Liberator operators in the Middle East and Asia, lacked the power-operated front turret of the later B.Mk VIs and B.Mk VIIIs. The application of nose art and mission symbols to the aircraft was commonplace.

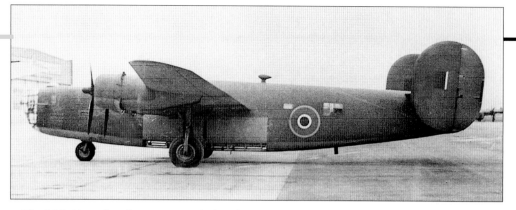

As the RAF's version of the B-24D, the Liberator B.Mk III was initially delivered with Martin A-3 top turrets and Consolidated A-6 tail turrets (below). Of the approximately 160 B.Mk IIIs delivered to the RAF, 15 were immediately allocated to Middle East Air Command, followed shortly after by 51 aircraft delivered directly to India. Operated by Nos 159, 160 and 355 Squadrons in Asia, the B.Mk III played a vital role in 1943 establishing the most effective way of attacking long-range targets, before the arrival of the definitive B.Mk VI in 1944. FK239 (left) is seen on its delivery flight through Dorval, Quebec, in 1942.

The Main Force bombers were guided to the aim point individually by their bomb-aimers, with the Bombing Leader assessing accuracy as he stood advising the Main Bomber Force as to any necessary corrections using the clock code relative to the marker fires.

There was a heavy emphasis on accuracy, and individual aircraft's bombing results were carefully assessed. For this purpose a camera was mounted in the tail of each aircraft, designed to photograph its bombs bursting on the target. As the bomb-aimer released the bomb train a flash flare was also released and the camera's shutter opened. The flash flare hung in a small parachute with its fuse mechanism set so as to explode with a bright flash exactly above the aircraft's bombs bursting below.

In addition to these conventional, medium-altitude bombing operations, No. 205 Group also flew large numbers of unconventional missions with its 'heavies', including some ultra low-level attacks, and including some low-level missions dropping supplies to partisans and resistance groups, rather than bombs on the enemy.

From 4 August 1944, No. 205 Group Liberators were heavily involved in operations in support of the ill-fated Warsaw rising. While the Russians (who had encouraged the rising) stood by a few kilometres away, waiting for the Germans to wipe out the hated Poles before advancing, General Bor Komorowski requested British and American assistance for the Home Army (the Polish resistance) in its struggle. Churchill rejected his request for the despatch of a British Airborne Brigade, although the First Polish Independent Parachute Brigade had been specifically formed to support just such a rising. It had originally been stipulated that the Parachute

Brigade would be the only element of the Polish Forces in Exile to come under the direct command of the Polish Government, but in the spring of 1944 General Montgomery had successfully lobbied for the brigade to be put under his command for the coming invasion of Europe. The Poles were forced to stand aside while the uprising was crushed, and the brigade (along with the rest of Britain's 1st Airborne Division) would be thrown away at Arnhem in September.

Polish supply-drops

Having denied the Poles their own paratroops, Churchill felt obliged to accede to their request for the air dropping of supplies, whatever the cost, and between 8 August and 22 September 1944, the RAF and SAAF mounted 196 sorties to drop weapons and supplies to the resistance in Warsaw.

Seven aircraft from No. 1586 Special Duties Flight made successful drops on 8 and 9 August, and 11 (including some Halifaxes from No. 148 Squadron) returned successfully on 12 August. The SAAF Liberator units and No. 178 Squadron joined the operation on 13/14 August, when 28 Liberators were despatched, 14 of which dropped successfully, 11 of which failed to find their drop zones and two failed to return.

One of the latter aircraft landed at Emilchino near Kiev, after an unusual mission. Bracketed by flak, the nervous No. 31 Squadron captain jettisoned his load one mile short of the DZ, before being hit in the port outer engine. This was enough for the captain, who put on his parachute and bailed out through the bomb doors, without warning the crew. Co-pilot Second Lieutenant 'Bob' Burgess became the youngest recipient of the DSO (Distinguished Service Order) when he took command of the crippled aircraft and flew it eastwards to safety, before pulling off his first night landing!

On 14/15 August only 12 aircraft of 26 found the drop zones, and eight failed to return. With a loss rate of 15 per cent, three times higher than the rate Bomber Command normally accepted, Jimmy Durrant (the SAAF CO of No. 205 Group) persuaded Air Marshal Sir John Slessor, (the RAF C-in-C in the Mediterranean) that the drops over Warsaw were prohibitively costly in precious aircraft and crews, and were of no military value. Unfortunately, neither he nor Slessor were able to convince Churchill to end the airlift, even with the support of Bomber Command's C-in-C, Sir Arthur Harris.

From 15 August the South African squadrons alternated their nightly drops with No. 178 Squadron. Slessor briefly halted the drops on 17 August (following the loss of 17 aircraft in just five days). No. 1586 Special Duties Flight continued the airlift from 20 August to 2 September, but lost eight aircraft in 16 sorties, and No. 34 Squadron, SAAF, and No. 178 Squadron rejoined the fray, losing five aircraft on 11 September alone. The last drops to the Polish Home Army were made on 22 September 1944. Of 196 sorties despatched (186 using Liberators), just 85 reached the target area, and

The RAF's Liberator B.Mk VI designation did not correspond to a single USAAF version, but was applied to a range of B-24Js and B-24Ls from a variety of production factories. Although the new power-operated front turret provided a much more effective deterrent to enemy fighters, and could be used at low level to strafe ground or maritime targets, the increased drag and corresponding degradation in speed ensured that the version was not universally popular with crews.

39 aircraft (31 Liberators) were lost. Despite dropping from low altitude (about 150 ft/46 m) and at low speed (130 mph/209 km/h with flap) much of the material dropped fell straight into German hands, and the Polish Home Army surrendered on 2 October 1944, after 63 days of fighting. The SAAF Wing lost 24 aircraft in six weeks, while No. 1586 Flight suffered almost 100 per cent casualties during the operation.

Supply-dropping and bombing missions continued to provide the SAAF and RAF Liberators in Italy with their 'bread and butter' for the remainder of the war, sometimes occasioning heavy losses.

On 12 October 1944, for example, 20 No. 31 Squadron SAAF Liberators set out from Celone in southern Italy to supply partisans in the Turin area, but six crashed in the Italian Alps with the loss of all 48 aircrew, due to an unpredicted change in the weather.

The dedicated supply-droppers of the Balkan Air Force SOE gained another Liberator unit when No. 301 Squadron reformed at Brindisi on 7 November 1944, while No. 205 Group's remaining bomber squadrons re-equipped with Liberator Mk VIs from October 1944.

No. 37 Squadron at Tortorella re-equipped from late October 1944, while No. 70 Squadron at Cerignola converted in January 1945 and Nos 104 and 40 Squadrons at Foggia began conversion in February and March, respectively.

Following the end of the war in Europe, the Liberator squadrons moved back to Egypt, where they disbanded or re-equipped with Lancasters. The surviving aircraft were gathered at the Maintenance Unit at K40, where German PoWs cut off the tailplanes and drove iron spikes into the engines. The aircraft were then abandoned to the elements and the locals, who quickly stripped them for scrap.

West Africa

Probably the smallest force of RAF Liberators overseas was based in the Gambia in west Africa, where there was a need to provide ASW cover for convoys from South Africa, which were threatened by small numbers of U-boats operating off the coast. This was mostly provided by flying boats (Sunderlands and Catalinas) operating from Sierra Leone and Senegal, and with a 'hub' at Bathurst (now Banjul) in the Gambia. The flying boat bases included the charmingly named Half Die Camp at Bathurst.

No. 200 Squadron, which had flown out to the Gambia in June 1941, re-equipped with Liberator GR.Mk Vs in July 1943. Although the West African maritime force made few contacts with enemy submarines, it had a massive deterrent effect and no ships were lost within range of the Sunderlands and Liberators while it was on patrol.

Far East bomber operations

When Japan joined the war in December 1941, it quickly captured Malaya, Singapore and most of Burma, and soon stood poised at the borders of India. It was immediately obvious that Air Command South East Asia (ACSEA) would need a heavy bomber force to attack enemy strategic targets at long ranges, and to counter enemy naval forces and submarines in the Indian Ocean. With forces in Europe being given the highest priority, all Lancaster, Halifax, and Stirling production was set aside for Bomber Command's offensive against the Reich, and the decision was taken that ACSEA's bomber units would be given Lend-Lease Liberators.

Even as the dust settled on Pearl Harbor, the RAF immediately began preparing to build up its forces in India, and in January 1942 two new Liberator squadrons, Nos 159 and 160, were established at Molesworth and Thurleigh. No. 159 was held in the Middle East between January and September 1942, while No. 160 flew ASW patrols from Northern Ireland in May and June before following No. 159 to the Middle East.

RAF Liberator operations in Asia and the Middle East were often conducted in primitive conditions, and it is a testament to the durability of the type that serviceability remained superior to many British-built types throughout its service period in-theatre. Goats and sheep are seen here using the shade cast by B.Mk VIII KG946 to shelter from the savage midday sun.

No. 159 Squadron's Liberators finally arrived at Salbani in October 1942, and soon settled into a routine of relatively small scale but very long-range night-bombing missions, involving between two and five aircraft. A second Indian-based Liberator squadron, No. 355, formed at Salbani on 18 August 1943, and flew its first operation on 20 November 1943. The third, No. 356, stood up at Salbani on 15 January 1944 and flew its first bombing operation on 27 July.

Aircrew for further ACSEA Liberator squadrons came mainly from No. 5 Operational Training Unit (OTU) that had formed at RCAF Boundary Bay on 1 April 1944. This Canadian base provided good flying weather, uncrowded airspace and the vast Pacific Ocean was adjacent, allowing SEAC-type long over-ocean missions to be accurately simulated. The OTU course provided approximately 35 hours of Mitchell flying and 70 hours on Liberators for pilots, engineers and navigators, with gunners and wireless operators flying only in the

African action earns Victoria Cross

Although vital in deterring enemy surface ships and U-boats from operating off the coast of west Africa, No. 200 Squadron's deployment to the Gambia saw only sporadic combat engagements with the enemy. However, despite the relative inactivity, one of the squadron's Liberators was involved in a ferocious engagement that would lead to the destruction of a U-boat, but also cost the lives of the aircrew.

On 11 August 1943, Flying Officer Lloyd Trigg, RNZAF, flying BZ832/D sighted U-486 running on the surface about 240 miles (386 km) south-west of Dakar. The U-boat was on route home to La Pallice, but its Captain, Oberleutnant Zur See Clemens Schamong, had ensured that its guns were manned and its crew alert. Trigg's Liberator was hit as it attacked, and the centre section fuel tanks were set ablaze. Trigg pressed home his attack, however, dropping a stick of six depth charges with unerring accuracy. Though his aircraft was then shot down, with the loss of all aboard, the U-boat had been mortally wounded, and sank 20 minutes later. Twenty of the crew escaped, but most were killed by sharks, and Schamong, his first lieutenant and five crew escaped by clambering aboard a dinghy from the crashed Liberator. The submariners were picked up by a Sunderland the next day and their testimony led to the award of a posthumous Victoria Cross to Trigg.

With a declining U-boat threat, No. 200 Squadron transferred to India in March 1944, joining what would be the RAF's largest force of Liberators.

FLYING OFFICER L. A. TRIGG, V.C.

Left: Following the successful invasion of Italy the RAF quickly established this newly-held territory as a base for Liberator B.Mk VI operations against enemy-held targets throughout southern and central Europe. With airfield space at a premium, facilities were often shared with USAAF units, as seen here as a pair of co-located Fifteenth Air Force P-38 Lightnings pass low over a No. 614 Squadron Liberator B.Mk VI at Foggia.

Below: Liberator operations conducted from Italy by the RAF's No. 205 Group and Balkan Special Operations Executive (SOE), combined with the two Liberator squadrons from the South African Air Force's No. 2 Wing, provided the Allies with a powerful force for attacking retreating German forces and supplying partisan fighters. No. 70 Squadron re-equipped with Liberator B.Mk VIs in January 1945, and missions conducted included minelaying of the Danube River. KL527 'W' is seen here while operating with the unit from Foggia.

Liberators. By contrast, USAAF B-24 pilots received 250 hours instruction

Graduating crews from No. 5 OTU were usually posted to No. 6 Ferry Unit, RAF Transport Command, whose task was to ferry new Liberators for ACSEA from Dorval, Canada, to the Far East. This proved to be an excellent means of getting Liberators to India, giving new crews further useful experience on type, and also releasing experienced transport crews for more important duties.

The destination for the Liberators from Dorval was Karachi's Drigh Road airfield, where aircraft were formally taken on charge before being flown on to Bamrauli (Allahbad), which was the reception park for aircraft and crews posted to the Liberator squadrons of No. 231 Group.

The next Liberator bomber squadrons to form were both existing Wellington bomber units. No. 215 Squadron at Digri began converting in July 1944, re-entering the lists in October, while No. 99 Squadron traded its Wellingtons for Liberator Mk VIs in September 1944 at Dhubalia.

Boosting the range

While the new units worked up to fully operational status, No. 159 Squadron developed and refined new operating procedures to increase its aircraft's range and endurance. Before this work began, a Liberator bomber had an effective range of about 1,200 miles (1931 km), carrying a modest 3,000-lb (1361-kg) bombload. Under the guidance of its CO, Wg Cdr J. Blackburn (an experienced Liberator hand from the Middle East) the squadron was able to increase range to about 3,000 miles (4828 km), while almost trebling the bombload to 8,000-lb (3629-kg).

Blackburn increased the distances flown gradually, first attacking the Kra Isthmus (2,300 miles/3701 km from base), then later the Malay Peninsula (2,800 miles/4506 km away) and finally the approaches to Penang harbour, 3,000 miles (4828 km) away.

This was achieved by increasing fuel tankage, and by stripping the aircraft of their ball turrets, nose guns, beam guns and armour, and by flying with skeleton crews, sometime including only a single pilot, a wireless operator, a navigator, a bomb-aimer and a single gunner.

The first 17-hour mission to mine Penang harbour was flown by 12 No. 159 Squadron Liberators on the night of 27 October 1944. The aircraft were modified and manned as detailed above, with fuel tanks in the forward bomb bay and about six unsweepable Bakelite mines in the rear bay. The day before the mission the aircraft deployed from Digri to an advanced landing ground near Calcutta where a longer runway was available – a useful precaution for fuel-laden aircraft that would be taking off 5,000 lb (2268 kg) overweight.

Blackburn took off first, having arranged to fire a green Very light to signal to the waiting aircraft to follow if he judged it safe. Once airborne the aircraft climbed very slowly to about 20,000 ft (6096 m), then almost immediately began an equally gentle descent to the target area, where the mines were laid from about 50 ft (15 m), with a complete lack of enemy opposition. No. 159 Squadron flew further long-range mining missions on 26 November, 10 January and, with 16 aircraft, on 11 January 1945.

Blackburn received an American DFC and a commendation for this work, and the USAAF quickly adopted his techniques for increasing the bombload and range of its own B-24s.

By this time No. 159 Squadron had been joined by new Liberator units. No. 215 Squadron rejoined the fight in October, and No. 99 flew its first Liberator operation (against railway targets at Pyininana in Burma) on 26 November, just beaten to this milestone by the newly formed No. 355 Squadron, which flew its first mission on 19/20 November, when three aircraft bombed the railway station at Mandalay. The squadrons initially concentrated to some extent on particular types of mission, with No. 159 serving as the mining experts, and with No. 99 and No. 215 often hitting the Burmese railway system, occasionally using radio-controlled Azon bombs, but usually relying on conventional precision bombing.

Apart from No. 159 Squadron, the SEAC Liberator squadrons used distinctive symbols on their fins and rudders to identify them to Allied fighters. No. 355 Squadron used black rudders on which were painted two vertical white stripes, while No. 356 used a large white diagonal cross on each face of the black rudders. No. 215 Squadron also used black rudders, but with two horizontal white bands, while No. 99 Squadron painted the top half of its aircraft's tail fins black and applied a white disc above the fin flash. Though these markings were all extremely distinctive, they did not entirely prevent 'friendly fire' incidents. In January 1945, for example, US Navy F4F Wildcats shot down six RAF Liberators dropping supplies to insurgents in French Indo-China.

Seen in desert camouflage, this B.Mk VIII was another of the aircraft assigned to operations in Italy. Italy-based No. 614 Squadron re-equipped with the Liberator between August 1944 and March 1945, replacing Halifax Mk IIs, and remained in-theatre until the end of the war.

Above: With the ventral ball turret much in evidence, this Consolidated-built B.Mk VI is seen soon after delivery to the UK, with its US serial number still present on the fin. The aircraft went on to serve with No. 357 Squadron in India between September 1944 and November 1945.

Top: No stranger to the sub-continent, having operated de Havilland DH.9As in India in 1919-20, No. 99 Squadron returned in June 1942 equipped with Vickers Wellingtons. The squadron is seen here at its home base of Dhubalia, Bengal, having re-equipped with the Liberator B.Mk VI, which it used from September 1944 on long-range day- and night-bombing missions against Japanese targets in Burma.

During the final months of the war the Liberators flew a wider variety of missions, including close air support for the 14th Army as it steadily pushed the Japanese out of Burma, bombing and strafing command posts, gun positions, troop concentrations, airfields, and road, rail and river traffic. Although aircrew and aircraft continued to arrive in theatre, the Liberator bomber force actually dwindled in size slightly.

No. 358 Squadron, for example, had been intended to join as another bomber unit, but flew only a single bombing raid on 13 January 1945 when eight aircraft bombed Mandalay, after which it was immediately transferred to the special duties role, dropping supplies and agents behind enemy lines. No. 215 Squadron was also released from bombing operations in April 1945. Because of the urgent need to keep the army supplied as it advanced through Burma, No. 215 Squadron converted to Dakotas and transferred to the transport role.

Maritime, Special Duties and Elint

While No. 159 Squadron remained dedicated to the bomber role when it finally arrived in India, its sister unit, No. 160 Squadron did not join No. 231 Group when it arrived in January 1943, but instead became a general reconnaissance (maritime patrol and ASW) unit reporting to No. 222 Group and based on the island of Ceylon. A second Liberator general reconnaissance unit, No. 354 Squadron, formed at Drigh Road, Karachi, on 10 May 1943 and moved to Cuttack in August 1943, finally moving to Ceylon in October 1944. Three further Liberator general reconnaissance units served in ACSEA. No. 200 Squadron moved from West Africa to St Thomas's Mount in March 1944, while No. 203 Squadron at Madura converted from Wellingtons to Liberators in October 1944, and then moved to Kankesanturai in Ceylon where it commenced operations in February 1945. Finally, the Dutch-manned No. 321 Squadron at China Bay, Ceylon, added Liberator Mk VIs to its Catalinas from December 1944.

Because those in command of the Liberator Heavy Bomber force in the Far East were less single-minded in their pursuit of an uncompromised strategic bombing campaign than senior officers in Bomber Command, there was a much more blurred line between the responsibilities of bomber and GR units.

On 18 January 1945, for example, a No. 159 Squadron Liberator B.Mk VI, flown by Flt Lt F. Barrett, spotted a Japanese ship and requested and received permission to attack the vessel instead of its original target. The bombs scored several direct hits and the vessel, a Japanese munitions ship, exploded violently and sank. On 15 June 1945, No. 356 Squadron attacked a Japanese tanker in the Gulf of Siam and succeeded in setting it ablaze. No. 159 Squadron then finished off the tanker, the two units sharing the biggest single anti-shipping success in Southeast Asia. Episodes like these added to the tally achieved by the dedicated maritime units.

Similarly, while the front-line Liberator bomber units flew their share of supply-dropping missions behind enemy lines, a number of dedicated special duties units were also equipped with the Liberator. No. 1576 Special Duties Flight at Digri, equipped with seven Hudsons and three Liberator Mk IIIs, became 'A' Flight of a new No. 357 Squadron on 1 February 1944, while No. 358 Squadron formed on 8 November 1944 at Kolar from the remnants of the disbanding No. 1673 HCU. The squadron flew a single bombing mission on 13 January 1945 before being permanently assigned to the special duties role.

The long-range supply-dropping operations flown by these units were no less hazardous than the missions flown by the bomber units. On

Right: The RAF Liberators' biggest single anti-shipping success of the Far East campaign occurred on 15 June 1945 in the Gulf of Siam when this 16,000-ton Japanese fuel tanker, already crippled by No. 356 Squadron B.Mk VIs, was attacked, set ablaze and finally sunk by Flt Lt Borthwick of No. 159 Squadron.

22 January 1945, No. 358 Squadron lost three out of 11 aircraft on a single mission. Although the ACSEA Liberators faced a generally more benevolent environment than bombers operating in Europe, with lighter fighter opposition and less sophisticated ground defences, when things did go wrong they tended to be very bad indeed.

Most operations involved long over-water legs far beyond the reach of Allied search and rescue assets, and in an aircraft whose ditching characteristics were at best problematic, over waters teeming with sharks. Overland operations were scarcely any less hazardous, often being flown at heights that made bailing out impossible, in a theatre of war in which being taken prisoner was often no more than a delayed death sentence. This was demonstrated on many occasions, but perhaps most starkly by an unfortunate crew from No. 1341 Flight, flying an Elint mission. On 31 January 1945 one of the flight's aircraft (BZ938/W) developed engine trouble as it returned from a mission to pinpoint the location of certain Japanese radars around Bangkok, Mandalay and Rangoon. Sqn Ldr Bradley, the Flight's CO and aircraft captain, ordered the crew to bail out and six did so successfully, linking up on the ground. Three more crewmembers were posted missing, presumed killed in the aircraft or in the jungle.

The six survivors started to make their way towards the coast, where a four-day SAR effort

As the war in the Pacific reached its conclusion Nos 99, 321 and 356 Squadron moved forward to the Cocos Islands in June/July 1945. One of the latter unit's aircraft is seen here prior to one of many supply-dropping missions supporting guerrillas operating in Malaya, in preparation for the planned, but never needed, invasion. At the end of hostilities No. 356 continued to operate from this base, flying supplies to Malaya and troops to Ceylon, before disbanding in November.

was being mounted, but were betrayed to the Japanese by a village headman. The skipper and navigator (the only officers in the crew) were taken to Japanese headquarters in Rangoon for detailed interrogation, while the NCOs remained in the hands of the Japanese 55th Engineering Regiment. Three of them were beaten and beheaded immediately, while the wireless operator, Flt Sgt Stanley James Woodbridge, was tortured and interrogated. When he refused to answer any questions he too was beheaded. He was later awarded a posthumous George Cross. After the war, three of the Japanese officers responsible, and a corporal, were tried and executed.

The final phase

With the collapse of Japanese resistance in Burma between March and May 1945, the RAF Liberator squadrons increasingly switched to special duties missions, dropping arms, equipment and agents to resistance movements and guerrillas in Malaya and Sumatra. No. 200 Squadron re-designated as No. 8 Squadron, and then moved to Minneriya in Ceylon, where it joined the other Ceylon-based GR units (Nos 160 and 203 Squadrons) in supply-dropping operations. No. 354 Squadron did not join the same effort, disbanding instead on 18 May. In order to stretch range to the maximum, the aircraft routinely flew without dedicated gunners (the wireless operator was a trained gunner and could man a turret in emergency) and during June 1945 a No. 8 Squadron Liberator clocked up an astonishing 24-hour and ten-minute sortie.

Nos 99, 321 and 356 Squadrons moved to the Cocos Islands in June and July, from where they flew missions to Sumatra and Malaya. The drops to guerrillas in Malaya were particularly challenging, having to cross Sumatra's mountains and then passing the Malacca Straits to the Cameron highlands. The drops had to be timed for about 16:00, giving the guerrillas just enough time to recover the weapons and return to their camps before nightfall, and before the Japanese could react. The Liberators then had to return to their remote island base in darkness, climbing over the massive tropical cumulonimbus clouds that built up during the afternoon.

Before the planned invasion of Malaya could be launched, America dropped the first atomic bomb on Hiroshima on 6 August, and another was dropped on Nagasaki three days later. Japan sued for peace on 10 August.

In the wake of the Japanese surrender there was chaos. Supplies to the PoW camps (often isolated in dense jungle far behind the former front-line, and days away from relief or liberation by ground forces) stopped, even where the Japanese guards did not simply abandon their former prisoners. This made dropping food and supplies to the camps an urgent priority. Even these missions were not without danger, and during a drop at Sungei Ron Camp at Palembang in Sumatra No. 99 Squadron lost a Liberator and its crew.

The RAF did not just look after its own, and as well as Operation Birdcage the Liberators participated in Operation Hunger, dropping 1.5 million lb (680 tonnes) of rice to villages in the hills of south Burma.

Most of the remaining Liberator squadrons (8, 99, 356, 357, and 358) disbanded in November 1945, while No. 321 returned to Dutch control in December. This left Nos 159 and 355 of the bomber squadrons, Nos 160 and 203 in the GR role and No. 232 in the transport role.

Australian service

Australian aircrew served with RAF Liberator units in all theatres, but the Royal Australian Air Force (RAAF) also had some seven Liberator squadrons (and sundry other units) of its own, though their acquisition was far from straightforward.

As a Dominion partner and a firm and loyal ally to Britain, Australia immediately threw its might behind Britain's war effort, which was naturally focused on the war in Europe and the Middle East. Under the Z Plan, it agreed to expand the RAAF to 36 squadrons, and to form 18 RAAF squadrons for service in Europe and the Middle East under RAF control. Twelve units were formed in the United Kingdom and five more in the Middle East, before Japan's entry into the war altered priorities. With the Japanese pushing south and west, in March 1942 the Australian Government was forced to approve a plan to increase the expansion to 72 squadrons, although the USA and Britain agreed that the Allies should continue to prioritise the war against Germany and Italy. With the enemy at its door, Australia had different priorities.

When Australia asked America for new aircraft its requests were referred by General Arnold to the British Chief of the Air Staff, who urged that nothing should be done that might detract from the effort in Europe. By strictly controlling the supply of aircraft to the RAAF in Australia, Portal

No. 159 Squadron's 'C' Flight was the only RAF Liberator unit in the Far East dedicated to the Elint role. Fitted with dedicated role-specific equipment, the flight monitored enemy radio traffic and intercepted emissions to plot Japanese radar positions. This 'C' Flight example, preparing to depart its Indian base, is fitted with exhaust flame-dampers beneath the nacelles.

RAF Liberators: Part 2

In the chaotic aftermath of war, Bomber Command's Liberators played a vital role in providing the newly-liberated Allied prisoners with much needed food, water and medical supplies. Here, Flying Officer Schmoyer of No. 356 Squadron releases supplies to a PoW camp in Burma. The day after the drops two-man liaison teams were parachuted in to assist with recovery.

hoped to maintain the flow of RAAF aircrew to the European theatre, and to avoid demands for the return of RAAF units from the UK to Australia.

There was seen to be a need for some re-balancing, however, and one of the UK-based RAAF squadrons was sent to the Middle East, and two more returned to Australia in 1942 to help fight the Japanese advance.

The policy of limiting the RAAF's growth in Australia was reversed after a visit to Australia by senior British officers in early 1943 that reported back to Portal "that the lack of decent aircraft was inhibiting the RAAF's development and having a most unfortunate effect on morale". Particular emphasis was placed on acquiring heavy bombers with which the RAAF could play a fuller part in taking the war to the Japanese.

The Australian government began to consider manufacturing heavy bombers in Australia in 1942, and the Liberator was considered carefully, its long range seeming well suited to the vast expanses covered by the RAAF in the Pacific theatre. In the end, however, the Lancaster was chosen for local production, although the war ended before this could start, and Australia eventually ended up building some 73 Avro Lincolns.

Despite Australia choosing not to build the Liberator under licence, the type was selected to form the basis of a new bomber force to operate alongside USAAF Fifth Air Force Liberator units in the South West Pacific Area, coming under the command of General George C. Kenney. These promised to replace American squadrons stationed in Darwin, releasing the USAAF aircraft to take part in the push against the Japanese homeland and leaving the new Australian squadrons to look after

Displaying the squadron's three black and two white vertical-striped rudder markings, this No. 355 Squadron Liberator B.Mk VI is seen en-route to another bombing raid over Burma. The word SNAKE, visible forward of the serial number on the rear fuselage, served the dual purpose of informing personnel in the Middle East, as the aircraft transitted through on delivery, that the aircraft was fitted with equipment for use in the Far East and also that it must not, under any circumstances, be appropriated by resident squadrons in that theatre.

'mopping-up' enemy pockets bypassed by the main Allied thrust.

The USAAF assisted the RAAF with obtaining aircraft and with training crews. Fifty-two RAAF aircrews were temporarily attached to the 380th Bomb Group to convert to type and gain 100 combat flying hours experience. The Australians attached to the 380th flew both as complete Australian-only crews and within mixed crews. At the end of their service with the 380th, the newly qualified RAAF Liberator aircrew then formed the backbone of the Australian-only B-24 squadrons.

The first RAAF units to re-equip with Liberators were Nos 21, 23 and 24 Squadrons, which had operated Vultee Vengeance dive-bombers before their withdrawal from New Guinea. At much the same time that these units returned home, in February and March 1944, the first nine Liberators (of an initial 12 B-24Ds) for the RAAF were delivered to Australia from surplus USAAF stocks, equipping No. 7 Operational Training Unit at Tocumwal, Victoria. These aircraft were used for operational and conversion training only, and were the first of 287 Liberators for Australia, 168 of which were supplied under the Lend-Lease terms, and which eventually equipped seven front-line squadrons and two flights.

It was August 1944 before the first RAAF Liberator squadron (No. 24) was ready for operations, but by the time the war ended in August 1945 the seven RAAF Liberator squadrons formed the backbone of the long-range heavy bomber force in the Netherlands East Indies (NEI), operating from bases in Northern Territory, Western Australia, Morotai (NEI) and Palawan (Philippines).

These front-line RAAF units, equipped with B-24Js, B-24Ls and B-24Ms, attacked Japanese targets in the Netherlands East Indies, especially on islands by-passed by the main Allied advance. In April 1945 the RAAF B-24s spear-headed operations against Japanese targets in Borneo, prior to the Australian landings at Balikpapan, Labuan Island and Tarakan.

A total of 82 aircraft from seven USAAF and three RAAF squadrons formed the bombing force during the landings at Balikpapan. Twenty-four B-24s from the US 13th Air Force bombed first, followed by 20 Liberators from the RAAF's 1st Tactical Air Force. These comprised seven aircraft from No. 21 Squadron, seven from No. 24 Squadron and six from No. 23 Squadron. They were followed by 38 B-24s from the 30th Bomb Group and the 380th Bomb Group of the US Fifth Air Force.

Although the squadron-strength units operated primarily in the bomber role, the two numbered flights undertook more specialised duties. No. 201 Flight flew radio and radar countermeasures and Elint missions, while No. 200 Flight was dedicated to the special duties role. After trials at Richmond in late December 1943, using a borrowed USAAF 380th Bomb Group aircraft, No. 200 Flight formed with two B-24Js from No. 24 Squadron, developing tactics and techniques and training for operations. By the time these began in February 1944 the unit had five Liberators and nine 11-man crews. The flight dropped supplies and agents in so-called 'storpedoes' 6-ft (1.83-m) long cardboard cylinders reinforced by metal straps.

Some of the RAAF's Liberator squadrons disbanded during 1945, and the remainder

Air Combat

Right: No. 356 Squadron became the RAF's third Liberator squadron in India in January 1944. The unit was heavily involved in the bombing of Japanese bases in Burma, Sumatra and Malaya having been declared operational that July. The Indian Liberator squadrons wore distinctive fin markings (No. 356's being a white cross on a black background) to identify them to Allied fighters.

Below: The Liberator B.Mk VI was easily distinguishable from earlier bomber versions by the addition of the twin 0.5-in (12.7-mm) nose turret. A new optical bombing station was built under the turret and, to house the 1,200 rounds of ammunition for the guns, the nose was extended by 10 in (25 cm).

effectively followed during 1946. No. 12 Squadron moved to Amberley during March 1946, followed by Nos 21 and 23 Squadrons in April, though all three were immediately reduced to a care and maintenance unit status, and flying virtually ceased, with some limited training and a handful of aircraft retained for VIP transport duties.

The last RAAF Liberators were struck off charge *en masse* in December 1952, although the last flying units had actually disbanded or re-equipped with Avro Lincolns by the end of 1947, apart from a handful of VIP transports. Of these A72-193 made the last flight by an RAAF Liberator on 21 January 1949. Most of the aircraft were scrapped at Tocumwal, but one aircraft has survived and is now being restored to static display standards.

In the Far East, large numbers of ex-RAF Liberators were gathered in a graveyard at No. 322 MU at Chakeri, near Cawnpore (now Kanpur).

The US had ended Lend-Lease in October 1945 and immediately ordered that all armament should be removed from the aircraft, before directing that all Liberators in SEAC should be demolished.

Nos 159 and 355 Squadrons were given a brief stay of execution while they undertook a photographic survey of Bengal for the then Bengali government, but their near-new Liberator Mk VIIIs were sent to Chakeri for demolition in March 1946, once the survey task was completed. The transports of No. 232 served on until August 1946, while No. 160 returned to the UK in June 1946, and No. 203 followed in May 1947, swelling the numbers of redundant Liberators gathered at Chakeri and other Indian MUs.

The first stage of the scrapping process was supposed to entail the removal of all equipment 'useful to the aircraft' including magnetos (thereby rendering all engines unserviceable), instruments, first aid kits and radios, while control cables were cut, sand was poured into engines, and ammunition chutes and turret motors were smashed. The army then drove tractors into the aircraft, smashing the turret Perspex and making holes in the Alclad skin with pickaxes and often smashing one main undercarriage leg.

Just as damagingly, the aircraft sat out unprotected and unserviced, without covers or control locks, in the worst Monsoon conditions India had to offer. By the time the British Forces left,

In 1942, with the Japanese invasion gathering pace in the Pacific, Australia desperately sought options in re-equipping its air force with modern four-engined bombers. After rejecting licence-building Liberators in favour of the Avro Lancaster, the RAAF instead acquired B-24J, L (above) and M variants direct from USAAF stocks. Used by seven operational squadrons, under USAAF command, the B-24s were used to 'mop-up' remaining Japanese resistance as the Allies advanced through 1944-45. The RAAF crews had notable successes against ground and sea targets including this Japanese installation at Burma Armapure (right) in January 1945. A total of 287 Liberators was eventually taken on charge by the RAAF.

Despite being supposedly irreversibly decommissioned under the terms of the Lend-Lease agreement, India returned over 40 ex-RAF Liberator Mk VIs and VIIIs to service. No. 6 Sqn, IAF, operated the type in the maritime reconnaissance role (right). A number survive today, including HE877, now owned by Pima County Air Museum, AZ, (above).

when India gained independence in August 1947, the aircraft were in a terrible condition.

Despite this, the new Indian Air Force decided to salvage some of the aircraft and return them to service, contracting HAL to do the work. The first problem was that the aircraft needed to be transported from Chakeri to HAL's factory at Bangalore. Making the aircraft capable of even a single ferry flight was a major undertaking and Mr Yelappa and his team surveyed the abandoned aircraft, identified those aircraft that could be made flyable and undertook temporary repairs by cannibalising parts from the remainder. B-24-qualified American pilots took one look at the aircraft and refused to fly them, despite the offer of generous bonuses.

Jamshed Kaikobad (Jimmy) Munshi, HAL's Chief Test Pilot, was made of sterner stuff and agreed to undertake the ferry flights refusing any extra payment over and above his normal HAL salary. Jimmy assembled a manual from loose pages left behind in some of the aircraft, and began flying the 42 aircraft (gear down) to Bangalore, despite having never flown a four-engined aircraft, and probably having never flown an aircraft with a nose gear. One aircraft caught fire en route, but the flames were doused using flasks of coffee! The exact number of aircraft returned to service is unknown, with different accounts suggesting totals of 36, 39, 40, 41, 42 or even 48. Forty-one may be the most likely number, with one further machine serving purely as a GI airframe.

Indian swansong

The aircraft were quickly refurbished and rebuilt by HAL, and No. 5 Squadron was formed on 2 November 1948 with the first six Liberators, later reaching its full strength of 16 aircraft. The Americans were horrified to discover that India had acquired serviceable Liberators, and especially to discover that it had done so without paying a single dollar. The RAF offered to help the Indians in returning the refurbished aircraft to operational service, however, and sent two experienced teams to Poona to help convert and train IAF crews to the B-24.

No. 6 Squadron re-equipped with 16 refurbished Liberators in January 1951, operating in the maritime reconnaissance role, while No. 16 Squadron was established with another two or three aircraft on 15 October 1951 as a Liberator training unit. Most of the No. 6 Squadron aircraft were fitted with the ASV-15 radar, with a retractable radome in place of the ball turret, and carried sonobuoys and depth charges. Some of them were GR.Mk VIIIs, but most were Mk VIs. Two more aircraft, modified to unarmed C-87 Liberator Express standards, were used as survey and reconnaissance aircraft by No. 102 Survey Flight.

Nos 5 and 16 Squadrons re-equipped with Canberra interdictors in 1957, but the Liberators of No. 6 Squadron continued in service, taking part in the takeover of the former Portuguese colonies in India in 1961, flying reconnaissance missions along the sea-lanes approaching Diu and Daman and dropping surrender leaflets over Goa. The Liberators also flew maritime patrol sorties during the 1965 Indo-Pakistani War, before retiring in 1968. This marked the end of the Liberator's military career – after 27 years in front-line service – nothing out of the ordinary, today, but an extraordinarily long service life for a wartime bomber.

When No. 6 Squadron finally retired its Liberators, one was presented to the RAF, ending up in the RAF Museum at Cosford, one was donated to the Royal Canadian Air Force, becoming part of the National Aeronautical Collection in Trenton, Ontario and three found homes in the US. Another (HE924, ex-KH342) was preserved in India, and two more remained in store at Poona. Of these, one ex-ground instructional aircraft (T-18 from the IAF Technical College at Jalahalli), originally delivered to the Blackbushe-based Warbirds of GB Ltd, was sold on to the Collings Foundation, who today operates the aircraft in USAAF markings as *All American* (port) and *The Dragon & His Tail* (starboard). Another, the former HE771 (ex-KH401), was delivered to Yesterdays Air Force, at Chino, CA, and has also flown, again in USAAF markings, as *Delectable Doris*, and is now in the hands of Kermit Weeks. Both airworthy Liberators saw extensive RAF service. Of the remaining ex-RAF Liberator survivors, one (HE877/A, ex-KH304) is on display at the Pima County Museum on the peripheries of Arizona's famous boneyard, wearing USAAF markings on one side, but retaining its IAF markings on the other, while another (HE807/K, the former KN751, with RAF SEAC colours to port and IAF markings to starboard), graces the RAF Museum at Cosford. The final ex-Indian Liberator (a B.Mk VIII, HE773/M, ex-KN820) is displayed by the Canadian National Aeronautical Collection in Coastal Command colours.

Jon Lake

RAF ground and aircrew on the Cocos Islands joyously celebrate in front of a No. 356 Squadron Liberator B.Mk VI, as news of the war's end comes through. With all British-built heavy bombers committed to the war in Europe, the Liberator, with its long range and excellent durability, had proved itself invaluable in Britain's ability to wage war against the Japanese.

RAF Liberator variants

LB-30A Liberator

Of the 165 LB-30s ordered by Britain, six (AM258-AM263) were diverted from an American order for seven YB-24s. The USAAC serials (40-696 to 40-701) for these aircraft were then re-used for six early B-24Ds. The six aircraft were ordered under contract A-5068, and were paid for in cash. The RAF LB-30A aircraft were to have been delivered with six 0.303-in (7.7-mm) machine-guns and Sperry 0-1 bombsights, but were then stripped and used as unarmed transports with 1,200-hp (895-kW) Pratt & Whitney R-1830-33C4-G engines driving A5/125 propellers. They had simple mechanical two-speed superchargers and lacked self-sealing fuel tanks, and were used by BOAC and RAF Ferry Command's Return Ferry Service. The first made its maiden flight on 17 January 1941, and deliveries began in March. One survived long enough to join No. 231 Squadron in July 1944, serving with the unit until November 1945. Others flew with No. 1425 Flight and No. 511 Squadron.

Delivered in August 1941, AM929 was one of 20 LB-30Bs taken on charge by the RAF. After service with No. 120 Sqn, Coastal Command, it was converted as a transport.

LB-30B Liberator I

As Britain faced Germany alone, desperate for modern aircraft and for symbols of US support, further Liberators were diverted from USAAC orders. Twenty (40-2349 to 40-2368) of the initial USAAC order for 38 B-24As were therefore delivered to the RAF as Liberator Is AM910-AM929. By the time they were built, the baseline Liberator design had matured, with the original XB-24 having been fitted with self-sealing fuel tanks and re-engined with R-1830-41 (and later -43) engines with turbosuperchargers. Some of these improvements were incorporated into the 20 Pratt and Whitney R-1830-33C4-G-engined RAF Liberator Is, which were also delivered with 0.303-in (7.7-mm) machine-guns. AM927 was diverted for conversion for 'special duties' but was damaged in an accident, and eventually entered USAAF service as a C-87.

Some of the LB-30Bs went to No. 120 Squadron, Coastal Command, while others (including AM911, AM913, AM914, AM918, AM920 and AM922) were used only for training Coastal Command aircrew (without radar or cannon armament) or augmented the LB-30As in the transport role. At least one of the radar- and cannon-equipped aircraft (AM929) was eventually modified back to transport configuration. Most of the transport Liberators were configured to carry 11 passengers, though some later conversions seated 20 or 21.

LB-30B Liberator 'GR.Mk I'

The first LB-30B to reach the UK was AM910, which arrived in April 1941 and was sent to Heston to serve as the Trial Installation aircraft, gaining a ventral gun pack containing four 20-mm Hispano cannon, a 1,000-watt generator in the port inboard nacelle, R.3003 IFF and ASV Mk II radar (with T.3040D transmitter and R.3039 receiver) with 'stickleback' antenna arrays on the top and sides of the rear fuselage, and Yagi arrays on the nose and underwing. The aircraft underwent trials at Boscombe Down in July and joined No. 120 Squadron at Nutts Corner in September 1941.

Nine further radar- and cannon-equipped Liberator Is were delivered to Coastal Command, comprising AM916, AM917, AM921, AM923, AM924, AM925, AM926, AM928 and AM929. Though sometimes referred to as GR.Mk Is, this designation is believed to have been unofficial. Five aircraft were converted to VLR (Very Long Range) standards by January 1942, with 2,650 Imp gal (12047 litres) of fuel in the wing tanks. This bestowed an astonishing 2,400-mile (3862-km) range – more than double that of the Sunderland flying boat. In RAF parlance VLR described an aircraft with a Prudent Limit of Endurance (PLE) in excess of 1,000-mile (1609-km) radius.

LB-30B Composite

Damaged in a crash at Dorval in June 1943, LB-30B AM920 was rebuilt with a lengthened C-87 type nose and a C-87 cabin 'canoe' in place of the bomb bays, and gained a unique half freight door. While retaining its LB-30B wing and fuel system, it was fitted with LB-30 engines and propellers. After RAF service, the aircraft was put onto the civil register as G-AHYB with BOAC, before being sold to France in April 1951. It was then converted to VVIP standards, becoming F-VNNP for Emperor Bao Dai of Vietnam, and later joining Acana as F-OASS.

LB-30 Liberator II 'B.Mk II'

Because they were diverted from US orders, and were ordered after the British took over the aircraft intended for the French, the RAF's first Liberators were designated LB-30A and LB-30B, while the Liberator IIs used the earlier LB-30 designation. As they were newly built, the LB-30s incorporated a host of improvements and refinements, some of them the result of service experience using the earlier versions.

Most obviously, the Liberator II introduced a 2-ft 7-in (79-cm) nose extension. Legend has it that designer Reuben Fleet looked at the XB-24 and pronounced it 'too stubby looking', adding the stretched nose for purely aesthetic reasons, though in doing so he made space for a proper navigator's compartment. The type used the same R-1830-33 (R-1830-33C4G) engines as the Liberator I, but with Curtiss Electric airscrews replacing the short-hubbed Hamilton Standard Hydromatic airscrews. All fuel tanks and lines were self-sealing.

Power-operated Boulton Paul gun turrets were added above the fuselage and in the tail, each containing four 0.303-in (7.7-mm) machine-guns.

A total of 140 was built (bringing the total to the required 165, including the six LB-30As and 20 LB-30Bs and a replacement for the first aircraft, which was lost during an early test flight) although 75 were taken over by the USAAC in the wake of Pearl Harbor. Two of the latter were lost, 23 eventually being returned to the British. Some of the returned aircraft may have retained their US Martin top turrets and open tail gun positions.

Production comprised AL503-AL642, plus FP685. AL503 was lost on its acceptance flight, and AL596 also crashed in the USA. AL508, 515, 521, 527, 532, 533, 539, 543, 567, 568, 570, 572, 573, 575, 576, 583, 586, 589, 594, 598, 601, 602, 604-609, 611-613, 615-618, 621-623, 626, 628, 629, 631-634, 637, 639-641 and FP685 were retained by the USA.

Of the 88 or 89 aircraft delivered to the RAF, most entered service in the bomber role, flying in the Middle East and India, though some were used as transports (with turrets removed) and by Coastal Command. Although sometimes referred to as B.Mk IIs, this designation is believed to have been unofficial.

Introducing the longer-nose, incorporated on all subsequent RAF Liberators, this is one of around 30 'B.Mk IIs' that saw combat service in North Africa during 1942.

LB-30 Liberator II 'GR.Mk II'

A very small number of Liberator IIs served with Coastal Command, mostly with No. 120 Squadron. The three Mk IIs in use with the squadron in late 1941 were used only for training and meteorological flights. Though sometimes referred to as GR.Mk IIs, this designation is believed to have been unofficial. Some may have had the same radar antenna array as the 'GR.Mk Is', but at least one (AL507) certainly had the later 'Dumbo' style undernose radome. This aircraft was the first with US Mk V (SCR 517) radar – equivalent to ASV.Mk III, and was used for trials by No. 120 Squadron between March and June 1942. None of the Coastal Command Liberator IIs was converted to VLR standard. The GR.Mk II had a range of 1,800 miles (2897 km) in standard fit.

LB-30 Liberator II Transport

One Liberator II transport, AL504, served with No. 511 Squadron in standard configuration between October 1942 and June 1944 and was used as Churchill's personal aircraft on several occasions. The aircraft was then converted, gaining a new interior, an electric galley and an R4Y-type single tailfin. In its new guise, and with the name *Commando*, the aircraft served with No. 231 Squadron. Other Liberator II transports included AL507 (G-AHYC), AL512 (G-AGEL), AL514 (G-AGJP), AL516 (G-AHZP), AL522 (G-AHYD), AL523, AL524 (G-AGTJ, later VH-EAJ), AL528 (G-AGEM), AL529 (G-AHYE), AL541 (G-AGTI, later VH-EAI), AL547 (G-AGKU), AL552 (G-AHZR), AL557 (G-AGZI), AL571 (G-AGZH), AL592 (G-AHYF), AL603 (G-AHYG), AL619 (G-AGKT) and AL627 (G-AHYJ).

Wartime Liberator transports flew in both military and civil (BOAC) guise, and small numbers were leased to Qantas. Post-war BOAC operated seven Liberator IIs until 1949, flying non-stop London-Montreal cargo services supported by Lancastrian tankers flying from Shannon and Goose Bay. Five Scottish Aviation Liberators participated in the Berlin Airlift, flying 497 hours and carrying 3,174 tons of cargo.

A total of 75 LB-30s was diverted from UK orders for service with the USAAC. This example, still wearing its RAF serial, was converted as a freighter for service with Air Transport Command's Pacific division.

Liberator B.Mk III

The Liberator B.Mk III was the RAF's version of the B-24D, with Pratt & Whitney R-1830-43 engines in slightly oval-section nacelles with intercooler intakes in the sides, driving A5/127 propellers. The Mk IIIs were delivered with Martin A-3 top turrets, mounted rather further forward than the Liberator II's top turrets had been, and containing twin 0.5-in (12.7-mm) machine-guns. The Mk IIIs were also delivered with similarly-armed Consolidated A-6 tail turrets. Most had the latter replaced by a Boulton Paul turret with four 0.303-in (12.7-mm) machine-guns. Early Liberator Mk IIIs had enclosed beam gun positions, with the single or twin guns using a small opening hatch in the bottom right part of the beam hatch's perspex panel. This was the arrangement used on the B-24H and B-24L.

Some 164 Liberator Mk IIIs were allocated, 159 of which actually arrived for RAF service. Three of the aircraft that crashed were not allocated RAF serials. It is difficult to determine exactly how many were B.Mk IIIs as opposed to GR.Mk IIIs.

The 15 aircraft delivered to MAAF, and the 51 delivered directly to India, were B.Mk IIIs, as were some of the 66 aircraft delivered to the UK (most of which were GR.Mk IIIs). Ten aircraft were delivered as transports, 11 were delivered as GR.Mk IIIAs and six were delivered to Nassau as GR.Mk IIIs.

Liberator III serials ran from FK214-245, FL906-937, FL939-940, FL943, FL968, FL992-995, BZ718, BZ748, BZ761-762, BZ833-842, BZ844-860, BZ890-909, BZ922-930, BZ932-936 BZ946-959.

Liberator GR.Mk III

The GR.Mk III was the Coastal Command ASW version of the Liberator Mk III. The aircraft had ASV.Mk II radar, but of a later type than had been fitted to the Liberator Is, and did not require the 'stickleback' antenna array on the spine. Whereas bomber Liberator Mk IIIs tended to retain their mid-upper turrets regardless, Coastal Command aircraft delivered to squadrons that operated over the Bay of Biscay

Liberator GR.Mk IIIA

There have been a number of explanations for the Liberator Mk IIIA designation, one being that the Mk III was applied to 'British purchased' aircraft, or alternatively the designation covered aircraft supplied under Lend-Lease. More accurate reports suggested that the designation covered 10 aircraft supplied with US armament, allowing them to enter operational service more quickly. This was true, insofar as it went, but, according to some reports, ignored the most significant feature of the Mk IIIA, which was that it was equipped with the 10 sets of Mk V radar released by the USA following the trials that No. 120 Squadron had undertaken using aircraft AL507.

The serials of the 10 Mk IIIAs were LV336-345, one of which was photographed in No.120 Squadron service without radar of any sort, but very clearly with US-type gun turrets.

VLR Liberator GR.Mk IIIAs carried their 2,650 Imp gal (12047 litres) of fuel in full main wing tanks and two full bomb cell auxiliary tanks.

Liberator GR.Mk V

When fitted with centimetric radar (ASV.Mk III), Coastal Command B-24D Liberators used the designation GR.Mk V. The new radar used either a ventral 'chin' radome, or a retractable 'dustbin' radome behind the bomb bays. Like the Liberator GR.Mk III, the GR.Mk V could be fitted with eight forward-firing rockets on stub wings on each side of the forward fuselage, though this was uncommon. Ten such aircraft were used by No. 311 Squadron from early 1944 until early 1945, and single RP-equipped aircraft were briefly used by Nos 224 and 53 Squadrons. The rocket-firing Liberators could still carry five 250-lb (113-kg) depth charges.

Some Mk Vs were equipped with underwing Leigh Lights, these initially being used by squadrons operating over the Bay of Biscay, and over the South Western approaches. Early Liberator Mk Vs had B-24H/B-24L-type enclosed beam gun positions, with the guns projecting through a small opening hatch in the bottom right part of the beam hatch. Later aircraft used the standard beam hatch, which swung inwards and upwards into the fuselage.

Liberator B.Mk VI

The Liberator Mk VI designation was applied to the first RAF Liberators delivered with a front gun turret. These were B-24Js and B-24Ls, from a variety of production blocks and factories, and thus displayed differences in turrets, nose glazing, engines and undercarriage door type, giving a number of RAF Liberator Mk VI configurations, all of which shared the same designation. Some (but not all) of the RAF's Ford-built B-24Js and B-24Ls were designated as Liberator Mk VIIIs, and these are described separately. The new turret knocked several knots off the Liberator's cruising speed, and the extra drag also reduced range.

Some 857 of the 1,144 RAF Liberator VIs were delivered as bombers, with a Sperry ball turret in the belly. They saw service in the Middle East, Mediterranean and Far East. Gaps in the following serial blocks comprised GR.Mk VIs and Mk VIIIs.

Wearing the unit's distinctive fin markings, this B.Mk VI belongs to No. 356 Squadron.

The GR.Mk V was not entirely restricted to operations over the Mediterranean and North Atlantic. BZ825 was one of a number of aircraft operated by No. 160 Squadron in the Far East and is seen here at Chakeri, India, in the immediate post-war period.

Liberator C.Mk III

Ten of the RAF Liberator Mk IIIs were delivered to Transport Command's No. 45 Group becoming C.Mk IIIs, comprising BZ718, BZ748, BZ907, BZ953, FK240, FK243, FL908, FL914, FL919 and FL995. BZ761 and 762 may have later been converted as transports.

retained the mid-upper turrets, while in other theatres they were removed. The GR.Mk III could be fitted with eight forward-firing 5-in (12.7-cm) HVAR rockets on stub wings on each side of the forward fuselage, though this was uncommon. From January 1944, some Mk IIIs were retrofitted with Leigh Lights underneath the starboard wing. The GR.Mk III had a range of 1,680 miles (2704 km) in standard fit.

The six aircraft delivered directly to No. 111 OTU were GR.Mk IIIs (FL968, FL992-994, BZ761 and BZ762), as were most of the 66 Mk IIIs delivered to the UK.

Liberator B.Mk IV

While some sources suggest that the B.Mk IV designation was allocated to the RAF's RCM Liberators, others insist that the designation was actually reserved for B-24Es, which were B-24Ds built by Ford at Willow Run under the Liberator Production Pool arrangement. The B-24E used Curtiss Electric propellers, and these were officially the differentiating feature of RAF Liberator Mk IVs, a handful of which may have been produced by conversion.

Liberator C.Mk V

One of the RAF's Liberator Mk Vs, BZ715, was delivered to No. 45 Group, Transport Command, as a transport, and others were converted to a similar standard after service with Coastal Command. These included BZ723, 744, 760, 769, 773, 781, 783, 786, 792-793, 804, 806, 862, 869, 871, 931, 941, FL941 970 and 979.

Other GR.Mk Vs were specially-equipped for VLR duties, initially with Nos 86 and 59 Squadrons of No. 15 Group, and with No. 120 Squadron in Iceland. These had all rear-facing armour removed, and usually lacked the mid-upper turret. Some had the rear turret removed, allowing two additional 250-lb (113-kg) depth charges to be carried. Early VLR GR.Mk Vs carried the same fuel tanks as VLR GR.Mk IIIAs, but thereafter carried full main tanks, full auxiliary wing tanks and one full bomb cell auxiliary tank. This gave the same 2,650 Imp gal (12047 litre) total, and the same 2,300 nm (4259 km) range. The standard GR.Mk V carried full main and outboard wing tanks auxiliary tanks, giving it a range of only 1,900 nm (3518 km).

The last GR.Mk Vs in service were withdrawn by No. 59 Squadron in April 1945. Later GR.Mk Vs had a single 0.5-in (12.7-mm) machine-gun with 400 rounds in each beam position, instead of twin 0.303-in (7.7-mm) guns with 500 or 1,000 rpg.

Some 205 GR.Mk Vs were allocated, but FL957 crashed before delivery and 15 were diverted to Canada (BZ725, 727-729, 732-739, 747 and 755-756). Serials of the GR.Mk Vs were BZ711-717, BZ719-747, BZ749-760, BZ763-832, BZ861-889, BZ910-921, BZ931, BZ937-945, FL938, FL941-942, FL944-967 and FL969-991.

B.Mk VI (Consolidated B-24J): BZ960, 962, 964-5, 973-4, 976-978, 980, 982, 983, 989-990, 992-993, 996-998, EV812-817, 820, 822, 825-826, 838-839, 841, 843-847, 849-852, 854-855, 857, 859-860, 862, 864-865, 867-868, 870, 875-876, 900-918, 920-932, 934, 937-938, 940-941, 944, 946, 949, 951-952, 957-971, 973-984, 989-991, 999, EW100-249, 250, 253-287, KG823-846, 871-894, 919-931, 933-942, 967-978, 993-999, KH100-122, 147-176, 201-218, 239-258, 269-288, 309-328, 359-368, 389-408, KK229-248, 269-288, 301-320, 343-362, KL352-367, KL368-388
B.Mk VI (Ford B-24J): KL392-393
B.Mk VI (Ford B-24L): KL390-391, 473, 475-476, 478-479, 481-489, 491-492, 494-495, 499, 501, 503-504, 507-508, 510, 512, 513, 515, 516, 521, 523-531, 534, 536-538, 540-541, 543, 545-549, 552, 556-557, 560, 563, 564, 569, 571-601, 607, 611-616, 618-630, 632-633, 635-639, 641-642, 641-652, 654, 657-658, 663-667, 669-670, 672-673, 676, 679, 681-682, 685-689, KN702-707, 744-752
Unknown variant, delivery unconfirmed: EW251-252, 932, 935, KL450-470, KL518-519, KL535, 539, 555, 602-606, KN708-719

Whereas the Liberator IIIs and Vs had their Consolidated tail turrets replaced by a Boulton Paul turret with four 0.303-in (7.7-mm) machine-guns, many early Liberator Mk VIs were delivered with four-gun Boulton Paul turrets, which were actually replaced by American turrets in-service.

In the MTO, Liberators used for night bombing often had their nose guns removed (because head-on attacks were not encountered). The beam guns and belly turrets were removed because the glow from the superchargers ruined gunners' night vision. Guns were often also removed from the mid-upper turrets, leaving only the tail turret.

Liberator GR.Mk VI

Although it was slower and had shorter range than the Liberator Mk V, the GR.Mk VI marked a real improvement over the earlier version. The new power-operated front gun turret with its twin 0.5-in (12.7-mm) machine-guns provided a better means of suppressing counter-fire from surfaced U-Boats, and was a more effective defensive weapon against enemy fighters. Some 287 GR.Mk VIs were delivered, most serving with home-based Coastal Command squadrons.

Like later examples of the GR.Mk V, the GR.Mk VI had a retractable 'dustbin' radome in place of the bomber's Sperry ball turret.

There was no VLR version of the Liberator GR.Mk VI, and all Coastal Command GR.Mk VIs had a range of 1,600 miles (2575 km) at 4,000 ft (1219 m) with fuel in the main and outboard wing auxiliary tanks.

A large number of GR.Mk VIs were fitted with, or had provision for, the Leigh Light, while aircraft flying from Ballykelly and the Azores were usually operated without mid-upper gun turrets. A handful may have been used with nose-mounted rocket projectiles.

GR.Mk VI (Consolidated B-24J): BZ961, 963, 966-972*, 975, 979, 981, 984-988, 991, 994-995, 999, EV818-819, 821, 823-824, 827-832, 833-837, 840, 842, 848, 853, 856, 858, 861, 863, 866, 869, 871-874, 877-899, 919, 933, 935-936, 939, 942-943, 945, 947-948, 950, 953-956, 972, 985-988, 992, 994-998, EW288-322, KG821-822, 847, 849-870, 895-918, 986, 990-992, KH123-124, 127, 134, 137-142, 144-145,

The GR.Mk VI, with its power-operated front turret and improved ASV radar, was the RAF's most important and capable long-range submarine hunter of the late-war period.

185-188, 190-200, 219-220, 267, 289, 295-297, 299-301, 303-305, 307, 331-332, 335-336, 339, 342-345, 348, 377-386, 409, 419-420, KK221-228, 251-252, 254-255, 257-258, 260, 265-267, 337, 340-342, 368, 371-378, KL348-351
*BZ970 was described as a GR.Mk IV (B-24H)

Liberator Mk VI (RCM) and transfers

About 28-30 ex-USAAF Eighth Air Force Liberators were transferred to the RAF mainly for use by No. 223 Squadron in the bomber support/radio countermeasures role. These consisted of 13 B-24H-15-FOs, three B-24H-20-FOs, two B-24J-65-CFs, two B-24J-70-CFs, and single examples of the B-24H-15-CF, B-24H-15-DT, B-24J-1-DT, B-24J-1-FO, B-24J-5-DT, B-24J-5-FO, B-24J-50-CF, B-24J-60-CF, B-24J-160-CO and B-24J-165-CO.

All had nose guns removed with nose turrets locked in the fore and aft position and faired with doped canvas, the astrodomes were faired over, new role equipment (including 'Window', 'Carpet', 'Dina', 'Piperack' and 'Jostle IV') was added in and behind the former bomb bay, and a host of associated antennas were incorporated on the fuselage and above the wing.

Liberator B.Mk VIII

The Liberator Mk VIII was not, as has often been said, an RAF B-24L or B-24M, though most of the Liberator VIIIs were Ford-built, and most were B-24Ls. Some Ford B-24Js and Ls were delivered as Liberator VIs, however. The B.Mk VIII was basically a B.Mk VI delivered to a higher modification standard, with a re-located wireless operator's station above the rear bomb bay. This gave more space in the nose for a better-equipped and roomier navigator's station.

From July 1944, 36 B.Mk VIIIs were delivered to the MTO with centimetric radar in place of the bomber's usual Sperry ball turret. They were used for Pathfinder operations. Most of the RAF's 108 B.Mk VIIIs were delivered to SEAC, and are believed to have retained their Sperry ball turrets. Nearly three times as many Mk VIIIs were delivered to Coastal Command, and these are described separately.

B.MK VIII (Consolidated B-24J): KG943-958, KH227-238, KH369-376
B.Mk VIII (Ford B-24L): KL608-610, 617, 631, 634, 640, 643, 653, 655-656, 659-662, 668, 671, 674-675, 677-678, 680, 683-684, KN759-760, 762, 764, 766, 768, 771-772, 774, 780-784, 790-794, 796, 798, 801-802, 806-808, 812, 814-816, 818, 820, 822, 824, 826, 828, 830, 832, 834, 836, KP126, KP128, KP130, KP132, KP134, KP136, KP138, KP140.

Liberator GR.Mk VIII

The GR.Mk VIII was a GR.Mk VI delivered to a higher modification standard, with a re-located wireless operator's station above the rear bomb bay and with improved 3-cm Mk X or Mk XV radar. About 318 of the 426 RAF Mk VIIIs were delivered as GR.Mk VIIIs, including Consolidated- and Ford-built B-24Js and Ford-built B-24Ls. Many (especially among the aircraft allocated to Nos 59, 206 and 547 Squadrons) had radar bomb sights and most were capable of carrying Leigh Lights. The radar bomb sight was known as LABS (Low Altitude Bomb Sight) for security reasons, and linked the main radar, the attack radar (in the bomb-aimer's position), the Fluxgate Master Compass and the auto-pilot. The GR.Mk VIIIs were the last RAF Liberators in service, some of them having been converted to C.Mk VIII standard.

With its distinctive single tail fin and stretched fuselage, the C.Mk IX was to have been procured in large numbers for use in Europe and the Far East. The end of the war brought about large-scale order cancellations and only 21 examples were delivered.

About 13 further USAAF Liberators (TW758-769) were transferred in-theatre in the MTO, but these were not converted to RCM configuration. The RAF serials of all known theatre transfers are given here: TS519-539, TT336-343*, TW758-769, VB852*, VB904*, VD245 and VD249*.
*TT336 and TT340 were converted as B.Mk IVs; TT343, VB852, VB904 and VD249 became C.Mk IVs.

Liberator C.Mk VII

The Liberator Mk VII designation was assigned to some 24 C-87-CF/RY-1 Liberator Express transport aircraft supplied to the RAF. They were broadly equivalent to the B-24D, but built as transports, with streamlined fairings replacing the turrets and with a full transport interior with windows along the fuselage. A double cargo door was fitted in the port rear fuselage. The nose fairing followed the same broad outline as the nose of the original Liberator Mks II/III/V, but was not glazed and was blunter in planform. The tip of the fairing formed a door, hinging open to starboard and, whereas this incorporated a small window on USAF C-87s, the door on most RAF Liberator C.Mk VIIs was 'solid'. The tail fairing formed a longer 'boat tail'. The aircraft were powered by R-1830-65 engines, driving A5/127 propellers. Two of the aircraft went to No. 300 Wing in Canada, and the rest to No. 45 Group and to No. 232 Squadron in India. They were serialled EW611-634.

Liberator C.Mk VIII

At the end of the war, large numbers of Liberator GR.Mk VIIIs were converted for transport duties. All guns were removed and the rear turret was faired over, while the bomb bays were sealed and fitted with wooden floors and seating for up to 26 personnel. Although the bomb aimer's glazing under the nose was retained, the nose turret was replaced by a streamlined metal fairing. The modifications gave the C.Mk VII a similar overall airframe configuration to the USAAF's C-109 fuel tankers.

A number of the C.Mk VIIs were taken over by BOAC at the end of the war and the type was used for supply flights during the Berlin airlift.

GR.Mk VIII (Consolidated B-24J): KG848, 959-966, 979-985, KH125-126, 128-133, 135-136, 143, 146, 177-184, 189, 221-226, 259-266, 268, 290-294, 298, 302, 306, 308, 329-330, 333-334, 337-338, 340-341, 346-347, 387-388, 410-418, KK249-250, 253, 256, 259, 261-264, 268, 289-300, 321-336. 338-339, 363-367, 369-370
GR.Mk VIII (Ford B-24J): KL394-449* (*Only KL394 exported, KL395-449 were retained in the USA)
GR.Mk VIII (Ford B-24L): KL471-472, 474, 477, 480, 490, 493, 496-498, 500, 502, 505, 509, 511, 514, 517, 520, 522, 532-533, 542, 544, 550-551, 553-554, 558-559, 561-562, 565-568, 570, KN719-743, 753-758, 761, 763, 765, 769-770, 773, 775-779, 785-789, 795, 797, 799-800, 803-805, 809-811, 813, 817, 819, 821, 823, 825, 827, 829, 831, 833, 835, KP125, 127, 129, 131, 133, 135, 137, 139, 141-156.

Liberator C.Mk IX

The Liberator C.Mk IX was the RAF designation for the C-87C/RY-3, and was the first RAF Liberator version built with a Privateer-type single tailfin and 7-ft (2.13 m) fuselage stretch. As purpose-built transports they used the same nose and tail fairings as the C.Mk VII. Serials were allocated for 111 RAF Liberator C.Mk IXs, JT973-999, JV936-999 and KE266-285. Only 21 (JT973, JT975-6, 978-9, 981-993, 997 and JV936) were actually delivered, two (JT982 and JT985) crashing after export. The variant served with Nos 231 and 232 Squadrons.

Liberator *Commando*

Winston Churchill's former personal aircraft, AL504, was later rebuilt with a Privateer-type lengthened fuselage and single tailfin, although it is believed to have retained its LB-30 wing and engines. The aircraft, named *Commando*, disappeared over the South Atlantic on 27 March 1945.

RAF Liberator units

ROYAL AIR FORCE COASTAL COMMAND

NO. 53 SQUADRON
No. 53 Squadron converted to the Liberator at Thorney Island from May 1943, initially operating the Liberator Mks V and VA. The squadron moved to Beaulieu in September, and to St Eval in January 1944, using the code '2-'. It augmented its GR.Mk Vs with GR.Mk VIs from June 1944, taking both versions to Reykjavik in September 1944. It used the codes 'PH-' from July 1944. Liberator GR.Mk VIIIs arrived in January 1945, and the earlier marks were withdrawn in March. The squadron moved back to St Davids on 1 June, and joined Transport Command on 25 June.

NO. 59 SQUADRON
No. 59 Squadron flew Liberator Mk IIIs from Thorney Island between August and December 1942, before converting to Fortresses. Liberator Mk Vs arrived in March 1943, and the squadron took these to Aldergrove on 11 May. In Northern Ireland the unit moved to Ballykelly on 14 September 1943, occasionally using the code letters 'WE-'. Liberator Mk Vs gave way to Mk VIIIs in March 1945. The squadron transferred to Transport Command and RAF Waterbeach after the war, using the code 'BY-'.

NO. 86 SQUADRON
No. 86 Squadron converted to the Liberator Mk IIIA from October 1942 at Thorney Island, conducting its first patrol on 16 February 1943. The 'BX-'coded Mk IIIAs were augmented by GR.Mk Vs from March, when the unit moved to Aldergrove. The unit moved to Ballykelly in September and to Reykjavik in March 1944, before settling at Tain on 1 July 1944. The GR.Mk IIIAs were withdrawn altogether in August 1944, and the Mk Vs gave way to GR.Mk VIIIs in February 1945. The code letters 'XQ-' were used occasionally. The unit joined Transport Command on 10 June 1945.

NO. 120 SQUADRON
No. 120 Squadron reformed at Nutts Corner on 2 June 1941 as Coastal Command's first Liberator squadron, beginning patrols in September. The squadron's Liberator Is served until February 1943, moving to Ballykelly in July 1942 and mounting a detachment to Reykjavik between September 1942 and April 1943. Liberator IIs served from December 1941 until December 1942, and Mk IIIs from June 1942 until January 1944. The squadron moved to Aldergrove in February 1943 and to Reykjavik on 13 April 1943. When the unit returned to Ballykelly on 23 March 1944 it operated only Liberator Mk Vs (which it used from December 1943 until January 1945). GR.Mk VIIIs were on charge from December 1944 until disbandment on 4 June 1945. The code letters 'OH-' were allocated, but rarely used. No. 120 reformed on 1 October 1946 by the renumbering of No. 160 Squadron at Leuchars, which was then in the process of converting from Liberator GR.Mk VIIIs to Lancaster GR.Mk 3s. The last Liberators continued until June 1947.

NO. 160 SQUADRON
No. 160 Squadron formed at Thurleigh on 16 January 1942, and on 7 May moved briefly to Nutts Corner for ASW patrols, finally leaving for the Far East on 8 June. No. 160 Squadron returned to Leuchars and Coastal Command in June 1946, operating six Liberator Mk VIIIs, beginning conversion to the Lancaster in August. It renumbered as No. 120 Squadron in October.

NO. 206 SQUADRON
No. 206 Squadron converted to the GR.Mk VI in April 1944, on its return from the Azores. Initially based at St Eval, using the code '4-', the unit moved to Leuchars on 11 July 1944, taking the code 'PQ-' and converted to the GR.Mk VIII between March and April 1945. The squadron transferred to Transport Command on 10 June 1945.

NO. 220 SQUADRON
No. 220 Squadron at Lagens (Azores) augmented its Boeing Fortresses with Liberator Mk VIs from December 1944, using the code letters 'KK-'. The last Fortresses were withdrawn from service in April 1945, and the squadron moved to St Davids, Wales, at the end of May before transferring to Transport Command that June. The unit disbanded a year later in May 1946.

No. 111 OTU was based in the Bahamas and fulfilled the majority of Coastal Command's operational training needs for the Liberator. The relatively safe operating environment allowed crews to hone their skills before allocation to an operational squadron.

Fitted with a Leigh Light beneath the starboard wing, this Liberator GR.Mk VIII is seen in service with No. 220 Squadron during late 1944.

NO. 224 SQUADRON
No. 224 Squadron converted to Liberators at Tiree in July 1942, moving to Beaulieu in September 1942, and to St Eval in April 1943. Liberator Mk Vs augmented Mk IIIs from March 1943, using 'XB-' codes from July 1944, but the unit took both types to Scotland, moving to Milltown in September 1944. 'QA-' codes were also used. The unit returned to St Eval in July 1945 and began converting to the Lancaster in October 1946.

NO. 228 SQUADRON
A short-lived Liberator operator, No. 228 Squadron was formed by the redesignation of No. 224Y Sqn at St Eval, Cornwall, on 1 June 1946. The unit flew Liberator GR.Mk VIIIs on transport, reconnaissance, air-sea rescue and meteorological tasks until 30 September 1946. It reformed as an Avro Shackleton MR.Mk 2 operator in July 1954.

NO. 311 SQUADRON
The Czech-manned No. 311 Squadron converted to Liberator Mk Vs in July 1943, operating from Beaulieu and, from February 1944, from Predannack. The squadron moved to Tain on 7 August 1944, where it re-equipped with 'PP-' coded Liberator GR.Mk VIs in July 1945, before transferring to Transport Command in June 1945.

NO. 547 SQUADRON
No. 547 Squadron converted to the Liberator Mk III in October 1943, operating from Thorney Island before moving to St Eval in January 1944, using the code '3-'. When this numerical station code system was stopped, in July 1944, the codes '2V-' were adopted. The last GR.Mk IIIs served on until September 1944, when the squadron moved to Leuchars, though GR.Mk VIs had been on charge since June, and GR.Mk VIIIs arrived in March 1945. The squadron disbanded on 4 June 1945.

AIR SEA WARFARE DEVELOPMENT UNIT
The Air Sea Warfare Development Unit was formed from the Coastal Command Development Unit at Thorney Island on 1 January 1945, briefly operating a number of Liberators.

COASTAL COMMAND DEVELOPMENT UNIT
The Coastal Command Development Unit naturally operated examples of most of the types on charge with the Command, from 22 November 1940 (when it formed at Carew Cheriton) to 1 January 1945 (when it became the Air Sea Warfare Development Unit). The unit operated from Ballykelly, Tain, Dale and Thorney Island.

COASTAL COMMAND FLYING INSTRUCTORS SCHOOL
Liberators were among the types operated by the Coastal Command Flying Instructors School which formed from No. 12 FIS at St Angelo on 23 February 1945. The unit moved to Turnberry in June and to Tain in November, before becoming the Coastal Command Instructors School on 29 October 1945. It disbanded on 1 April 1946.

NO. 1(COASTAL) OTU
No. 1(C) OTU at Thornaby gained Liberators in March 1943. The Liberator Flight detached to Beaulieu on 24 March, then to Aldergrove in September, merging into No. 1674 HCU on 10 October.

NO. 4 REFRESHER FLYING UNIT
No. 4 RFU at Haverfordwest added Liberators to its complement in June 1944 and moved to Mullaghmore in September. The unit disbanded on 5 October 1944.

NO. 1653 CONVERSION UNIT
No. 1653 Conversion Unit formed at Polebrook on 9 January 1942 by absorbing No. 108 Conversion Unit and re-equipping with Liberator IIs. The unit continued to train front-line Liberator pilots after the unit moved to Burn in June 1942, before disbanding in October 1942.

NO. 1674(HEAVY) CU
No. 1674(H) Conversion Unit formed at Aldergrove on 10 October 1943 to train No. 17 Group crews on the Fortress, Liberator and Halifax. The Liberator Flight remained at Aldergrove when the rest of the unit moved to Longtown. The unit disbanded on 30 November 1945. The unit used Liberator GR.Mk IIIs, Vs, VIs and VIIIs.

NO. 111 OPERATIONAL TRAINING UNIT
No. 111 OTU was the only one of 11 OTUs (Nos 110-120) intended to train aircrew on US types in the USA, Cuba and the Bahamas to become operational. At least four of these were to have been equipped wholly or partly with Liberators. No. 111 OTU formed at Oakes Field, Nassau, on 20 August 1942 with 35 B-25s and 29 Liberators, and had a satellite field at Windsor Field. The Mitchells were used to tow targets and to familiarise pilots and engineers with modern US landplanes. The unit flew operational patrols (giving graduating crews invaluable role experience in a 'permissive' air environment). No. 111 OTU and its Liberators moved back to Lossiemouth from July 1945, vacating Oakes Field by 12 September. The unit disbanded on 21 May 1946.

Air Combat

ROYAL AIR FORCE TRANSPORT COMMAND

No. 53 Squadron
As a transport unit (from 25 June 1945), No. 53 operated Liberator Mk VIs and VIIIs from Merryfield and then, from 17 September 1945, from Gransden Lodge until it disbanded in June 1946.

No. 59 Squadron
The squadron transferred to Transport Command and flew trooping flights to India from Waterbeach, where it was stationed between September 1945 and 15 June 1946.

No. 86 Squadron
The Squadron moved to Oakington on 14 August 1945, forming part of No. 301 Wing. It continued in the transport role until disbanded on 25 April 1946.

No. 102 Squadron
The Halifax-equipped No. 102 Squadron moved to Bassingbourn on 8 September 1945 where it converted to Liberators, which it used for trooping flights to India until 28 February 1946, when it disbanded.

No. 206 Squadron
No. 206 Squadron and its GR.Mk VIIIs transferred to Transport Command on 10 June 1945, moving from Leuchars to Oakington on 1 August, and continued in the transport role until disbanded on 25 April 1946.

No. 220 Squadron
No. 220 Squadron transferred to Transport Command in June 1945, moving to Waterbeach in September. The unit gained Liberator Mk VIIIs in July 1945, but operated a mix of VIs and VIIIs until disbanded on 25 May 1946.

No. 231 Squadron
The former No. 45 Group Communications Squadron, No. 231 Squadron formed at Dorval on 8 September 1944, and flew transatlantic services using a wide variety of aircraft types, including the Liberator II and Liberator C.Mk IX. The unit moved to Bermuda in September 1945, where it disbanded in January 1946.

No. 232 Squadron
No. 232 Sqn at Stoney Cross traded Wellingtons for Liberator C.Mk IXs in February 1945, before moving to India.

JT978 was one of a number of Liberator C.Mk IXs to see service with No. 232 Squadron in late 1945.

No. 246 Squadron
No. 246 was a dedicated transport unit which formed at Lyneham in October 1944 with a variety of Liberator variants inherited from No. 511 Squadron. These were operated until November 1945, though they were augmented by Halifaxes and Yorks from November 1945, and by Skymasters from April 1945.

No. 311 Squadron
No. 311 Squadron transferred to RAF Transport Command in June 1945, and flew between the UK and Czechoslovakia until disbanded as an RAF unit on 15 February 1946.

No. 422 Squadron
No.422 Sqn, a former Coastal Command Sunderland unit, moved to Bassingbourn on 5 June 1945, re-equipping with Liberator VIs and VIIIs in August 1945, before disbanding on 3 September 1945.

No. 426 Squadron
A Canadian Halifax bomber unit, No. 426 Squadron transferred to Transport Command in June 1945, re-equipping with Liberators. It disbanded on 31 December 1945.

No. 466 Squadron
A Canadian Halifax unit, No. 426 Sqn transferred to Transport Command on 7 May 1945, re-equipping with Liberators in October. It disbanded on 26 October, before conversion was complete.

No. 511 Squadron
Formed from No. 1425 Flight on 10 October 1942, No. 511's Liberators flew between Lyneham and Gibraltar, and, from October 1943, on to India. By July 1944 the large squadron had 13 Avro Yorks in 'A' Flight, and 12 Liberators in 'B' Flight, the latter being passed to No. 246 Squadron in October 1944.

No. 1332 Conversion Unit
No. 1332 Conversion Unit formed at Longtown on 5 September 1944, and trained Transport Command Stirling, Liberator and York crews, operating from Nutts Corner from October, Riccall from April 1945 and Dishforth from November 1945. Strength included the Liberator II, III, VI, VII, VIII and IX. It used the unit code prefix 'YY-'.

One of many Liberator IIs converted for the transport role, AL578 took the name Marco Polo when serving with No. 231 Squadron within No. 45 Group. Important passengers included Earl Mountbatten of Burma, who was flown over the 'hump' from India to China in August 1943.

No. 1409 Long Range Meteorological Flight
No. 1409 Meteorological Flight added 'Long Range' to its designation in October 1945, when it partially re-equipped with Liberators at Lyneham. It disbanded on 13 May 1946.

No. 1425 Communications Flight
No. 1425 Flight formed at Prestwick on 30 October 1941 with three Liberators, and developed a long-range ferry service to the Middle East. It was successively based at Honeybourne, Hurn and Lyneham, gaining further aircraft and finally re-designating as No. 511 Squadron on 10 October 1942. The flight operated a Liberator service between the UK and Gibraltar from November 1941 until October 1942.

No. 1427(FT) Flight
No. 1427(Ferry Training) Flight was established to give ATA pilots flying experience on four-engined aircraft and occasionally used a Liberator. The unit was operational from December 1941 until April 1943.

No. 1445 Flight
The flight formed on 27 February 1942 at Lyneham, tasked with preparing and despatching 32 Liberators to the Middle East for Nos 159 and 160 Squadrons. The flight became No. 301 Ferry Training Unit's 'C' Flight on 1 November 1942.

ROYAL AIR FORCE BOMBER COMMAND

No. 223 Squadron
No. 223 Squadron reformed at Oulton on 12 August 1944, and operated a variety of ex-USAAF B-24Hs and B-24Js (sometimes known as RCM Liberators or Liberator RCM.Mk VIs) in the bomber support role. These night camouflage-painted aircraft carried the code '6G-'. The squadron disbanded on 29 July 1945, having started to convert to the B-17 in April.

No. 1699 Flight
Originally formed at Sculthorpe on 24 April 1944 to role-train No. 100 Group's B-17 crews, moving to Oulton on 16 May. No. 1699 Flight gained a Liberator training role from September 1944. Its aircraft used the code '4Z-'. The flight was redesignated as No. 1699 (Bomber Support) Conversion Unit on 24 October 1944.

No. 1699(Bomber Support) Conversion Unit
No. 1699 Flight was redesignated as No. 1699(Bomber Support) Conversion Unit on 24 October 1944, and provided conversion training for No. 100 Group Fortress and Liberator aircrew until the end of the war, disbanding on 29 June 1945.

Named Commando, and serving as the personal transport for the British Prime Minister Winston Churchill, AL504 underwent conversion to incorporate an R4Y-type tail fin during its period of service with No. 231 Squadron.

RAF Liberator Units

ROYAL AIR FORCE MIDDLE EAST AND MEDITERRANEAN COMMANDS

No. 37 Squadron
No. 37 Squadron converted to the Liberator Mk VI at Tortorella, Italy, from late October 1944, and operated the type in the bombing and resupply roles until the end of the war, dropping supplies to Yugoslav partisans and attacking targets throughout Italy and the Balkans. The squadron moved to Aqir in October 1945, then to Shallufa in December, before disbanding on 31 March 1946.

No. 40 Squadron
No. 40 Squadron converted to the Liberator B.Mk VI in March 1945 at Foggia, moving to Abu Sueir in October. The Liberators were phased out in January 1946.

No. 70 Squadron
No. 70 Squadron converted to the Liberator VI in January 1945 at Cerignola, moving to Aqir in October and to Shallufa in December. The squadron disbanded on 31 March 1946.

No. 104 Squadron
No. 104 Squadron began conversion to the Liberator B.Mk VI in February 1945 at Foggia, moving to Abu Sueir in late spring. The squadron converted to Lancasters during November.

Seen at Gianaclis, Italy, in late 1945, this B.Mk VI belongs to No. 148 Squadron.

No. 108 Squadron
Liberator Mk IIs supplemented No. 108's Wellingtons between November 1941 and June 1942, operating from Kabrit. During November and December 1942 a remnant of No. 108 Squadron operated the Liberators of the Special Operations (Liberator) Flight, before disbanding on 25 December.

Liberator 'B.Mk II' AL579 served with No. 159 Squadron in North Africa during 1942.

No. 614 Squadron
No. 614 Squadron at Amendola began receiving Liberator B.Mk VIIIs in August 1944, and these had entirely replaced the remaining Halifaxes by March 1945. The squadron disbanded on 27 July, immediately reforming as No. 214 Squadron.

Heavy Bomber Conv. Unit
A Heavy Bomber Conversion Unit was formed to meet MEAF requirements for Liberator aircrew. Based at Lydda, the unit became No. 5 Heavy Bomber Conversion Unit before re-designating as No. 1675 Heavy Conversion Unit.

No. 108 Conversion Unit
No. 108 Conversion Unit began forming at Polebrook in December 1941, intending to convert crews to the Liberator for Middle East-based squadrons. However, the unit was absorbed by No. 1653 Heavy Conversion Unit before any aircraft arrived.

No. 1(ME) C&C Unit
No. 1(Middle East) Check & Conversion Unit formed at Bilbeis on 1 June 1943 from a number of refresher training and check and conversion flights. The unit was redesignated as No. 1330 Conversion Unit on 29 June 1944, when its complement included Liberator VIs.

No. 1330 Conversion Unit
No. 1330 CU formed from No.1(ME) Check & Conversion Unit and performed type conversion for in-theatre pilots in the Middle East. It was based at Bilbeis between 29 June 1944 and 1 March 1946. Strength included the Liberator VI.

No. 148 Squadron
The Special Liberator Flight (X Flight) at Gambut became No. 148 Squadron on 14 March 1943, operating Liberator IIs and IIIs and Halifaxes in the supply-dropping role, supplying arms and equipment to resistance groups in Greece, Yugoslavia and Albania. The Liberator B.Mk VI was introduced in March 1945.

No. 159 Squadron
No. 159 Squadron's Liberator IIs arrived in the Middle East, at Fayid, in April 1942, and flew bombing operations against targets in North Africa, Italy and Greece until flying on to India, leaving on 10 May.

No. 160 Squadron
No. 160 Squadron formed at Thurleigh on 16 January 1942, and on 7 May moved briefly to Nutts Corner for ASW patrols, finally leaving for the Far East on 8 June. Though the squadron's groundcrew had left for India in February, the aircraft and aircrew were held in the Middle East until after the Battle of El Alamein. Individual Liberator IIIs then began leaving for India, though a handful of Liberator IIs remained in the Middle East, forming No. 178 Squadron.

No. 178 Squadron
No. 178 Squadron formed at Shandur on 15 January 1943 from a No. 160 Squadron detachment. Liberator IIs were augmented by Halifaxes between May and September 1943, and then by Liberator IIIs from September 1943, but both types were replaced by Liberator B.Mk VIs from December. The squadron moved from Hosc Raui to Terria, El Adem, Celone and Amendola, where it ended the war. The unit moved to Ein Shemer in August 1945 and converted to Lancasters in November.

No. 214 Squadron
No. 614 Squadron at Amendola, Italy, renumbered as No. 214 Squadron on 27 July 1945, and moved to Palestine (Ein Shemer and then on to Fayid in Egypt) in August. The squadron flew Liberator Mk VIs until December, when conversion to Lancasters was completed.

No. 1586 Special Duties Flight
No. 138 Squadron's Polish Flight at Derna redesignated as No. 1586 Special Duties Flight on 3 November 1943. The unit flew a mix of Halifaxes and Liberators dropping supplies to partisans in Poland, Yugoslavia and northern Italy. The flight moved to Brindisi in January 1944, and between April and May was known as No. 1586 Special Duties (Operations) Flight. The unit redesignated as No. 301 Squadron on 7 November 1944.

No. 1675(Heavy) Conversion Unit
The Heavy Bomber Conversion Unit at Lydda was redesignated as No. 1675(Heavy) Conversion Unit on 15 October 1943, providing conversion training for MEAF Liberator crews. The unit moved to Abu Sueir in August 1944 and disbanded on 12 October 1945. The unit used Liberator IIs, IIIs, Vs, and VIs.

No. 31 Squadron, SAAF
No. 31 Heavy Bomber Squadron was re-established in January 1944 at Zwartkop Air Station, departing for Egypt on 30 January 1944. The unit arrived at Almaza from 19 February 1944 and the aircrews converted at No. 1675 Conversion Unit at Lydda, Palestine. The squadron moved to Kilo 40 on 27 April, forming the basis of No. 2 Wing, SAAF, and started flying operations one month later. The squadron moved to Foggia, Italy, from 16 June. After the German surrender the squadron adopted a trooping role until it was withdrawn from operations on 5 December 1945 and disbanded the following day.

No. 34 Squadron, SAAF
No. 34 Squadron formed after No. 31, but joined it at Kilo 40 during May 1944. Thereafter, both South African units operated together as No. 2 Wing, SAAF.

No. 301 Squadron
No. 301 Squadron reformed at Brindisi, Italy, on 7 November 1944 with the redesignation of No. 1586(Special Duties) Flight. The squadron flew Halifax Mk IIs, Mk Vs and Liberator Mk VIs on secretive supply-dropping missions to Poland, northern Italy and Yugoslavia. The squadron left its Liberators behind in April 1945 when it returned to Blackbushe in the UK to become a Vickers Warwick-equipped transport squadron.

An important element of the Mediterranean Allied Strategic Air Force in Italy was the Liberators of No. 2 Wing, South African Air Force. This B.Mk VI, operated by No. 34 Squadron, is seen during a mission in late 1944.

Air Combat

Royal Air Force South East Asia and West Africa Commands

No. 8 Squadron
No. 8 Squadron enjoyed a brief six-month stint as a Liberator unit when No. 200 Squadron at Jessore re-designated. No. 8 then moved to Minneriya, in Ceylon, from where it flew mainly supply-dropping missions to resistance groups in Malaya. It disbanded on 15 November 1945.

No. 99 Squadron
No. 99 Squadron traded Wellingtons for Liberator B.Mk VIs in September 1944 at Dhubalia. The squadron moved to the Cocos Islands in July ready for the liberation of Malaya, then disbanded on 15 November. The squadron's aircraft wore a small white disc on their black-painted rudders.

No. 159 Squadron
No. 159 Squadron finally arrived at Salbani on 30 September 1942, then flew from Digri (from October 1943) with a brief deployment to Dhubalia in March 1944. Liberator Mk IIIs gave way to Mk VIs in March 1944, and Mk VIIIs arrived in June. From 30 June 1944 'C' Flight took over the Elint flight previously parented by No. 160 Squadron. This flight was detached to Ceylon in April 1945, and became No. 1341 Flight in May 1945. The remainder of the squadron switched to transport duties after the Japanese surrender, moving back to Salbani in October 1945 and disbanding on 1 June 1946.

No. 160 Squadron
After the Battle of El Alamein No. 160 Squadron began to trickle its Mk III aircraft out to Shandur and Salbani, before moving to Ratmalana (Ceylon) in February 1943. The squadron operated in the shipping protection, minelaying and reconnaissance roles, moving between Sigiriya, Minneriya and Kankesanturai, augmenting its Mk IIIs with Mk Vs from June 1943 and Mk VIs from June 1944 until January 1945. The unit switched to supply-dropping for resistance movements in Malaya, and Sumatra in June 1945, and then to transport duties after the Japanese surrender. The squadron used the codes 'BS-'. The squadron moved back to Leuchars in Scotland in June 1946, joining Coastal Command.

No. 200 Squadron
No. 200 Squadron at Yundum in the Gambia traded Hudsons for Liberator Mk Vs in June 1943. The unit moved to St Thomas's Mount in March 1944, flying ASW patrols. The unit switched to supply-dropping in April 1945, moving to Jessore. The unit renumbered as No. 8 Squadron on 15 May 1945.

No. 203 Squadron
No. 203 Squadron at Madura converted from Wellingtons to Liberator GR.Mk VIs in October 1944, and moved to Kankesanturai in Ceylon to start operations in February 1945. The unit switched to the transport role after the Japanese surrender and gained Mk VIIIs in January 1946. The unit returned to the UK in May 1947, leaving behind its Liberator Mk VIIIs.

No. 215 Squadron
No. 215 Squadron flew its final mission with Wellingtons on 23 June 1944, and moved to Kolar to convert to the Liberator Mk VI. The squadron resumed bomber operations in October, operating from Digri and then, from December, from Dhubalia. The squadron's aircraft wore two horizontal white stripes on their black-painted rudders, before they were replaced by Dakotas in April 1945.

No. 232 Squadron
No. 232 Squadron moved to India in February 1945, setting up shop at Palam as a transport unit with Liberator Mk IIIs, VIs and VIIIs, and Skymasters. The aircraft maintained a service between Ceylon and Australia, moving base to Poona in May 1946 and disbanding on 15 August 1946.

No. 292 Squadron
A Jessore-based air-sea rescue unit equipped with Walruses, Warwicks and Sea Otters, No. 292 Sqn gained Liberator Mk VIs in December 1944. The squadron moved to Agartala in February 1945, and disbanded in June, its task taken over by Nos 1347, 1348 and 1349 Flights.

Wearing the unit's allocated tail marking of two horizontal white stripes on black-painted fins, this Liberator B.Mk VI belongs to No. 215 Squadron, an important unit of Strategic Air Force, Eastern Air Command. The aircraft is seen here approaching its base after a successful bombing attack on Japanese facilities at Rangoon.

No. 321 Squadron
A unit formed from Royal Netherlands Naval Air Service personnel, No. 321 Squadron at China Bay, Ceylon, augmented its Catalinas with Liberator Mk VIs from December 1944. The squadron moved to Cocos Island in July 1945 and to Batavia in October, transferring back to Royal Netherlands Air Force command on 8 December 1945. The Catalinas were retained, but the Liberators were returned to the USAAF and replaced by Dakotas.

No. 354 Squadron
No. 354 Squadron formed as a General Reconnaissance Squadron at Drigh Road, but did not receive its first aircraft until after a move to Cuttack on 17 August 1943. The unit was equipped with Liberator Mk Vs until May 1944, augmenting these with Liberator IIIAs between December 1943 and April 1944, and replacing both versions with the Liberator VI from February 1944. The unit moved to Minneriya in October 1944, but returned to Cuttack on 5 January 1945, disbanding there on 18 May 1945.

No. 355 Squadron
No. 355 formed as a heavy bomber unit at Salbani on 18 August 1943, initially equipping with Liberator IIIs. The unit flew its first operation on 20 November 1943. Liberator VIs arrived in March 1944, and the last IIIs bowed out in July. The Mk VIs themselves gave way to Mk VIIIs from August 1945, flying survey and transport missions after the Japanese surrender, before disbanding on 31 May 1946. Apart from a brief stay at Digri between January and April 1946, the unit remained at Salbani. The squadron's aircraft wore two vertical white stripes on their black-painted rudders.

No. 356 Squadron
No. 356 Squadron joined No. 355 at Salbani when it formed there on 15 January 1944. The squadron used only the Liberator Mk VI throughout its career, flying its first bombing operation on 27 July 1944. Bombing and mining sorties continued until July 1945, when the squadron moved to Cocos Island to prepare for the liberation of Malaya. After a period of supply-dropping and transport operations, the squadron was disbanded on 15 November 1945. The squadron's Liberators wore a white diagonal cross on their black-painted rudders.

No. 357 Squadron
No. 357 Squadron was formed at Digri on 1 February 1944, 'A' Flight taking over the seven Hudsons and three Liberator Mk IIIs of No. 1576 Special Duties Flight, and with a new 'B' Flight forming with Catalinas at Redhills Lake. The Liberators flew long-range supply and agent-dropping flights to Malaya and Sumatra, with Liberator Mk VIs arriving in September 1944. The Catalinas broke away to form No. 628 Squadron in March 1944, and in January 1945 the last Liberator Mk IIIs left and the Hudsons were replaced by a mix of Dakotas and Lysanders. The unit moved to Jessore in September 1944 and to Mingaladon in September 1945, although it maintained numerous detachments including China Bay, Cox's Bazaar, Don Muang, Drigh Road, Dum Dum, Meiktela, Minneriya and Toungoo. The squadron disbanded on 15 November 1945.

No. 358 Squadron
No. 358 Sqn was formed on 8 November 1944 at Kolar from the remnants of the disbanding No. 1673 Heavy Conversion Unit. The squadron moved to Digri on 3 January, and flew just one bombing mission on 13 January 1945 before being permanently assigned to 'special duties', dropping supplies to resistance movements and, after VJ-Day, to PoW camps while operating from Jessore (from 10 February). The unit moved to Bishnupur on 19 November 1945, where it disbanded two days later.

No. 160(Special) Flight
Two aircraft were assigned to the detection of Japanese radio and radar signals in December 1943, operated by hand-picked and specially trained crews. These were attached to No. 160 Squadron at Sigiriya. The flight moved to Salbani on 1 May, but found insufficient space and moved on to Digri three days later, still under the administrative control of No. 160 Squadron. The flight ceased to be controlled by No. 160 Squadron on 30 June, when it became 'C' Flight of No. 159 Squadron, with both aircraft requiring major overhauls and both crews 'tour-expired'.

No. 356 Squadron formed at Salbani, India, in January 1944, equipped with newly arrived Liberator B.Mk VIs. Declared operational in July of that year, the unit advanced through the Pacific attacking Japanese ground installations and shipping. One of the squadron's aircraft is seen here while operating from the Cocos Islands in the final days of the war.

RAF Liberator Units

Liberator B.Mk VIII KP136, seen here cruising over the Indian Ocean, served with No. 355 Squadron from August 1945 to May 1946.

No. 1673(Heavy) Conversion Unit
No. 1673(Heavy) Conversion Unit formed from No. 1584(HB) Conversion Unit at Salbani on 1 February 1944. The unit moved to Kolar in April 1944 and then disbanded in November, passing its aircraft, personnel and role to No. 358 Squadron and No. 6 Refresher Flying Unit. The unit used Liberator Mk IIIs and VIs.

No. 1341 Radio Countermeasures Flight
No. 1341 Flight took over the Liberators of No. 159 Squadron's 'C' Flight at Digri on 13 May 1945, adding these to its Elint Halifaxes. The Liberators operated from China Bay during July, but then transferred to transport operations after VJ day, flying from Pegu and Raipur before the flight disbanded on 30 October 1945.

ASW Development Unit
The Air Sea Warfare Development Unit (ACSEA) formed at Ratmalana, India, on 15 December 1944 to develop GR and ASW tactics and equipment in the Far East. The unit included a number of Liberators on its strength before disbanding in November 1945.

Heavy Bomber Conv. Unit
The Heavy Bomber Conversion Unit in India was formed at Salbani on 1 September 1942 to provide Liberator crews for SEAC bomber units. It became No. 1584 Heavy Bomber Conversion Flight on 23 July 1943.

No. 1346 Air-Sea Rescue Flt
No. 1346 Flight formed from No. 1 ASR Flight at Kankesunterai on 15 June 1945, operating Warwicks while awaiting Liberator Mk VIs, which arrived in October. The flight maintained a detachment in the Cocos Islands, but disbanded on 20 April 1946.

No. 1347 Air-Sea Rescue Flt
No. 1347 Flight formed from No. 2 ASR Flight at Agartala on 15 June 1945, taking over a flight of Liberator Mk VIs from the disbanding No. 292 Squadron. The flight moved to Akyab in September, adding Warwicks, and to Chittagong in October. The flight maintained a detachment at Cochin but disbanded on 1 June 1946.

No. 1 AGS (India)
The complement of No.1 Air Gunnery School (India), based at Bairagarh from May 1943 until July 1945, included a number of Liberators, alongside various other types.

No. 6 Refresher Flying Unit
No. 6 RFU formed from No. 1673 HCU at Kolar on 8 November 1944 to provide refresher training for returning Liberator aircrews in India and acclimatisation for crews joining from Canadian OTUs. It disbanded on 31 December 1945.

No. 1348 Air-Sea Rescue Flt
No. 1348 Flight formed from No. 3 ASR Flight at Agartala on 15 June 1945, taking over a flight of Liberator Mk VIs from the disbanding No. 292 Sqn. The unit moved to Pegu in January 1946, and began converting to the Lancaster from February before disbanding on 15 May 1946.

No. 1349 Air-Sea Rescue Flt
No. 1349 Flight formed from No. 4 ASR Flight at Agartala on 15 June 1945. The unit moved to Mauripur in January 1946, and disbanded on 15 May 1946.

No. 1331 Conversion Unit
No. 1331 Conversion Unit formed from a Check & Conversion Unit at Mauripur on 1 September 1944, quickly moving to Risalpur. It performed conversion training for ferry crews, before disbanding on 15 January 1946. Aircraft on strength included Liberator Mk IIIs and Mk VIs.

No. 1584(HB) Conv. Flight
No. 1584(Heavy Bomber) Conversion Flight formed at Kolar on 1 September 1942 to convert aircrew from Wellingtons and other types to the Liberator. It became a full conversion unit in November 1943 before redesignation as No. 1673 HCU on 1 February 1944. The unit used Liberator Mk IIIs, Vs and VIs.

No. 1576(Special Duties) Flt
No. 1576(SD) Flight augmented its Hudsons with three Liberator Mk IIIs, but was almost immediately absorbed in the new No. 357 Squadron, which formed at Digri on 1 February 1944, becoming the new unit's 'A' Flight.

No. 1354 DDT Spraying Flt
No. 1354 DDT Spraying Flight formed at Digri on 1 July 1945 from the based DDT Flight (formed on 10 March), equipped with five Liberator Mk VIs and two Austers. The unit moved to Mingaladon in November, then on to Pegu, where it was re-designated as No. 1354 Spray (Insecticide) Flight before disbanding on 15 February 1946.

Royal Australian Air Force

No. 12 Squadron, RAAF
No. 12(Heavy Bomber) Squadron began to reform at Cecil Plains on 14 December 1944, receiving its first Liberators on 5 February 1945. No. 12 Squadron's HQ at Cecil Plains closed down on 30 April 1945 and reopened at McMillan's Road campsite, Darwin, on the following day. No. 12 Squadron's Liberators began operations from Darwin on 26 May 1945 and flew 108 operational sorties before the cessation of hostilities. The squadron then operated in the transport role. It left Darwin for Amberley on 1 March 1946 and was reduced to a care and maintenance unit, with limited aircrew training controlled by No. 82 Wing. Three Liberators were retained on unit establishment for VIP duties until 1947, when the squadron completed conversion to the Lincoln. The unit's aircraft wore 'AU-' or 'NH-' codes.

No. 21 Squadron, RAAF
No. 21 Squadron at Camden, New South Wales, converted to the Liberator In July 1944 after its Vengeances were withdrawn from New Guinea. The squadron moved to Fenton, and participated in a number of attacks on Japanese convoys during the final days of the war. After some supply-dropping and transport work, ferrying former PoWs and returning personnel to Australia, No. 21 Squadron moved to Amberley on 18 April 1946. It was immediately reduced to care and maintenance status and effectively ceased to exist. The squadron used the code 'MJ-'.

No. 23 Squadron, RAAF
No. 23(City of Brisbane) Squadron was originally formed in February 1939. Along with other RAAF Vengeance squadrons, it was withdrawn to the Australian mainland in February 1944, and was reduced to a cadre unit pending the allocation of Liberators. After conversion, No. 23 Squadron's Liberators deployed to Fenton, Northern Territory, in April 1945, flying reconnaissance and anti-shipping operations until the end of the war. The squadron then dropped supplies to Allied PoWs still in Japanese hands before evacuating them to Australia and flying troops home from the war. No. 23 Squadron moved to Amberley on 18 April 1946, but was immediately reduced to care and maintenance status effectively ceasing to exist. The squadron used 'NV-' codes.

No. 24 Squadron, RAAF
No. 24 Squadron was originally formed in June 1940, and received Liberators at Manbulloo in March 1944. After conversion the unit moved to Fenton, Northern Territory, on 1 September 1944, and forward-deployed to Truscott on 9 November. The next day A72-38 flew the first RAAF B-24 operation from Truscott, a shipping sweep between Bali-Lombok-Soembawa-Soemba. No. 24 flew anti-shipping strikes, reconnaissance missions and bomber attacks and operated briefly from Corunna Downs in November 1944. Strikes continued until Japan surrendered, after which the squadron's aircraft ferried PoWs from Moratai to Australia. The unit used 'GR-' codes.

No. 25 Squadron, RAAF
No. 25(City of Perth) Squadron served as an air defence unit on Australia's west coast for most of the war, and did not convert to the Liberator until early 1945. The squadron re-equipped at RAAF Cunderdin, Western Australia, and from March 1945 flew bombing missions in the Netherlands East-Indies, Timor and Java. After the Japanese surrender No. 25 Squadron returned PoWs and troops to Australia and disbanded in July 1946. Its Liberators wore 'SJ-' codes.

No. 99 Squadron, RAAF
No. 99 Squadron formed with Liberators at Leyburn, Queensland, on 1 February 1945, moving to Jondaryan in March, and to Darwin from May (an advance party, with the bulk of the squadron following as the war ended). The squadron ferried former PoWs and troops home, before moving to New South Wales in November 1945, where it disbanded in June 1946. The squadron used 'AU-' and 'UX-' codes.

No. 102 Squadron, RAAF
No. 102 Sqn formed at Cecil Plains as a Liberator unit on 31 May 1944, remaining there until it disbanded on 18 March 1946. The aircraft (which may have embraced a radio/radar countermeasures trials role) wore 'BV-' codes.

No. 201 Flight, RAAF
The shadowy No. 201 Flight at Darwin operated in the radio and radar countermeasures and Elint roles, and its aircraft wore 'LV-' codes.

Among Royal Australian Air Force Liberator units was No. 21 Squadron, seen here operating from Tarakan in June 1945.

No. 200 Flight, RAAF
No. 200 Flight at Leyburn, Queensland, began clandestine operations (including dropping Commandos behind enemy lines) in February 1944. The aircraft sometimes used McGuire Field on Mindoro Island, Philippines, as a forward base. The unit disbanded in December 1945. Its aircraft wore 'NX-' codes.

No. 7 OTU, RAAF
No. 7 OTU at Tocumwal, New South Wales, received its first Liberators in February 1944, and thereafter took over responsibility for training all RAAF Liberator aircrew. The OTU's aircraft wore yellow numerical codes.

INDEX

Page numbers in **bold** refer to illustration captions.

11 (see PZL)
63 (see Potez)
180 (see Aero Boero)
182 Skylane (see Cessna)
206 Stationair (see Cessna)
210 (see Bell)
210 Centurion (see Cessna)
404 Titan (see Cessna)
407/430 (see Bell)
500C (see Hughes)
550 Citation II (see Cessna)
707 (see Boeing)
707 Phalcon/Condor (see Boeing/IAI/Elta)
717/720/737/738/747 (see Boeing)
748 (see HAL/Avro)
757-200 (see Boeing)
767 (see Boeing)
2707 SST (see Boeing)

A

A-1/A-1A (see Alenia/EMBRAER)
A-4 (see McDonnell Douglas)
A-10 Warthog (see Fairchild)
A-29A/B Super Tucano (see EMBRAER)
A36 Hofit (see Beech)
A-50 Golden Eagle (see KAI/Lockheed Martin)
A-50 'Mainstay' (see Beriev/Ilyushin)
A109E (see Agusta)
A160 Hummingbird VTUAV (see Boeing/Frontier Systems)
A300 *et seq* (see Airbus Industrie)
A400M (see Airbus Military Company)
AAI
RQ-7 Shadow UAV: 18, 28
RQ-7B UAV: 12
AB-139/AB139 (see Agusta-Bell; AgustaWestland)
Abraham Lincoln, USS: 16
AC-47 (see Douglas)
AC-47T Fantasma (see Basler)
AC-130 (see Lockheed Martin)
Advanced Light Helicopter, Dhruv (see HAL)
Aermacchi M-346: 4
Aero Boero 180: **10**
Aerocomp Comp Air 7SL: 20
Aeromot Super Ximango: **10**
Aérospatiale
HH-65A Dolphin: 86, 88, 89, 90, **91**, 92
HH-65B: 88, 89, 90, **91**, 92
HH-65C: 89, 90, 92
HH-65X project: 92
MH-65C: 92
Aérospatiale/Westland Gazelle AH.Mk 1: 21
Agusta
A109E: 92
MH-68A Sting Ray: 88, 89, **89**, 90, 92
Agusta-Bell AB-139: 12, 92
Agusta/Sikorsky SH-3D/H: **29**
AgustaWestland
AB139: 10
EH101 Merlin: 6, 11
EH101 Merlin Joint Supporter: 9
Super Lynx 300: **12**
AH-1 (see Bell)
AH-64A/D (see Boeing/McDonnell Douglas Helicopters)
Airborne Laser, YAL-1A (see Boeing)
Air Breathing Satellite (ABS) UAV project: 59
Airbus Industrie
A300: 21, 28-29
A319CJ (Corporate Jet): 9, 111
A330 MRTT: 113
A330-200 MRTT: 9
Airbus Military Company A400M: 4, 6
Air Combat – RAF Liberators at war – Part 2: Service abroad, variants and operators: 156-173, **156-173**
Air Defence Ship: 69
Air Power Analysis – US Department of Homeland Security: 78-93, **78-93**
Air Power Intelligence: 4-15, **4-15**
Airtech CN-235M: 12
Alenia
C-27J: 6
G-222: 6
Alenia/EMBRAER
A-1/A-1A (AMX): **36**
RA-1A/B: **36**
Alenia/Lockheed Martin
C-27J Spartan: **13**
Alizé (see Breguet)
Altair UAV (see General Atomics)
Altus I/II UAV (see General Atomics)
AMX International AMX-T: 13
Antonov
An-12: 13
An-32: 8
An-74T-200A: 12
An-74TK-200S: 12
An-124: 31
Apache, AH-64A/D (see Boeing/McDonnell Douglas Helicopters)
Argonaut, HMS: 103
Argus, P-2H (see Canadair)
AS 350 *et seq* (see Eurocopter)
AStar, AS 350B/B2/B3 (see Eurocopter)
AT-26 Xavante (see EMBRAER)
AT-27 Tucano (see EMBRAER)
AT-63 Pampa (see Lockheed Martin Argentina)
Atlantic (see Breguet; Dassault)
ATR 72MP-500 (see Avions de Transport Regional)
Aurora Flight Sciences Perseus/Theseus UAVs: **49**
Australia
Royal Australian Air Force: 162-164, 173
AV-8B Harrier II (see McDonnell Douglas/BAe)
Avions de Transport Regional ATR 72MP-500: 12
Avro
Lancaster: 147, 150, 159, 163
Lincoln: 163, 164

B

B-1B (see Rockwell)
B-2 (see Northrop Grumman)
B-17 (see Boeing)
B-24 Liberator (see Consolidated)
B-52 (see Boeing)
'Badger' (see Tupolev Tu-16)
BAe Strikemaster: 13
BAE Systems
Harrier: 103, 119
Harrier GR.Mk 7/7A/9: 61
Harrier T.Mk 4(I): 68, **69**
Harrier T.Mk 10/12: 61
Harrier T.Mk 60: 68, **69**
Hawk Mk 120: 6
Hawk Mk 128 AJT (advanced jet trainer): 6
Nimrod: 21
Nimrod MR.Mk 2: 14, 27
Nimrod MRA.Mk 4: **6**, 14
Sea Harrier: 60-69, **60-69**, 103
Sea Harrier FA.Mk 2: 61-67, **61**, **63**, **65-66**, 69
Sea Harrier FRS.Mk 51: 68, **68**, 69
Sea Harrier T.Mk 8: 61, 62, **66**, **67**
Bandeirante, EMB-110 (see EMBRAER)
Barboor, RC-707 (see Boeing)
Basler AC-47T Fantasma: 9
Bateleur UAV (see Denel)
'Bear'-class USCG cutter: 92
Beech
A36 Hofit: **14**
C-12: 18
C-12C King Air: 82, 83, **83**, 84
C-12M: 83, **83**, 84
King Air 200: 83, **83**
T-6A: 9
Bell
210/407: **8**
430: 80
AH-1: 82
AH-1W Cobra: 18, 25, 26, 27, **27**, 28
HV-911 Eagle Eye VUAV: 92, **93**
JetRanger: 20
Kiowa: **31**
OH-58D Kiowa Warrior: 14-15, **16**, 18, **18**, 21, 25, 26, 28-29
TH-67A+: 12
UH-1: 79, 82
UH-1B: 125
UH-1H: 8, **9**, 11, 13, 14
UH-1H Iroquois: **31**
UH-1H Super Huey: 80, **80**
UH-1N Huey: 18, 25, 27, 28, **28**
Beriev/Ilyushin
A-50 'Mainstay': 117
A-50EhI: 10
Bf 109 *et seq* (see Messerschmitt)
BGM-34A (see Teledyne Ryan)
Blackburn Skua: 132
Black Hawk/Blackhawk (see Sikorsky)
'Blackjack' (see Tupolev Tu-160)
Blenheim (see Bristol)
Boeing
707: 98-117, **98-117**
707-1: 100
707-3F5C: 109
707-3F9C: 105
707-3J9C: 108, 115, **115**
707-3J9C Peace Station: 110, **110**
707-3K1C/-3L5C/-3L6B/-3L6C: 109
707-3M1C: 108
707-3P1C/-3W6C: 109
707-6: 98
707-7-27/-39: 99
707-120: 100, 101, 106, 108, 115, **115**
707-120B: 100
707-124: 108, 112
707-125: 108
707-126: 109
707-131: 102, 108
707-138: 100
707-138B: 108, 109
707-153: **98**, 106
707-226: 109
707-227: 100
707-307C: 107, **107**
707-320: 100, 101, 108, 115
707-320B: 100, 105, 106, 117, **117**
707-320C: 100, 110
707-321: 108
707-321B: 107, 108, 109, 111-112, **112**, 116
707-323C: 107, 109, 113
707-324C: 109
707-326: 109
707-327C: 107
707-328: 108
707-329: 108, 116
707-330B: 108, 109
707-330C: 109
707-331C: 110
707-336B: 108
707-337C: 108, 116
707-338: 113
707-338C: 108
707-340C: 109
707-344C: 117
707-346C: 113
707-347C: 107
707-351B: 109
707-351C/SCD: 108, 109
707-353B: 106
707-355C: 107
707-366C: 108
707-368: 113
707-368C: 108, 109
707-370C: 108
707-373C: 113
707-382B: 106, 109, 111
707-385C: 106, 110
707-387B: 108
707-387C: 116
707-396C: 106
707-420: 100
707-700: 101, 111
707-720B: 100
707 IAI EW conversions: 115-117, **115**, **116**, **117**
707 IAI tanker conversions: 113, **113**
707 Iranian EW conversion: 115, **115**
707 Paul Revere MC^{2}A-X testbed: 105, 107, 114, **114**
707 Re'em: 102, 115
717: 100
720: 100, 101, 109, **109**
720-023B: 109
720-025: 117, **117**
720-047B/-051B/-060B: 109
720 LRPA ASW test aircraft: 117, **117**
737 AEW&C: 11
737 MMA: 50-51
738: 109
747: 19, **100**, 102, **103**, 105, 107
757-200: 8
767: 105, 111
2707 SST: 105
B-17: 151, 153
B-52: 114
C-18A: 107, **107**, 114
C-40A/C: 12
C-135: 98, 100, 101
C-135FR: **37**
C-137: 107, **107**
C-137B/C: 106, **106**
CC-137 Husky: **102**, 107, **107**, 110, **110**
CT-49: **103**
E-3 Sentry: 24, 98, **100**, 101, 103-104, 105, 111
E-3A: **100**
E-3B Block 30: **98-99**
E-3D: 105, **105**
E-6: 101, 105
E-6A: **100**
E-10A MC^{2}A (Multi-Mission Command and Control Aircraft): 114
EC-18: **103**
EC-18B: 113-114, **113**, **114**
EC-18D ACMMCA: 107, 114, **114**
EC-137D: **100**, **101**, 107, **107**
'EC-707' Airborne Command Post: 112, 116
EC-707 Chasidah: 115, **115**
KC-135 Stratotanker: 17, 25, 101
KC-135A/D/E: 101
KC-135T: **20**
KC-137: **102**, 110, **110**, 111, **111**
KC-707: 5
KC-707 Saknayee: 112, **112**, 115
KC-767A: 10-11
KE-3A: 111, **111**, 115
Model 367-80: 98, 99-100, 109
Model 367-138B: 100
RC-135: 101, 102
RC-707 Barboor: **104**, 115, 116, **116**
RE-3A: 111, 115, **115**
RE-3B: 115
Sentry AEW.Mk 1 (E-3D): 105, **105**
TC-18E: **103**, 107
VC-25A: **106**
VC-137A: 106, **106**
VC-137B: 106
VC-137C: **99**, 106, **106**
X-45A UAV: 56, **56**
X-45C UCAV: 54, 55, **56**
YAL-1A Airborne Laser: **6**
Boeing 707 military variants – Part 1: 98-117, **98-117**
Boeing/Frontier Systems A160 Hummingbird VTUAV: 57-58, 59, **59**
Boeing Helicopters
CH-46: 18
CH-46E Sea Knight: 18, 25, 26, 27, 28-29
CH-47 Chinook: **17**, 18, 21, 22, **22**
CH-47C: 13
CH-47D: 12, 14, 18, 21, 22, 28-29
CH-47F: 12
Chinook HC.Mk 2: 28-29
MH-47D/E/G Chinook: 12
Boeing/IAI/Elta 707 Phalcon/Condor: **105**, 117, **117**
Boeing/McDonnell Douglas
C-17 Globemaster III: 18, 21, **22**
C-17A: 14, 28-29
CF-18 Hornet: **102**
F-15 Eagle: 62, 102
F-15C: 17, 25
F-15E Strike Eagle: 17, **19**, **20**, 25-26, 28, 44
F-15K: **12**
F/A-18 Hornet: 25, **57**, **70**, **71**
F/A-18A: **58**
F/A-18C: **24**, 26, **73**, **74**, 76
F/A-18D: **16**, **26**, 27, 28
F/A-18E Super Hornet: 72, **72**, **73**, 76
F/A-18E/F Block II upgrade: 7
F/A-18F: **15**, **72**, 76
Boeing/McDonnell Douglas Helicopters
AH-64A Apache: **17**
AH-64D: 12, **17**, 18, 25, 26, 28-29
AH-64D Longbow Apache: 21
AH-64D Saraf: **14**
Unmanned Little Bird VTUAV: 7
Boeing/Northrop Grumman
E-8 J-STARS: 40, 98, 101, 104, 105
E-8C J-STARS: **46**, **101**
Bombardier
Dash 8Q-202 MPA: 85-86, **87**
Global 5000: 13
Bonhomme Richard, USS: 14
BQM-34 Firebee drone (see Ryan)
Brazil
Força Aérea Brasileira (air force): 34-39, **34-39**
Breguet
Alizé: 68
Atlantic: 48, 51
Bristol, HMS: 103
Bristol Blenheim: 128
Bronco, OV-10A (see Rockwell)

C

C-2A Greyhound (see Northrop Grumman)
C-5 Galaxy (see Lockheed Martin)
C-12 (see Beech)
C-17 Globemaster III (see Boeing/McDonnell Douglas)
C-18A (see Boeing)
C-27J (see Alenia)
C-27J Spartan (see Alenia/Lockheed Martin)
C-37A (see Gulfstream Aerospace)
C-40A/C (see Boeing)
C-87 Liberator Express (see Consolidated)
C-130 Hercules (see Lockheed Martin)
C-135/C-137 (see Boeing)
C-141 StarLifter (see Lockheed Martin)
Canadair P-2H Argus: 117
Canberra PR.Mk 9 (see English Electric)
Cardiff, HMS: 103
Carl Vinson, USS: 15
Catalina (see Consolidated)
Cayuse, OH-6A (see Hughes)
CC-137 Husky (see Boeing)
Centurion, 210 (see Cessna)
Cessna
182 Skylane: 79, 80
206 Stationair: 79, 80, 81, 82, 83, 84
210 Centurion: 79, 80, 81, 82, 83, 84
210M: **83**
404 Titan (C-28A): 79, 81
550 Citation II: 82, 83, **83**, 84
OA-37B: 9
UC-35: 18
CF-18 Hornet (see Boeing/McDonnell Douglas)
CH-21 (see Piasecki)
CH-46 (see Boeing Helicopters)
CH-47 Chinook (see Boeing Helicopters)
CH-53 (see Sikorsky)
CH2000 (see SAMA)
Charles de Gaulle: 62
Chasidah, EC-707 (see Boeing)
Chengdu
F-7: **20**
J-10: 9
Chetak (see HAL)
Cheyenne III, PA-42 (see Piper)
Chinook (see Boeing Helicopters)
Citation II, 550 (see Cessna)
CN235-300M (see EADS/CASA)
CN-235M (see Airtech)
Cobra, AH-1W (see Bell)
Colibri, EC 120B (see Eurocopter)
Comet (see de Havilland)
Comp Air 7SL (see Aerocomp)
Condor, 707 (see Boeing/IAI/Elta)
Consolidated
B-24E Liberator: 167
B-24H: 170
B-24J: 163, **164**, 167, 168, 170
B-24L: 163, **164**, 167, 168
B-24M: 163, **164**
C-87 Liberator Express: 165
Catalina: 172
LB-30 Liberator: **166**
LB-30 Liberator II 'B.Mk II': 157, **157**, 166, **166**, **171**
LB-30 Liberator II 'GR.Mk II': 166
LB-30 Liberator II Transport: 166
LB-30A Liberator: 166
LB-30B Composite: 166
LB-30B Liberator I: 166, **166**
LB-30B Liberator 'GR.Mk I': 166
Liberator: 156-173, **162**, **163**, **164**, **173**
Liberator B.Mk III: 157, **157**, **158**, 161, 167
Liberator B.Mk IV: 167
Liberator B.Mk VI: **156**, 157, **158**, 159-161, **160**, **161**, **163-165**, 167, **167**, 171, **171**, 172, **172**
Liberator B.Mk VIII: **156**, 157, **159**, **160**, 164, 171, **173**
Liberator C.Mk III/V: 167
Liberator C.Mk VII/VIII: 168
Liberator C.Mk IX: 168, **168**, 170, **170**
Liberator *Commando*: 168, **170**
Liberator GR.Mk III/IIIA: 167, 169
Liberator GR.Mk V: 159, 167, **167**, 169
Liberator GR.Mk VI: 168, **168**, 169, 172
Liberator GR.Mk VIII: 165, **165**, 168, 169, **169**, 170
Liberator Mk II: 156-157, 169, 170, **170**, 171
Liberator Mk III/V: 169, 171, 172
Liberator Mk VA: 169
Liberator Mk VI: 165, **165**, 168, 170, 171, 172, 173
Liberator Mk VIII: 170, 172
XB-24: 166
Cougar (see Eurocopter)
CP-140 (see Lockheed)
Cruzex 2004: 34-39, **34-39**
CT-49 (see Boeing)
CVW-5 – At home at Atsugi: 70-77, **70-77**

D

D.XXI (see Fokker)
DarkStar UAV, RQ-3 (see Lockheed Martin/Boeing)
Dash 8Q-202 MPA (see Bombardier)
Dassault
Atlantic: 24
F-103E (Mirage IIIEBR): **34**, **35**
HU-25 Guardian: **87**, 90
HU-25A: **87**, 88, 89, 90
HU-25B: 90
HU-25C: **87**, 90
HU-25C+: **87**, 89, 90
HU-25D: 89, 90
Mirage IIIEBR (F-103E): **34**, **35**
Mirage 50DV/EV: **39**
Mirage 50M Pantera: 9
Mirage 2000: **7**
Mirage 2000-5: 62-63
Mirage 2000-5DDA/EDA: 13-14
Mirage 2000-9: 10
Mirage 2000C/N: **37**
Dassault/EADS/Thales EuroMALE UAV: 49, 56, **59**
DC-8 (see Douglas)
DC-10-40 (see McDonnell Douglas)
DC-130E (see Lockheed Martin)
Debrief: 30-33, **30-33**
de Havilland
Comet: 101
Mosquito: 154
Denel Bateleur/Seeker UAVs: 57
Dhruv Advanced Light Helicopter (see HAL)
Distant Star UAV, P-610 (see Lockheed Martin)
Dolphin, HH-65 (see Aérospatiale)
Douglas (see also McDonnell Douglas)
AC-47: 9
DC-8: **37**, 98, 99, 100, 105
VC-118A: 106

E

E-2C Hawkeye (see Northrop Grumman)
E-3 Sentry (see Boeing)
E-6 (see Boeing)
E-8 J-STARS (see Boeing/Northrop Grumman)
E-10A MC^{2}A (see Boeing)
EA-6B Prowler (see Northrop Grumman)
EADS
Eagle 1 UAV: 49, **52**, 53, **53**, 56
Eagle 2 UAV: 53, 56
EADS/CASA CN235-300M: 92, **93**
Eagle, F-15 (see Boeing/McDonnell Douglas)
Eagle 1/2 UAVs (see EADS)
Eagle Eye VUAV, HV-911 (see Bell)
EC-18 (see Boeing)
EC 120B Colibri (see Eurocopter)
EC-130J (see Lockheed Martin)
EC 135 (see Eurocopter)
EC-137D (see Boeing)
EC 225 (see Eurocopter)
EC-707 (see Boeing)
EC 725 Cougar Mk II+ (see Eurocopter)
EH101 Merlin (see AgustaWestland)
Elbit/Silver Arrow
Hermes 450 UAV: 50, **54**, 81, **81**
Hermes 450S UAV: 54, **54**
Hermes 1500 UAV: 50, **54**
EMBRAER
A-29A/B Super Tucano: **37**
AT-26 Xavante: **35**, **37**
AT-27 Tucano: 13, **35**
EMB-110 Bandeirante: **11**
EMB-145 Legacy: 10
EMB-201R: **10**
EMB-314 Super Tucano: 9, 13
R-99A/B: **36**
English Electric Canberra PR.Mk 9: **22**, 47
Enterprise, USS: 22, 24
EP-3E (see Lockheed)
Eurocopter
AS 350B AStar (Ecureuil): 79, 80, 82
AS 350B2: 83, 84, 86, **87**
AS 350B3: 79, 80, **81**
AS 532 Cougar: **38**
AS 532AL: 8
AS 532UE (HM-3): **11**
AS 532UL: 11
AS 550A2 Fennec: **11**
AS 565AA Panther (HM-1): **11**
AS 565MB Panther: 8
EC 120B Colibri: 80
EC 135: 10
EC 225 Super Puma/Cougar: 9
EC 725 Cougar Mk II+: 9-10
Tiger ARH: 31, **31**
Eurofighter
Typhoon: 5, **5**
Typhoon T.Mk 1: **5**
Typhoon Tranche 2: 5
EuroHawk UAV (see Northrop Grumman)
EuroMALE UAV (see Dassault/EADS/Thales)
E-X AEW programme: 11
Exercise
Cruzeiro do Sul (Cruzex) 2004: 34-39, **34-39**
Linked Seas 00: 42
Red Flag 3-04: 30, **30**
Explorer, MD-900 (see McDonnell Douglas Helicopters)
Extender, KC-10A (see McDonnell Douglas)

F

F4F Wildcat (see Grumman)
F-5B/E (see Northrop)
F-7 (see Chengdu)
F-14 Tomcat (see Northrop Grumman)
F-15 (see Boeing/McDonnell Douglas)
F-16 Fighting Falcon (see Lockheed Martin)
F-35A/B/C Joint Strike Fighter (see Lockheed Martin)
F-86 Sabre (see North American)
F-103E (see Dassault)
F-105 Thunderchief (see Republic)
F/A-18 (see Boeing/McDonnell Douglas)
F/A-22A Raptor (see Lockheed Martin)
Fairchild
A-10 Warthog: **19**, 41, 95, 96
A-10A: 17
A-10C: 4, **4**
'Famous'-class USCG cutter: 92
Fantasma, AC-47T (see Basler)
FCA (Future Cargo Aircraft): 6
Fennec, AS 550A2 (see Eurocopter)
Fighting Falcon, F-16 (see Lockheed Martin)
Fightinghawk, A-4AR (see McDonnell Douglas)
FINDER UAV (see General Atomics)
Firebee drone, BQM-34 (see Ryan)
Fire Scout UAV, RQ-8 (see Northrop Grumman)
'Fitter-F' (see Sukhoi Su-20)
'Fitter-G' (see Sukhoi Su-22UM3)
'Fitter-J' (see Sukhoi Su-22M2K)
FMA IA-63 Pampa: 8
Focke-Wulf Fw 190: 151, 153
Fokker
D.XXI: 131, **133**
G.1: 128
France
Armée de l'Air (air force): **37**
'Fulcrum-C' (see Mikoyan MiG-29S/SE)
Furious, HMS: 132

Future Cargo Aircraft (FCA): 6
Fw 190 (see Focke-Wulf)
F-X combat aircraft: 13

G

G.1 (see Fokker)
G-222 (see Alenia)
G550 (see Gulfstream Aerospace)
Galaxy, C-5 (see Lockheed Martin)
Gazelle AH.Mk 1 (see Aérospatiale/ Westland)
General Atomics
Altair UAV: 42, 46, **49**, 52, 55
Altus I/II UAV: **49**
FINDER UAV: **42**
I-Gnat ER UAV: **44**, 45
Mariner UAV: 51-52, 55, **55**, 92
MQ-1 Predator UAV: 23, 25, 28, **43**, 44
MQ-1L Predator UAV: 15, 42, **43**, **44**
MQ-9A Predator B UAV: 42, 44, 46-47, 48, **48**, **49**, 50, **50**, 51, 52, 54, 55
Predator C UAV: 56
Prowler II UAV: **42**
RQ-1 Predator UAV: 23, 25, 28, 40-42, **40**, **42**, **43**, 44, 45-46, 53, 86
RQ-1K Predator UAV: **42**
RQ-1L Predator UAV: 15, 28-29, **29**, **41**
Warrior UAV: 12, 52-53
George Washington, USS: 24, 25, 26
Germany
Luftwaffe
4.(F)/14: 134-135, **134**, **135**
Aufklärungsgruppe 11: 136, **140**
Aufklärungsgruppe 14: 136, **139**
1./Aufklärungsgruppe 22: 136
Ergänzungs (Zerstörer) Staffel: **131**
Erprobungsgruppe 210: 130 131, 133-134. 136, **136**, 139
13.(Z)/JG 5: **146**, 147
1.(Z)/JG 77: 140
Stab./KG 55: **137**
LG 1: 129, **131**, **132**, **135**, 136, **137**
LG 2: 136
Nachtjagdstaffel Norwegen: **154**
Nahaufklärungsgruppen 2/4/5/6/8/14: 147
NJG 1: 136, 146, **146**, 147, **147**, **148**, **149**, **152**, **154**, **155**
NJG 3: **135**, **137**, 147, 149-150, **152**, **155**
NJG 4: 137, 139, **146**, 147, **147**, **153**
NJG 5: **144**, 147, **148**, 151, **153**
NJG 6: 143, **143**, 153
5./NJG 200: **149**
SKG 210: 139-140, **140**
StG 1: 147
ZG 1: 129, 131, **131**, 132, **132**, **133**, 139, 140, **145**, 146, 147, **148**
ZG 2: **128**, **129**, 135, **136**
ZG 26: 133, **135-142**, 136-140, **145**, 147, **150**, 151, **151**, 153-154
I./ZG 52: **128**, **132**
ZG 76: 129-132, **129**, **131**, **133**, 135-139, **136**, **150**, 154
Sonderkommando Junck: 138
Gladiator, (see Gloster)
Global 5000 (see Bombardier)
Global Hawk UAV, RQ-4 (see Northrop Grumman)
Global Military Aircraft Systems (GMAS): 6
Globemaster III, C-17 (see Boeing/ McDonnell Douglas)
Gloster Gladiator: 131, 132, 138
GMAS (Global Military Aircraft Systems): 6
Golden Eagle, T-50/A-50 (see KAI/ Lockheed Martin)
Greyhound, C-2A (see Northrop Grumman)
Gripen (see Saab)
Grumman (see also Northrop Grumman)
F4F Wildcat: 131
OV-1 Mohawk: 82
S-2 Tracker: 82
Guardian, HU-25 (see Dassault)
Gulfstream I, VC-4A (see Northrop Grumman)
Gulfstream Aerospace
C-37A (Gulfstream V): 89, 90, **93**
G550: 11

H

HAL
Chetak: **68**
Dhruv Advanced Light Helicopter: **7**, 11
HJT-36 Intermediate Jet Trainer (IJT): **7**, 10
HJT-39: 10
Tejas Light Combat Aircraft: **7**, 10
HAL/Avro 748: 10
HALE/MALE Unmanned Air Vehicles – Part 2: 21st Century warfighters: 40-59, **40-59**
Halifax (see Handley Page)
'Hamilton'-class USGC cutter: 92
Handley Page Halifax: 147, 151, 159, 171
Harrier (see BAE Systems)
Harrier II, AV-8B (see McDonnell Douglas/BAe)
Harry S Truman, USS: 24, 25
Hawk (see BAE Systems)
Hawker Hurricane: 128, 132, 133, 134-135, 155
Hawker Siddeley
HS 125: 21, **21**
HS-748 (C-91): **10**
P.1127 Kestrel: 118, 119, 120
Hawkeye, E-2C (see Northrop Grumman)
HB 350L1 (see Helibras)
HC-130H/J/P (see Lockheed Martin)
Heinkel He 111: 135, 138
Helibras HB 350L1 (HA-1): **11**
Hercules (see Lockheed Martin)
Hermes, HMS: 103
Hermes 450/1500 UAVs (see Elbit/ Silver Arrow)
Heron UAV (see IAI Malat)
HH-60 (see Sikorsky)
HH-65 (see Aérospatiale)
Hiller UH-12: **127**
HJT-36/-39 (see HAL)
Hofit, A36 (see Beech)
Hornet (see Boeing/McDonnell Douglas)
HS 125 (see Hawker Siddeley)
HS-748 (see Hawker Siddeley)
HU-25 Guardian (see Dassault)
Huey, UH-1 (see Bell)
Hughes
500C: 80
OH-6A Cayuse: 79, **79**, 80
Hummingbird, Model 330 (see Lockheed)
Hummingbird VTUAV, A160 (see Boeing/Frontier Systems)
Hunter UAV, RQ-5 (see IAI/Malat/TRW)
Hunter II UAV, Model 397 (see Northrop Grumman/IAI/Aurora Flight Sciences)
Hurricane (see Hawker)
Husky, CC-137 (see Boeing)
HV-911 Eagle Eye VUAV (see Bell)

I

IA-63 Pampa (see FMA)
IAI
Kfir CE: **15**
Lavi: 9
IAI/AAI RQ-2 Pioneer UAV: 18, 25, 27, 29
IAI Malat
Heron UAV: 49, **52**, **53**
Heron TJ UAV: 50
Heron TP UAV: 49, 56
IAI/Malat/TRW
RQ-5 Hunter UAV: 18, 28
RQ-5A Hunter UAV: 81
IAR-Aérospatiale IAR-330 Puma: 13
I-Gnat ER UAV (see General Atomics)
Ilyushin
Il-76: 13, 117
Il-78MKI: **7**
India
Air Force: 165, **165**
Navy: 68-69, **68**, **69**
Intermediate Jet Trainer, HJT-36 (see HAL)
'Invincible'-class carrier: 67
Iraq: Airpower in the counter-insurgency war: 16-29, **16-29**
Iroquois, UH-1H (see Bell)

J

J-10 (see Chengdu)
Japan
Japanese Maritime Self-Defence Force (NAF Atsugi): 71-76, **71**, **72**, **74**, **76**
JAS 39 Gripen (see Saab)
Jayhawk, HH-60J (see Sikorsky)
JetRanger (see Bell)
John C. Stennis, USS: 15
John F. Kennedy, USS: 15, 24, **24**, 25, 28
Joint STARS, E-8 (see Boeing/ Northrop Grumman)
Joint Strike Fighter, F-35A/B/C (see Lockheed Martin)
Joint Unmanned Combat Air System (J-UCAS): 54, 55, 56, **56**, 57, **57**
Jordan
Royal Jordanian Air Force
Special Operations Flying Unit: 13
J-STARS, E-8 (see Boeing/Northrop Grumman)
Ju 52 *et seq* (see Junkers)
J-UCAS (Joint Unmanned Combat Air System): 54, 55, 56, **56**, 57, **57**
Junkers
Ju 52: 131, 138
Ju 88: 137, 153
Ju 90: 138

K

Ka-28 *et seq* (see Kamov)
KAI (Korean Aerospace) KO-1: 9
KAI/Lockheed Martin T-50/A-50 Golden Eagle: 6
Kamov
Ka-28: **68**
Ka-31: **7**, **68**
Ka-60: 14
Karine A: 102
KC-10 (see McDonnell Douglas)
KC-130 Hercules (see Lockheed Martin)
KC-135 *et seq* (see Boeing)
KE-3A (see Boeing)
Kestrel, P.1127 (see Hawker Siddeley)
Kfir CE (see IAI)
King Air (see Beech)
Kiowa (see Bell)
Kiowa Warrior, OH-58D (see Bell)
Kitty Hawk, USS: 43, 71, **73**
Knighthawk, MH-60S (see Sikorsky)
KO-1 (see KAI)
Korean Multi-purpose Helicopter (KMH) programme: 11

L

LALEE UAV (see Singapore Technologies)
Lancaster (see Avro)
Lavi (see IAI)
LB-30 Liberator (see Consolidated)
LC-130 (see Lockheed Martin)
Legacy, EMB-145 (see EMBRAER)
Liberator (see Consolidated)
Liberator Express, C-87 (see Consolidated)
Light Combat Aircraft, Tejas (see HAL)
Lightning, P-38 (see Lockheed)
Lincoln (see Avro)
Lockheed
augmented jet test rig: **121**
CP-140: 117
EP-3E: 53
Model 330 Hummingbird (VZ-10, later XV-4A): **118-119**, 119-122, **122**, 124-127, **124**
NP-3D Orion: 14
P-3 Orion: 50, 51, 82
P-3A/B: 82, 83, 84, **85**
P-3B(AEW): 83, 84, **85**
P-3C: 11, 24, 71
P-3C Lot II: 11
P-3K: 8
P-38 Lightning: **160**
TriStar: **22**
U-2: 43
VC-121E: 106
VTOL design studies: 119-120, **120**
XV-4A Hummingbird (VZ-10): **118-119**, 119-122, **122**, 124-127, **124**
XV-4B Hummingbird II: **126**, 127
Lockheed Martin
AC-130: 22, 25, 27, 28, 45
AC-130H/U: 25-26
C-5 Galaxy: 18
C-5B: 28-29
C-130 Hercules: 8, 18, 20, **20**, 25, 27, 45, 102
C-130B: 6, 8
C-130E: **9**, 15, 20
C-130F: 6
C-130H: 8, 9
C-130H3: 15
C-130H-30: 8, 9
C-130J: 7, 88, 90
C-130J-30: 10
C-141 StarLifter: 104, 107, 109
C-141C: 14
DC-130E: 41, **41**
EC-130J: 7
F-16 Fighting Falcon: 9, 10, **16**, 17, **19**, **20**, 22, 25-26, 28, **34**, 44, 95, 96, 102
F-16 Block 40/50: 30, **30**
F-16 MLU: 64
F-16A/B: 11
F-16A/B M3 MLU: 30
F-16B: **38**, **39**
F-16C: 15, 94, **94**, 95, **97**
F-16C+ Block 25: 95, **95**, **96**, 97
F-16C/D Block 50: 9
F-16CG: 28-29
F-16CJ: 25
F-16D: 95
F-35A Joint Strike Fighter: 7
F-35B: 61
F-35C: 56
F/A-22A Raptor: **12**, 15
HC-130H Hercules: 88, 89, **89**, 90, 92
HC-130J: 89, **89**, 92
HC-130P: 25
Hercules: **22**
Hercules C.Mk 1/3/5: 21
Hercules C.Mk 4: 21, **21**
KC-130 Hercules: 25, **26**, 27, 102
KC-130H: **38**
KC-130J: 7, 10, 14
KC-130T: 25
LC-130: 14
MC-130/-130P: 25
P-610 Distant Star UAV: 54
Penetrating High-Altitude Endurance (PHAE) UAV: 54
S-3B Viking: 16, 25
Team US101: 6-7
Lockheed Martin Argentina AT-63 Pampa: 8
Lockheed Martin/Boeing RQ-3 DarkStar UAV: 53, 54
Longbow Apache, AH-64D (see Boeing/McDonnell Douglas Helicopters)
Lynx (see Westland)

M

M-346 (see Aermacchi)
'Mainstay' (see Beriev/Ilyushin A-50)
'Majestic'-class carrier: 68
Mariner UAV (see General Atomics)
MC^2A-X testbed, 707 Paul Revere (see Boeing)
MC-130 (see Lockheed Martin)
McDonnell Douglas (see also Boeing/ McDonnell Douglas)
A-4 Skyhawk: **34**
A-4AR Fightinghawk: **38**
DC-10-40: 112
KC-10: 25
KC-10A Extender: 17
QF-4 Phantom: 63
QF-4 Lot 11: 12-13
RF-4E: **103**
McDonnell Douglas/BAe AV-8B Harrier II: 18, 25, 26, **26**, 27, 28
McDonnell Douglas Helicopters
MD500: 13
MD500E: 79, **79**, 80, 82, 83, 84, **87**
MD530F: 7
MD600N: 79, **79**, 80
MD-900 Explorer: 92
Me 210 *et seq* (see Messerschmitt)
Merlin, EH101 (see AgustaWestland)
Messerschmitt
Bf 109: 129, 130, 132, 133, 134, 138, 151, 153
Bf 109E: 129, **139**
Bf 109G-6: **153**
Bf 110: 128-155, **128-129**, **132**, **133**, **135-140**, **144**, **146**, **148**, **149**, **151**, **154**, **155**
Bf 110A: **130**, 145
Bf 110A-0: 129, 145
Bf 110B: 145
Bf 110B-0: 129, **130**, 145, **145**
Bf 110B-1: 129, **130**, **131**, 145
Bf 110B-2: 145
Bf 110B-3: **131**, 145
Bf 110C: **128**, 129-132, **129**, **131-133**, **135**, 136-138, **136**, **144**, 145, 146, **147**, **148**
Bf 110C-0: 129, **131**, 145
Bf 110C-1: 129, **131**, **132**, 145
Bf 110C-1/U1: 145
Bf 110C-2/C-3: 145
Bf 110C-4: 130, **136**, 134-135, **134**, 136, **136**, 145, **146**, **147**
Bf 110C-4/B: **144**, 145
Bf 110C-5: 131, 145
Bf 110C-5/N: 145
Bf 110C-6: 130-131, **144**, 145
Bf 110C-7: 145
Bf 110D: 145
Bf 110D-0: **133**, 145
Bf 110D-1: **133**, 135, **137**
Bf 110D-1/R1: 132, **133**, 145
Bf 110D-1/R2: 145
Bf 110D-1/U1: 145
Bf 110D-2: 145
Bf 110D-3: **129**, **138**, **140**, 145, **145**
Bf 110E: 137-138, **138**, **139**, 145, **146**
Bf 110E-0/E-1: 145
Bf 110E-1/R2: 145
Bf 110E-1/U1: 145
Bf 110E-1/U2: 145, **145**
Bf 110E-2: **141-142**, 145
Bf 110E-3: 145
Bf 110F: 139, 145, 146
Bf 110F-1: 145, **145**
Bf 110F-2: **144**, 145, **146**
Bf 110F-3: 145
Bf 110F-4: 145, 146, **146**, 147, **147**
Bf 110F-4a: 145, **148**
Bf 110F-4/U1: 145
Bf 110G: **144**, 145, 146, **149**, **154**, **155**
Bf 110G-0/G-1: 145
Bf 110G-2: 145, **149**, **150**, 151, 152, **152**, 153, 155
Bf 110G-2/R1 and Bf 110G-2/R3: 145
Bf 110G-2/R4: **150**
Bf 110G-2/R5: **144**
Bf 110G-3: 145
Bf 110G-4: 145, 147, 148, **148**, **151**, **152**, **154**
Bf 110G-4b: **149**
Bf 110G-4b/R3: **153**
Bf 110G-4c/R3: **153**, **154**, **155**
Bf 110G-4d/R3: 143, **143**
Bf 110G-4/R3: **135**
Bf 110G-4/U6: **149**
Bf 110H: 145
Bf 110 V1: **130**, 145
Bf 110 V2/V3: 145
Me 210: 139, 140, 146
Me 323: **149**
Me 410: 154
Messerschmitt Bf 110 – Underrated Warrior: 128-155, **128-155**
MH-47D/E/G Chinook (see Boeing Helicopters)
MH-53M (see Sikorsky)
MH-60J/S (see Sikorsky)
MH-65C (see Aérospatiale)
MH-68A Sting Ray (see Agusta)
Mi-17 *et seq* (see Mil)
Midway, USS: **75**
Mikoyan
MiG-21: 68
MiG-25R: **20**
MiG-29: 102
MiG-29K: 10, 69
MiG-29S/SE 'Fulcrum-C': 8
MiG-29SMT: 10, 13
MiG-29UB: 8
MiG-29UBT: 13
Mil
Mi-17: 8, 13, **29**
Mi-24: 8
Mi-24V: 13
Mi-28N: 14
Mirage (see Dassault)
MMA (Multi-mission Maritime Aircraft): 50-51
Model 234 (see Teledyne Ryan)
Model 281 Proteus (see Scaled Composites)
Model 330 Hummingbird (see Lockheed)
Model 367-80 (see Boeing)
Model 367-138B (see Boeing)
Model 395/396 UAVs (see Northrop Grumman)
Model 397 Hunter II UAV (see Northrop Grumman/IAI/Aurora Flight Sciences)
Mohawk, OV-1 (see Grumman)
Mosquito (see de Havilland)
MQ-1 Predator UAV (see General Atomics)
MQ-9A Predator B UAV (see General Atomics)
MRE (Multi-Role Endurance) UAV: 54
MRH-90 (see NH Industries)
Multi-mission Maritime Aircraft (MMA): 50-51
Multi-Role Endurance (MRE) UAV: 54
Mustang, P-51 (see North American)

N

'Narasuan'-class frigate: **12**
News Report – Iraq: Airpower in the counter-insurgency war: 16-29, **16-29**
NH Industries MRH-90: **31**
Nimitz, USS: 24
Nimrod (see BAE Systems)
North American
F-86 Sabre: **110**
P-51 Mustang: 154
Northrop
F-5B: **103**
F-5E: **34**, **35**
Northrop Grumman
B-2: 56
C-2A Greyhound: 25, 76, **76**
E-2C Hawkeye: **7**, 14, 24, 25
E-2C Hawkeye Group II: **70**, **71**, **75**, 76, **76**
EA-6B Prowler: 24, 25, 28-29, **70**, **75**, 76
EuroHawk UAV: 48, 51
F-14 Tomcat: **15**
F-14A: **103**
F-14B: **24**, 25
Model 395/396 UAVs: 56
RQ-4 Global Hawk UAV: 92
RQ-4A Global Hawk UAV: **40**, 42-44, **45**, 46, **46**, **47**, 51, **51**, 52, **55**, 58
RQ-8B Global Hawk UAV: 48, **55**
RQ-8 Fire Scout UAV: **40**
VC-4A Gulfstream I (G-159): 89, 90, **93**
X-47B UCAV: 54, 55, 56, **57**
Northrop Grumman/IAI/Aurora Flight Sciences Model 397 Hunter II UAV: 12, 53
Northrop Grumman/Scaled Composites X-47A Pegasus UCAV: **40**, 54-55, 57, **57**
NP-3D Orion (see Lockheed)

O

OA-37B (see Cessna)
'Ocean Hawk' (see Sikorsky SH-60F)
OH-6A Cayuse (see Hughes)
OH-58D Kiowa Warrior (see Bell)
Omega Air, Inc.: 111-112
Orion (see Lockheed)
OV-1 Mohawk (see Grumman)
OV-10A Bronco (see Rockwell)

P

P-2H Argus (see Canadair)
P-3 Orion (see Lockheed)
P-38 Lightning (see Lockheed)
P-47 Thunderbolt (see Republic)
P-51 Mustang (see North American)
P-610 Distant Star UAV (see Lockheed Martin)
P.1127 Kestrel (see Hawker Siddeley)
PA-18 *et seq* (see Piper)
PAK FA (see Sukhoi)
Pampa, AT-63 (see Lockheed Martin Argentina)
Pampa, IA-63 (see FMA)
Panavia
Tornado F.Mk 3: 8, 62
Tornado GR.Mk 4: **13**, **20**, 21, **21**, 27, 47
Tornado GR.Mk 4A: **13**
Pantera, Mirage 50M (see Dassault)
Panther, AS 565AA/MB (see Eurocopter)
Paul Revere MC^2A-X testbed, 707 (see Boeing)
Pave Low, MH-53M (see Sikorsky)
PC-12/45 (see Pilatus)
PC-21 (see Pilatus)
Peace Station, 707-3J9C (see Boeing)
Pegasus UCAV, X-47A (see Northrop Grumman/Scaled Composites)
Peleliu, USS: 25
Penetrating High-Altitude Endurance UAV (see Lockheed Martin)
Perseus UAV (see Aurora Flight Sciences)
PHAE UAV (see Lockheed Martin)
Phalcon, 707 (see Boeing/IAI/Elta)
Phantom, QF-4 (see McDonnell Douglas)
Photo Report – Cruzex 2004: 34-39, **34-39**
Piasecki CH-21: **125**
Pilatus
PC-12/45: **83**, 84-85
PC-21: 9
Pioneer UAV, RQ-2 (see IAI/AAI)
Pioneers & Prototypes – VTOL flat-risers – Lockheed XV-4 and Ryan/GE XV-5: 118-127, **118-127**
Piper
PA-18 Super Cub: 79, 80-81, **81**
PA-42 Cheyenne III: 82, 83, 84, **85**
Potez 63: 128
Predator UAV (see General Atomics)
Prince of Wales, HMS: 61
Proteus, Model 281 (see Scaled Composites)
Prowler, EA-6B (see Northrop Grumman)
Prowler II UAV (see General Atomics)
Puma (see Westland/Aérospatiale)
Puma, IAR-330 (see IAR-Aérospatiale)
PZL 11: 129
PZL-Swidnik W-3 Sokol: 26

Q

QF-4 Phantom (see McDonnell Douglas)
Queen Elizabeth, HMS: 61

R

R-99A/B (see EMBRAER)
RA-1A/B (see Alenia/EMBRAER)
RAF Liberators at war – Part 2: Service abroad, variants and operators: 156-173, **156-173**
Raptor, F/A-22A (see Lockheed Martin)
RC-135 (see Boeing)
RC-707 Barboor (see Boeing)
RE-3A/B (see Boeing)
Re'em, 707 (see Boeing)
Republic
F-105 Thunderchief: **110**
P-47 Thunderbolt: 153, 154, 155
'Rescue Hawk' (see Sikorsky HH-60H)
RF-4E (see McDonnell Douglas)
Rockwell
B-1B: 25, 28-29
OV-10A Bronco: 9, 13
XFV-12A: 126
RQ-1 Predator UAV (see General Atomics)
RQ-2 Pioneer UAV (see IAI/AAI)
RQ-3 DarkStar UAV (see Lockheed Martin/Boeing)
RQ-4 Global Hawk UAV (see Northrop Grumman)
RQ-5 Hunter UAV (see IAI/Malat/TRW)
RQ-7 Shadow UAV (see AAI)
RQ-8 Fire Scout UAV (see Northrop Grumman)
Ryan
BQM-34 Firebee drone: 41
VZ-3RY Vertiplane: 121
X-13 Vertijet: 118, 120
Ryan/General Electric
VZ-11 (later XV-5A Vertifan): **118-121**, 120-121, 123-127, **123**, **124**, **125**
XV-5B: 127, **127**

S

S-2 Tracker (see Grumman)
S-3B Viking (see Lockheed Martin)
S-70A/B (see Sikorsky)
S-92 (see Sikorsky)
Saab
Gripen: 4, 32
JAS 39B Gripen: **5**
JAS 39C/D Gripen: 10
JAS 39EBS-HU Gripen: 4, **5**
SHARC UAV: **59**
SK 60: 32, **32**, 33, **33**
Sabre, F-86 (see North American)
Saipan, USS: 25
Saknayee, KC-707 (see Boeing)
SAMA CH2000: **9**, 20
Saraf, AH-64D (see Boeing/McDonnell Douglas Helicopters)
SB7L-30 Seeker (see Seabird Aviation)
Scaled Composites Model 281 Proteus: 57, 58, **58**
Scharnhorst: 132, **147**
Seabird Aviation SB7L-30 Seeker: 20
Sea Harrier (see BAE Systems)
Sea Harrier Farewell: 60-69, **60-69**
Seahawk, S-70B (see Sikorsky)
Sea King (see Westland)
Sea King, UH-3H (see Sikorsky)
Sea Knight, CH-46E (see Boeing Helicopters)
Sea Stallion, CH-53E (see Sikorsky)
Seeker, SB7L-30 (see Seabird Aviation)
Seeker UAV (see Denel)
Sentry (see Boeing)
SH-3D/H (see Agusta/Sikorsky)
SH-60F 'Ocean Hawk' (see Sikorsky)
Shaanxi Y-8: 13
Shadow UAV, RQ-7 (see AAI)
SHARC UAV (see Saab)
Short Stirling: 147, 148, 159
Sikorsky
Black Hawk: 102
CH-53 Yasur: 102
CH-53E Sea Stallion: 18, **20**, 24, 25, 27, **27**
HH-60G: **20**, 25
HH-60H 'Rescue Hawk': 25, 76, **77**
HH-60J Jayhawk: 88, 89, 90, **91**, 92, **92**
HH-60T: **91**, 92
MH-53M: 25
MH-53M Pave Low: 26
MH-60J: 89, 92
MH-60S Knighthawk: 7, 12, 14
S-70A Blackhawk: 11
S-70A-6 (HM-2): **11**
S-70B Seahawk: 11
S-92: 12
SH-60F 'Ocean Hawk': 25, 76, **77**
UH-3H Sea King: 24, **24**
UH-60 Black Hawk: **17**, 18, 21, 22, 25, 26, 27, 28-29
UH-60A: 14, 15, 28-29, 82, 83, 84, **86**, **87**
UH-60L: 12, 13, 28-29
UH-60M: 12
Singapore Technologies LALEE UAV: 57
SK 60 (see Saab)
Skua (see Blackburn)
Skyhawk, A-4 (see McDonnell Douglas)
Skylane, 182 (see Cessna)
Sokol, W-3 (see PZL-Swidnik)
South Africa
Air Force: 157, 158, 159, 171, **171**
Spartan, C-27J (see Alenia/Lockheed Martin)
Special Feature – Sea Harrier Farewell: 60-69, **60-69**
Special Feature – Tucson Testers – ANG/AFRC Test Center: 94-97, **94-97**
Special Report – CVW-5 – At home at Atsugi: 70-77, **70-77**
Spitfire (see Supermarine)
Stationair, 206 (see Cessna)
StarLifter, C-141 (see Lockheed Martin)
Sting Ray, MH-68A (see Agusta)
Stirling (see Short)
Stratotanker, KC-135 (see Boeing)
Strike Eagle, F-15E (see Boeing/ McDonnell Douglas)
Strikemaster (see BAe)
Su-30MKI (see Sukhoi/Hindustan)
Sukhoi
PAK FA: 5
Su-20 'Fitter-F': 8
Su-22M2K 'Fitter-J': 8
Su-22UM3 'Fitter-G': 8
Su-25/-25UB: 8
Su-25UBK: 13
Su-27: 9
Su-27SK/UBK (J-11/JJ-11): 9
Su-30: 9
Su-30K: 4-5
Su-30MK2: 9
Su-30MKK (J-13): 9
Sukhoi/Hindustan Su-30MKI: 4-5
Super Cub, PA-18 (see Piper)
Super Hornet, F/A-18E/F (see Boeing/McDonnell Douglas)
Super Huey, UH-1H (see Bell)
Super Lynx 300 (see AgustaWestland)
Supermarine
Spitfire: 128, 132, 133, 135, 155
Spitfire Mk VB: 134-135
Super Puma, EC 225 (see Eurocopter)
Super Tucano (see EMBRAER)
Super Ximango (see Aeromot)
Sweden - Air Force
'Team 60': 32-33, **32**, **33**

T

T-6A (see Beech)
T-50 Golden Eagle (see KAI/Lockheed Martin)
Taksin, HTMS: **12**
TC-18E (see Boeing)
Tejas Light Combat Aircraft (see HAL)
Teledyne Ryan BGM-34A/Model 234: 41, **41**
TH-67A+ (see Bell)
Theseus UAV (see Aurora Flight Sciences)
Thunderbolt, P-47 (see Republic)
Thunderchief, F-105 (see Republic)
Tiger ARH (see Eurocopter)
Titan, 404 (see Cessna)
Tomcat, F-14 (see Northrop Grumman)
Tornado (see Panavia)
Tracker, S-2 (see Grumman)
TriStar (see Lockheed)
Tu-16 *et seq* (see Tupolev)
Tucano, AT-27 (see EMBRAER)
Tucson Testers – ANG/AFRC Test Center: 94-97, **94-97**
Tupolev
Tu-16 'Badger': 10
Tu-95MS: 14
Tu-160 'Blackjack': 14
Typhoon (see Eurofighter)

U

U-2 (see Lockheed)
U-486: 159
UC-35 (see Cessna)
UH-1 (see Bell)
UH-3H Sea King (see Sikorsky)
UH-12 (see Hiller)
UH-60 Black Hawk (see Sikorsky)
United Kingdom
Fleet Air Arm: 23, **23**, **60**, 61, **61**, 62, 63, **63**, **64**, **67**
Joint Force Harrier: 61-62
Royal Air Force
Air Sea Warfare Development Unit: 169, 173
DDT Spraying Flight: 173
Long Range Meteorological Flight: 170
No. 1 Air Gunnery School (India): 173
No. 1(Coastal) OTU: 169
No. 1(ME) C&C Unit: 171
No. 4 Refresher Flying Unit: 169
No. 5 Heavy Bomber Conversion Unit: 171
No. 6 Refresher Flying Unit: 173
No. 108 Conversion Unit: 171
No. 111 OTU: 169, **169**
No. 160 (Special) Flight: 172
No. 205 Group: 157-158, 159
No. 231 Group: 160
No. 257 Wing: 157
No. 1330 Conversion Unit: 171
No. 1331 Conversion Unit: 173
No. 1341 Radio Countermeasures Flight: 161-162, 173
No. 1346/1347/1349 Air-Sea Rescue Flights: 173
No. 1425 Communications Flight: 170
No. 1427 (Ferry Training) Flight: 170
No. 1445 Flight: 170
No. 1576 Special Duties Flight: 161
No. 1584 (Heavy Bomber) Conversion Flight: 173
No. 1586 Special Duties Flight: 158, 159, 171
No. 1653 Conversion Unit (CU): 169
No. 1673 (Heavy) CU: 161, 173
No. 1674 (Heavy) CU: 169
No. 1675 (Heavy) CU: 157, 171
No. 1699 (Bomber Support) CU: 170
Royal Air Force: squadrons
No. 8: 162, 172
No. 31: 157, 158
No. 32 (The Royal): 21, **21**
No. 37/No. 40: 159, 171
No. 53/No. 59: 167, 169, 170
No. 70: 21, 159, **160**, 171
No. 86: 167, 169, 170
No. 99: 21, 160, **161**, 162, 172
No. 102: 170
No. 104: 159, 171
No. 108: 156-157, 171
No. 120: 21, 166, **166**, 167, 169
No. 148: 157, 171, **171**
No. 159: 157, **157**, 159, 160, 161, **161**, 162, **162**, 164, 171, **171**, 172
No. 160: 157, **157**, 159, 161, 162, 164, **167**, 169, 171, 172
No. 178: 157, 158, 171
No. 200: 159, 161, 162, 172
No. 203: 161, 162, 164, 172
No. 206: 14, 21, 169, 170
No. 214: 171
No. 215: 160, 161, 172, **172**
No. 220: 169, **169**, 170
No. 223: 168, 170
No. 224: 167, 169
No. 228: 169
No. 231: 166, 168, 170, **170**
No. 232: 162, 164, 168, 170, **170**, 172
No. 292: 172
No. 301: 159, 171
No. 311: 167, 169
No. 321/No. 354: 161, 162, 172
No. 355: 159, 160, 162, **163**, 164, 172, **173**
No. 356: **156**, 159, 160, 161, **161-165**, 162, **167**, 172, **172**
No. 357/No. 358: 161, 172
No. 511: 166
No. 547: 169
No. 614: 157, **160**, 171
United States
Air Force
332nd Air Expeditionary Wing: 25
335th FS 'Chiefs': **19**
379th Air Expeditionary Group: **20**
407th/447th/506th Air Expeditionary Groups: 25
447th Air Expeditionary Wing: 18
Air National Guard
162nd Fighter Wing: 95, **96**
ANG/AFRC Test Center: 94-97, **94-97**
Coast Guard: 78, 86-92, **87**, **89**, **91**, **92**
Aircraft Repair and Supply Center: 88, 89
C-130J Aircraft Project Office: 88, 89, **89**
Coast Guard Aviation Training Center: 88, 89, **92**
Commander Atlantic/Pacific Area: 88, 89
Helicopter Interdiction Tactical Squadron: 88, 89, **89**
Maritime Defense Zone Atlantic/Pacific: 88, 89
Customs and Border Protection: 78-86
Defense Advanced Research Projects Agency (DARPA): 54, 55, **56**, 57, 58
Department of Homeland Security: 78-93, **78-93**
Immigration and Naturalization Service: 78
Joint Interagency Task Force: 82, 86
Marine Corps
I Marine Expeditionary Force: 18, 25, 29
3rd Marine Air Wing: 18, 25, 26, 27, 28
11th Marine Expeditionary Unit (SOC): 25, 27-28
15th Marine Expeditionary Unit (SOC): 18
24th Marine Expeditionary Unit (SOC): 18, 25
HMH-465: **20**
HMLA-169: **27**
VMA-542: **26**
VMFA(AW)-224 'Bengals': **26**
VMFA(AW)-242 'Bats': **16**, **26**
Naval Reserve
VAW-78 'Fighting Escargots': 14
Navy
CVW-5: 70-77, **70-77**
CVW-17: **24** 28
HC-2 Det 2 'Desert Ducks': **24**
HS-14 'Chargers': 76, **77**
VAQ-136 'Gauntlets': **75**, 76
VAW-115 'Liberty Bells': **75**, 76, **76**
VF-103: **24**
VFA-27 'Royal Maces': **72**, **73**, 76
VFA-34: **24**
VFA-102 'Diamondbacks': **72**, 76
VFA-131: 26
VFA-192 'Golden Dragons': **73**, 76
VFA-195 'Dambusters': **74**, 76
VRC-30 Det 5: 76, **76**
Office of Air & Marine Operations: 78, 82-86, **83**, **85**, **86**, **87**
Office of Border Patrol: 78-81, **79**, **80**, **81**
Unmanned Little Bird VTUAV (see Boeing/McDonnell Douglas Helicopters)
US Department of Homeland Security: 78-93, **78-93**
US101 (see Lockheed Martin: Team US101)
U-X UAV: 54

V

Variant File – Boeing 707 military variants – Part 1: 98-117, **98-117**
VC-4A Gulfstream I (see Northrop Grumman)
VC10 (see Vickers)
VC-25A (see Boeing)
VC-118A (see Douglas)
VC-121E (see Lockheed)
VC-137A/B/C (see Boeing)
Venezuela
Força Aérea Venezolana (air force): **34**, **38**, **39**
Vertifan, XV-5A (see Ryan/General Electric)
Vertijet, X-13 (see Ryan)
Vertiplane, VZ-3RY (see Ryan)
Vickers
VC10: 21, **22**
Wellington: 130, 156
Viking, S-3B (see Lockheed Martin)
Vikramaditya, INS: 69
Vikrant, INS: 68
Viraat, INS: 68, **68**
VTOL flat-risers – Lockheed XV-4 and Ryan/GE XV-5: 118-127, **118-127**
VZ-3RY Vertiplane (see Ryan)
VZ-10 (see Lockheed XV-4A Hummingbird)
VZ-11 (see Ryan/General Electric)

W

W-3 Sokol (see PZL-Swidnik)
Warplane Classic – Messerschmitt Bf 110 – Underrated Warrior: 128-155, **128-155**
Warrior UAV (see General Atomics)
Warthog, A-10 (see Fairchild)
Wellington (see Vickers)
Westland (see also AgustaWestland)
Lynx AH.Mk 7: 21, **23**, 29
Lynx AH.Mk 9: 21, **23**
Lynx HMA.Mk 8: 22, **22**
Sea King HC.Mk 4: 21, 23, **23**
Sea King Mk 42B: **68**
Westland/Aérospatiale
Puma: 21, **22**
Puma HC.Mk 1: 29
Wildcat, F4F (see Grumman)

X

X-13 Vertijet (see Ryan)
X-45A/C UAV (see Boeing)
X-47A Pegasus UCAV (see Northrop Grumman/Scaled Composites)
X-47B UCAV (see Northrop Grumman)
Xavante, AT-26 (see EMBRAER)
XB-24 (see Consolidated)
XFV-12A (see Rockwell)
XV-4A/B Hummingbird/Hummingbird II (see Lockheed)
XV-5A/B (see Ryan/General Electric)

Y

Y-8 (see Shaanxi)
YAL-1A Airborne Laser (see Boeing)
Yasur, CH-53 (see Sikorsky)

Picture acknowledgments

Front cover: USAF via Tim Ripley, Cees-Jan van der Ende, MoD. **4:** USAF via David Donald. **5:** A.B. Ward, Saab via David Donald (two). **6:** A.B. Ward, USAF via David Donald. **7:** David Willis (five). **8:** Bell via David Donald (two). **9:** MNSTC-I via Tom Kaminski (three). **10-11:** Chris Knott/API (eight). **12:** AgustaWestland, Boeing via Tom Kaminski, USAF via Tom Kaminski. **13:** Peter R. Foster (two), Alenia via Tom Kaminski. **14:** Shlomo Aloni (two). **15:** Gert Kromhout (four), Fernando Jimenez. **16:** USAF via Tim Ripley (TR), USMC via TR, US Army via TR. **17:** US Army via TR (three). **18:** US Army via TR. **19:** USAF via TR (four). **20:** USAF via TR (four), USMC via TR. **21:** USAF via TR, via TR, RAF Deployed Operation Deatchment Basra via TR. **22:** RAF Deployed Operation Deatchment Basra via TR (two), via TR, Tim Ripley (three). **23:** Tim Ripley (three). **24:** US Navy via TR (three), Tim Ripley. **25:** US Navy via TR (two). **26-28:** USMC via TR. **29:** RAF Deployed Operation Detachment Basra via TR (two). **30:** Danny van der Molen (four). **31:** Nigel Pittaway (five). **32-33:** Robert Hewson (five). **34:** Cees-Jan van der Ende (four). **35:** Cees-Jan van der Ende, Chris Lofting (four), René van Woezik. **36:** Chris Lofting (four), René van Woezik. **37:** Chris Lofting (three), René van Woezik, Chris Knott/API. **38:** Chris Lofting (two), Cees-Jan van der Ende (two). **39:** Chris Lofting (four), Cees-Jan van der Ende, René van Woezik. **40:** General Atomics – Aeronautical Systems Inc. (GA-ASI), Northrop Grumman. **41:** Teledyne Ryan Aeronautical (two), GA-ASI. **42:** GA-ASI via David Donald (two), GA-ASI (two). **43:** USAF (two), USAF via David Donald, USAF via Tim Ripley. **44:** USAF, GA-ASI. **45-46:** Northrop Grumman. **47:** USAF via David Donald (two), Ted Carlson/Fotodynamics. **48:** GA-ASI, Ted Carlson/Fotodynamics. **49:** Ted Carlson/Fotodynamics, GA-ASI (two), NASA Dryden. **50:** GA-ASI, via Bill Sweetman (two), NASA Dryden. **51:** EuroHawk via David Donald (two), via Tim Ripley. **52:** Shlomo Aloni, via Bill Sweetman (two). **53:** EADS, Shlomo Aloni (two). **54:** Elbit, Office of Border Patrol via Tom Kaminski, Shlomo Aloni. **55:** GA-ASI, via Bill Sweetman (two), David Willis, USAF via David Donald. **56:** Boeing (two), Boeing via David Donald. **57:** Northrop Grumman (three). **58:** NASA Dryden (two). **59:** Saab via David Donald (two), Boeing via David Donald (two). **60:** Henri-Pierre Grolleau (two). **61:** Tony Holmes (two). **62:** Henri-Pierre Grolleau (two). **63:** Tony Holmes (two). **64:** Jamie Hunter/Aviacom, Tony Holmes. **66:** Jamie Hunter/Aviacom. **67:** Jamie Hunter/Aviacom, Tony Holmes. **68-69:** Simon Watson/Wingman Aviation. **70-77:** Ted Carlson/ Fotodynamics. **78:** via Tom Kaminski (TK). **79:** Tony Zeljeznjak via TK, OBP via TK (two), Tom Kaminski, via TK (two). **80:** Tony Zeljeznjak via TK. **81:** OBP via TK (two), Tom Kaminski, Ted Carlson/ Fotodynamics. **83:** AMO via TK (two), Tom Kaminski (three), Ted Carlson/Fotodynamics, Rafal Szczypek via TK. **85:** Ted Carlson/ Fotodynamics, AMO via TK (two), Sunil Gupta/Lockon Aviation Photography via TK, Jonathan Derden via TK. **86:** Sunil Gupta/ Lockon Aviation Photography via TK. **87:** Field Aviation via TK, AMO via TK (two), Tom Kaminski, Ted Carlson/Fotodynamics (three). **89:** Ted Carlson/Fotodynamics (two), Lockheed Martin via TK, Ray Rivard via TK, USCG via TK. **91:** Ted Carlson/Fotodynamics (three), USCG via TK (three). **92:** USCG via TK, Ted Carlson/Fotodynamics. **93:** Ted Carlson/Fotodynamics, Peter J. Cooper, EADS, Gulfstream. **94-97:** Andy Wolfe. **98:** Boeing. **99:** Ted Carlson/Fotodynamics, Boeing. **100:** Boeing (three). **101:** Northrop Grumman, Bob Archer/ Milslides. **102:** Boeing, David Donald (three). **103:** Boeing, Aerospace, David Donald, Peter R. Foster. **104:** Chris Knott/API, TRH, via Bob Archer/Milslides, Peter R. Foster. **105:** MoD, Ivan Siminic. **106:** Boeing (two), via Bob Archer/Milslides, Bob Archer/Milslides (two), David Donald. **107:** Chris Knott/API, Boeing, Paul Bigelow via Bob Archer/Milslides, Peter R. Foster, Bob Archer/Milslides. **108:** Cees-Jan van der Ende, David Donald, Chris Knott/API, Steve Miller via Bob Archer/Milslides, Boeing, Bob Archer/Milslides. **109:** Peter R. Foster, Chris Knott/API (two), Aerospace, Bob Archer/Milslides. **110:** Boeing via TRH, Boeing (two), Canadian Forces, TRH. **111:** Cees-Jan van der Ende (two), Ted Carlson/Fotodynamics, Peter R. Foster, Aerospace. **112:** Jim Winchester (three), IAI, Peter R. Foster. **113:** David Donald, Ivan Siminic, Peter R. Foster (two), Chris Knott/API. **114:** Bob Archer/Milslides, Craig Kaston via Tom Kaminski, via Tom Kaminski. **115:** Bob Archer/Milslides, via Bob Archer/Milslides, Peter R. Foster, Chris Knott/API (two). **116:** Bob Archer/Milslides (two), Chris Ryan, Chris Knott/API, Cees-Jan van der Ende. **117:** Cees-Jan van der Ende, David Donald, Boeing (two). **118:** Aerospace. **119:** Lockheed, Aerospace. **120:** Ryan, Lockheed (three). **121:** NASA via Daniel J. March, Aerospace, Lockheed. **122:** Aerospace (two), Lockheed (two). **123:** Ryan (three). **124:** Aerospace, Lockheed. **125:** Aerospace (two), Ryan (two). **126:** Aerospace, Lockheed. **127:** Ryan via Daniel J. March, TRH. **128:** TRH (two). **129:** TRH (two), Aerospace. **130:** Aerospace (four). **131:** John Weal (two), Aerospace (two). **132:** Dr Alfred Price (two), John Weal, Aerospace. **133:** John Weal (two), Dr Alfred Price (two). **134:** TRH, Aerospace, Dr Alfred Price. **135:** Aerospace (three), John Weal, TRH. **136:** John Weal, Dr Alfred Price (two), Aerospace. **137:** John Weal (two), TRH (two), Dr Alfred Price. **138:** John Weal (three), TRH. **139:** John Weal, Aerospace (three). **140:** John Weal (four), Dr Alfred Price. **144:** Dr Alfred Price (six), TRH (two), Aerospace (two), John Weal (two). **145:** Dr Alfred Price, Aerospace (three). **146:** John Weal (three). **147:** Aerospace, Dr Alfred Price (two), John Weal. **148:** Aerospace, Dr Alfred Price, John Weal. **149:** Dr Alfred Price, Aerospace (two), TRH, John Weal (two). **150:** Dr Alfred Price, Aerospace, John Weal. **151:** John Weal (three), Dr Alfred Price. **152:** Dr Alfred Price (three), Aerospace, John Weal. **153:** Dr Alfred Price (two). **154:** Aerospace, Dr Alfred Price (two). **155:** Dr Alfred Price, John Weal, Aerospace. **156:** Aerospace (two). **157:** via Martin Bowman (three). **158:** Air Ministry (two), via Martin Bowman. **159:** via Martin Bowman (two). **160-161:** via Martin Bowman. **162:** via Martin Bowman, Aerospace. **163:** via Martin Bowman, IWM. **164:** via Martin Bowman (two), RAAF (two). **165:** Harry Gann, IAF, Aerospace. **166:** MoD, via Martin Bowman, Harry Gann. **167:** via Martin Bowman, Aerospace. **168:** via Martin Bowman, MoD. **169:** via Martin Bowman, via Jon Lake. **170:** via Martin Bowman, MoD, Aerospace. **171:** Aerospace (three). **172:** IWM (two). **173:** via Martin Bowman, RAAF.